Co-Creativity and Engaged Scholarship

Alex Franklin
Editor

Co-Creativity and Engaged Scholarship

Transformative Methods in Social Sustainability Research

palgrave
macmillan

Editor
Alex Franklin
Centre for Agrecology
Water and Resilience (CAWR)
Coventry University
Coventry, UK

The majority of the chapters in this book result from the RECOMS project. This project has received funding from the European Union's Horizon 2020 research and innovation programme under the Marie Skłodowska-Curie grant agreement No 765389.

ISBN 978-3-030-84247-5 ISBN 978-3-030-84248-2 (eBook)
https://doi.org/10.1007/978-3-030-84248-2

This Palgrave Macmillan imprint is published by the registered company Springer Nature Switzerland AG
The registered company address is: Gewerbestrasse 11, 6330 Cham, Switzerland

Foreword

Over the past two decades, young scholars in Europe and beyond have developed a new understanding of the critical social, economic and environmental challenges that humankind is facing around the globe. They are highly aware that the unprecedented destruction of the planet, its natural resources and its species means that we cannot continue with 'business as usual'. Many scholars from the sustainable sciences to the newly established environmental humanities have demanded that the Academy rises to the occasion: A world that teeters on the brink of destruction does not need more conventional scholarship but new, more creative and more concerned forms of research and engagement. In order to imagine and to build a more sustainable world, we must rethink the very way in which we create knowledge, communicate across different disciplines and engage with various publics. To that effect, we need to navigate the multiple environmental and social challenges and negotiate—through language, art and action—between the discourses of politicians, planners, natural and social scientists, social justice activists and others.

This volume stems from the work of fifteen young scholars and some of their mentors who are participants in an Innovative Training Network (ITN) funded by the European Commission. The network is comprised of a unique consortium of scientists, practitioners and change agents from eleven public, private and non-profit organizations located in six European Union countries. All of these early career researchers are working towards building a better understanding of—and greater support for—resourceful environmental practices in communities in Europe and beyond. Their central goal has been to assess, re-think and re-work existing concepts and methodologies for the twenty-first century. Their project is an ambitious one. It comes out of a timely call and their own aspirations to make a difference in both social relationships and in human interactions with the environment. It is informed by three fundamental insights: First, the understanding that the problems that societies are facing in a globalized world need perspectives that reach beyond regional and national analysis. Second, the realization that co-creation is a powerful practice that can (and should) replace older notions of the 'creative genius'. And third, the insight that co-creativity can serve as an effective research tool and an instrument of social change.

On the most obvious level, *Co-Creativity and Engaged Scholarship* is an introduction into participatory and transdisciplinary methodologies. It is, however, more than that, not least because the young scholars present first-hand accounts of their own scholarly endeavours and social experiences. What shines through is their engagement with the public, guided by academic curiosity, resulting in a combination of literary lightness and a summons for creative action. The essays focus on such diverse topics as sustainability and arts collaborations in The Netherlands; improvisational theatre and deep mapping as creative forms of action and understanding; photography as a means to bridge the differences between Asian and European cultures; and many more. Together, these chapters remind us that creativity is a social practice. In an age marked by competition and individual tendencies, they call for a counter-hegemonic scholarship that is based on fully inclusive and truly democratic participation, and on alternatives to the predominant ideology of economic progress. Each essay demonstrates that research can and must comprise much more than the economized and industrialized production of knowledge. They illustrate that a critical approach can be creative, and that, in turn, creative research can promote critical thinking.

It is my hope that *Co-Creativity and Engaged Scholarship* will help a new generation of scholars to discover and explore new transformative research methods, take grounded, informed positions in political debates and work towards building a more resilient, ecological and sustainable world.

Christoph Georg Lichtenberg, an eighteenth-century German inventor and philosopher, insisted that his private library should consist only of books containing new ideas. Most books, he maintained, did not meet this criterion. With the following pages, editor Alex Franklin, together with the young scholars who contributed, have created precisely such a volume.

Tutzing, Germany Christof Mauch
April 2021

Acknowledgements

The majority of the research informing this edited collection has been funded by a Marie Skłodowska-Curie Innovative Training Network Grant (No. 765389) awarded through the European Union's Horizon 2020 research and innovation programme. It is also because of this award that it has been possible to publish this book in Open Access format. On behalf of all the book contributors, and also all those involved with the 'RECOMS' project more broadly, I take this moment to acknowledge the incredible opportunity which MSCA funding presents for engaging in truly transformative programmes of action-orientated research and learning. I also thank all the many individuals and institutions who have contributed to the making and delivery of 'RECOMS' across the last four years. I have a debt of gratitude that extends to every single one of you.

My particular thanks to Christof Mauch for kindly providing the foreword to this collection. His willingness and generosity in doing so serves also as a good reflection of the time and commitment given to mentoring of early stage researchers by all the members of the RECOMS consortium—academic and practitioner alike. It has been a deep privilege to collaborate with you all.

The cover illustration is by Imogen Humphris. I am grateful to Imogen for allowing us to use the image, not least because it serves as such a rich and evocative example of what can be realised through co-creative research practice.

My thanks to Palgrave for all their enthusiasm and support (as well as patience) during the writing of this book.

And finally, a personal thank you to Yasmin and Faisal: for helping me see the world in a whole new way, for reminding me that nothing is ever fixed, for demonstrating that with an open mind the possibilities and opportunities for creative transformation remain endless, and for making me appreciate once more the importance of play.

Contents

Notes on Contributors

Nohemi Ramirez Aranda is an architect, M.Sc. in Urban Management and Development, and Ph.D. student in Geography at Ghent University and ILVO as part of the RECOMS programme. Her research focuses on developing a Public Participatory GIS (PPGIS) tool that can map the social values that green open spaces provide to people. Moreover, the research addresses how these values can be further included in spatial planning and policy for better community management and preservation of green open spaces across Belgium.

Dr. Marco Armiero is Research Director at the Institute for Studies on the Mediterranean (CNR Italy) and director of the Environmental Humanities Laboratory (KTH Sweden). He has worked on the nationalization of nature, migrations and environment, and environmental justice. With his research, he has contributed to bridging environmental humanities and political ecology. His most recent publication is *Wasteocene: Stories from the Global Dump* (Cambridge University Press, 2021). He is the elected president of the European Society for Environmental History.

Prof. Dr. Jos Arts is full-professor Environment and Infrastructure Planning and Head of the Department of Spatial Planning and Environment, Faculty of Spatial Sciences, University of Groningen, The Netherlands. He is also extraordinary professor at Northwest University, Potchefstroom, South Africa. He has organized many international workshops, conferences and published widely about impact assessment, evaluation and environmental, spatial and infrastructure planning. His research focuses on institutional analysis and design for integrated planning approaches for sustainable infrastructure networks.

Dr. Lorena Axinte is currently working in mobility and transport, helping cities make their streets more human-friendly. She has a doctorate in Geography & Planning from the Sustainable Places Research Institute, Cardiff University, where her research focused on city-regional planning and development. Interested in governance and participation, Lorena looked for windows of opportunity to engage young people in conversations about the future of their areas.

Zhanna Baimukhamedova is based at the Rachel Carson Center, LMU Munich, Germany. She holds an M.Sc. degree in Urban Studies. For her graduate programme, she attended six universities in Brussels, Vienna, Copenhagen, and Madrid and wrote her thesis on how technology and social media affect tourism patterns in Berlin and Madrid. For her BA degree, Zhanna focused on the emigration of the Kazakhstan-born Germans after the fall of the Soviet Union. Zhanna's research experience includes projects on heritage and memory, nature-wildlife conflicts and environmental history/stories both in Europe and Kazakhstan. In March 2019, Zhanna joined the Rachel Carson Center as a RECOMS Early Stage Researcher and a PhD candidate. Her dissertation project is 'Transforming the Bavarian Forest: Socio-Ecological Crises, Community Resilience and Sustainability from a Historical Perspective'.

Dr. Iain Biggs RWA, is former Director of the PLaCE Research Centre, UWE, Bristol, and was elected to the Royal West of England Academy in 2012. He has held a Moore Institute Visiting Fellowship at the NUI, Galway (2014), and is currently an Honorary Research Fellow at the Duncan of Jordanstone College of Art and Design, University of Dundee, and a Visiting Researcher Fellow at the Environmental Humanities Research Centre, Bath Spa University. A proponent of deep mapping, he is currently involved in projects in Wales, Ireland, Scotland, Norway and The Netherlands. He regularly examines doctorates in the UK and Ireland. Recent publications include chapters in *The Routledge Companion to Art in the Public Domain*(Routledge, 2020) and *Walking Bodies: Papers, Provocations, Actions* (Triarchy Press, 2020). He is co-author, with Professor Mary Modeen, of *Creative Engagements with Ecologies of Place: Geopoetics, Deep Mapping and Slow Residency* (Routledge, 2020).

Scott Davis is a socio-spatial planning Phd candidate at the Faculty of Spatial Sciences in the University of Groningen. Scott's Ph.D. is part of the wider Marie Curie ITN RECOMS project. His research investigates community resilience, exploring how fostering a communities' cultural relationship with their immediate environment can play a role in encouraging local voices to secure a more powerful say over decisions relating to their place. Scott's previous background includes working for the British Red Cross as a social researcher on the EU FP7 'DRIVER+' project. Through this project he supported the development of community resilience crisis management scenarios by delivering community resilience workshops to flood-risk communities in Scotland and analysing their effectiveness. Scott has also worked at the Scottish Government in environmental policy (land-use, forestry) and as a social researcher in public health. He holds an undergraduate degree in International Policy and a postgraduate degree in Social Research Methods.

Sofia De la Rosa Solano is a Marie Skłodowska Curie fellow, based at Groningen University, The Netherlands, working in the RECOMS project. Her research interests include environmental history, political ecology and environmental justice. More specifically, she is interested in how inequalities are expressed in both society and space, particularly through water. Prior to joining RECOMS, she completed a bachelor degree in History at the National University of Colombia and an M.A. in Latin American Studies at the University of Amsterdam. She has worked in multiple NGOs, government institutions and academic centres in Colombia, The Netherlands and the UK on water-related research.

Dr. Alex Franklin is an Associate Professor at the Centre for Agroecology, Water and Resilience (CAWR), Coventry University, UK. Her research explores collaborative forms of environmental practice. Her interest in practices of care and co-creativity extends also to more-than-human relations. Originally trained as a human geographer, her involvement since 2005 in a series of inter- and transdisciplinary research centres sees her identifying more closely these days with the broader field of (social) sustainability science. She is currently involved with a number of national and international research projects, including coordinating the 'RECOMS' H2020 MSCA Innovative Training Network (2018–2022).

Dr. Jana Fried is an Assistant Professor at the Centre for Agroecology, Water and Resilience (CAWR) at Coventry University, UK, with expertise in community-based and community-focused research, particularly in the areas of science and society, health and development, environmental risk perception and community resilience. With a background in geography, sociology and urban planning, she has been working in various international and interdisciplinary settings on issues pertaining to community resilience and community well-being.

Dr. Gloria Giambartolomei holds a doctorate from the Centre for Agroecology Water and Resilience (CAWR) at Coventry University (UK). She holds a Research Master's from Utrecht University (NL) in Sustainable Development—Environmental Governance, and a degree in Politics, Philosophy and Economics from LUISS University, Rome (IT). She is particularly interested in exploring the personal, political and power dimensions of social-ecologically just sustainability transformations—what triggers processes of transformation of individual and collective values, beliefs and paradigms, both in everyday practices, and at the institutional level. For her Ph.D., she worked at the interface between communities, government and public and third sectors organizations, to understand how a feminist and democratic ethics of care can contribute to the development of meaningful and inclusive collaborative practices for the sustainable management of the natural resources in Wales.

Viola Hakkarainen is an interdisciplinary Ph.D. candidate based at LUKE and the University of Helsinki. Her research focuses on epistemic dimensions of human-nature relationships. She is interested in understanding how to enhance inclusivity of diverse knowledges, expertise and worldviews and enable plurality in different ecosystem governance settings for just sustainability transformations.

Prof. Dr. Lummina G. Horlings studied land- and water management and human (environmental) geography. She obtained a Ph.D. in Policy Science, and currently holds a chair in Socio-Spatial Planning at the Faculty of Spatial Sciences of the University of Groningen in The Netherlands. Her research work broadly deals with 'the human factor' in spatial planning. She is an expert in socio-spatial planning, place-based development, sustainability, collective citizenship, place leadership and values. She coordinated the Marie Curie ITN programme SUSPLACE and participates in the Marie Curie ITN project RECOMS. Within RECOMS she supervises three research fellows. She also participated in the COST Action IS1007 'Investigating Cultural Sustainability' and in the EU projects: Rural Alliances, DERREG, EUWelNet and GLAMUR. Lummina G. Horlings is a member of the Research Committee of the Regional Studies Association and the RSA Research Network 'The Place of Leadership in Urban and Regional Development'. She has edited several books, on passion, vital coalitions, leadership and the role of culture in regional development.

Imogen Humphris is based at the University of Groningen, The Netherlands. She is formally trained in architecture and is centrally concerned with the social landscape of urban contexts, particularly in marginalized settings. Her art-design-research practice explores notions of place and its becoming from a variety of perspectives and seeks to foster conversation around the multiplicitous nature of space. Over the last 10 years her work has become increasingly focused on the power and limitations of civic initiatives, both formal and informal, acting as a community planning facilitator, researcher and activist. During this time she has worked across a variety of international and local settings from Scotland and England to The Netherlands, Kenya and Palestine.

Prof. Dr. Wim Leendertse works as a senior advisor at Rijkswaterstaat, the executive organization of the Dutch Ministry of Infrastructure and Water Management, The Netherlands. He is also part-time professor Management in Infrastructure Planning at the University of Groningen, The Netherlands.

Stephen Leitheiser is a Ph.D. candidate in the Department of Spatial Planning and Environment at the University of Groningen, The Netherlands. He is interested in radical democracy, regenerative society, agroecology, food sovereignty and the commons, and his research engages with citizens and civic groups who put these concepts into practice.

Kei Yan Leung is a Marie Skłodowska Curie Ph.D. fellow with the RECOMS project, based at University of Natural Resources and Life Sciences, Vienna, Austria. Her Ph.D. project looks at how farmers relate to art and explore nature-cultural relations and mind-body experiences of farmers. Her research interest in small-scale, sustainable farming began in 2015 when she engaged in several farming communities in Hong Kong and started growing rice. After receiving a bachelor degree in Social Sciences from the University of Hong Kong (HKU) in 2015, the first-hand engagement in farming communities in Hong Kong inspired her to start a Master of Philosophy (M.Phil.) in sociology at HKU. In 2017, she completed the M.Phil. and worked as a part-time research assistant in HKU until joining RECOMS.

Dr. Angela Moriggi is a social scientist working for transdisciplinary sustainability projects since 2013, with extensive fieldwork experience in Finland, Italy and China. She is research fellow at the Department of Land, Environment, Agriculture & Forestry of the University of Padova, focusing on transformative social innovation, rural development and co-creative approaches. The red thread in Angela's work is the design, use and evaluation of participatory and action-oriented methods and processes. Her research, teaching and facilitation work is informed by care ethics & feminist approaches to transformative change. She co-founded 'Re.Imaginary', an open-access database of creative methods, and co-edited *Once Upon the Future*, an anthology of children stories to inspire sustainable change-makers. She holds a Ph.D. in Rural Sociology from Wageningen University (NE).

Dr. Luke Owen is an Assistant Professor based in the Centre for Agroecology, Water and Resilience, Coventry University. He is a social scientist with a background in human geography and specializes in a range of qualitative research methodologies and data analysis techniques. Luke's research interests include Alternative Food Networks and Short Food Chains and how they contribute to rural development and the livelihoods of small-scale food producers.

Kelli Rose Pearson is the co-founder of Re.imaginary Group which works in the arena of leadership, creative practices and storytelling for regenerative sustainability. In the past, she's been the owner of a successful cafe and community hub in Savannah, Georgia, a consultant in the field of international development and sustainability, and a research fellow focused on the topic of transformative mindsets and sustainable place-shaping. In her free time she writes middle grade adventures about the mythical Cosmos Mariners. Her formal education includes an M.Sc. in Environmental Governance from the University of Freiburg in Germany, a B.A. in Comparative Religion from Carleton College in Minnesota, a Certificate in Sustainability Leadership from the University of Oregon and a Certificate in Ecological Design from the Ecosa Institute in Arizona.

Maria Alina Rădulescu is an Early Stage Researcher affiliated with Rijkswaterstaat, the executive organization of the Dutch Ministry of Infrastructure and Water Management, and a Ph.D. candidate at Groningen University. Maria has extensive experience in the field of urban design and spatial planning and her research interests lie at the junction of public participation, spatial and infrastructure planning, and policy-making.

Talitta Reitz is a landscape architect, researcher and doctoral candidate. Growing up in Curitiba, the 'green capital' of Brazil, awakened her curiosity for sustainability, planning, design and landscapes. Moving to Los Angeles, Talitta acquired a master's degree in landscape architecture and urban planning from the University of Southern California as a part of a Brazilian scientific mobility scholarship programme. During her graduate studies, Talitta received a heritage conservation certificate and an ASLA merit award; she also took part in a participatory community project for a small city park. This project sparked her interest in participatory design and community engagement. For this reason, she joined RECOMS, a research consortium investigating community engagement across multiple European countries. With an Early Stage Researcher position and a Marie Skłodowska-Curie fellowship, she currently pursues a Ph.D. candidacy in environmental humanities at the Rachel Carson Center for Environment and Society, LMU, Munich. Her doctoral research focuses on the environmental history of bicycle cities, exploring the relationships between cycling cultures, urban landscapes and transportation planning.

Sergio Ruiz Cayuela is a militant researcher currently conducting his Ph.D. between the Rachel Carson Center in Munich and the Centre for Agroecology, Water and Resilience, Coventry University, Coventry, UK. He is a Marie Skłodowska Curie fellow within the RECOMS network. Sergio is interested in commoning practices as potentially emancipatory forms of social organization. He is working on strategies for commons expansion that simultaneously focus on increasing material autonomy and building commoning subjectivities. Sergio is involved in several self-organized community groups, political organizations and social movements working on diverse fronts such as migration, ecology, solidarity economy and housing.

Dr. Gwenda Van der Vaart is an Assistant Professor at the Department of Spatial Planning and Environment at the Faculty of Spatial Sciences, University of Groningen. Her research interests include arts and culture in communities (in terms of community development, resilience, liveability, community engagement, cultural infrastructure, coping with place change), the relationship between citizens and the government (community initiatives, community participation, giving an active role to citizens) and creative and arts-based research methods.

Yanthe van Nek is a prizewinning community artist residing in the north Netherlands who specializes in co-creating environments in which sustainable social movements can come to life. Yanthe is interested in chaos, anger and the perfect imperfect. Her field of study is the part that lies between perfection and destruction and believes that participatory art must be more than having nice conversations—art, research and the social context must coincide. Her most recent accolades include 'winner of the Klaas Dijkstra Art Academy Prize 2016' and co-winner of the Dutch 'Lang Leven Kunstprijs 2018'. Her academic achievements include a B.A. Hanze University Groningen Art Academy Minerva, HKU University of the Arts Fine Art Foundation Course and VU University Amsterdam Propaedeutic year English.

Rubén Vezzoni is an Early Stage Researcher at LUKE (Natural Resources Institute Finland) and a Ph.D. candidate in Political, Societal and Regional Change at the University of Helsinki. He is interested in heterodox economics, institutional change and the role of academia in society. In his Ph.D., he is exploring the ways in which climate change is shaping fiscal policy. He is involved in several initiatives concerning participatory (and economic) democracy, as well as environmental justice.

Agnes Zolyomi is an Honorary Research Fellow at Centre for Agroecology, Water and Resilience, Coventry University. She has worked with NGOs since 2005 at national and international levels on climate change and biodiversity. For over ten years she has worked specifically on EU relevant biodiversity policies focusing on advocacy work as well as on international processes such as the IPBES or the Convention of Biological Diversity. During working closely with the European Commission and EU Member States and leading various communication campaigns, she could collect personal experience about how advocacy can achieve impact. Agnes has also worked with academia, at the Centre for Agroecology, Water and Resilience, Coventry University working on the RECOMS—resilience and resourceful communities Horizon 2020 project. Currently, she is involved in ecosystem services assessments and the Horizon 2020 EU 1.5 Degree Lifestyles project that deals with behaviour change and communication on lifestyles transformations for achieving the EU climate goals.

List of Figures

List of Tables

1

Introduction: Sustainability Science as Co-Creative Research Praxis

Alex Franklin

Introduction

The conception of this book has been guided by three inter-related aims. The first is to encourage reflection and debate on the relationship between collaboration and creativity ('co-creativity') within sustainability science research. The second aim is to support researchers in actively promoting and nurturing, but also managing and responding to, the effects of co-creativity within their research. The third is to better understand the potential of engaged and co-creative scholarship in furthering transformative sustainability agendas. Clearly, these aims are underpinned by a number of beliefs, including: that creativity and collaboration are co-present within sustainability research, that the two do and 'should' go hand-in-hand in the design and practicing of transformative research, and that a transformative agenda is central to much

A. Franklin (✉)
Centre for Agroecology, Water and Resilience (CAWR), Coventry University, Coventry, UK
e-mail: alex.franklin@coventry.ac.uk

A. Franklin (ed.), *Co-Creativity and Engaged Scholarship*,
https://doi.org/10.1007/978-3-030-84248-2_1

of the sustainability sciences. Also, though, that the pursuit of engaged scholarship and co-creative research as a form of praxis requires further critical consideration within the context of the sustainability sciences.

This chapter is primarily concerned with introducing the concept of co-creativity. The term is applied here in reference to both individual methods and overarching research approaches that seek to engender collaborative and creative forms of action and reflection. In the case of collaboration (i.e., the 'co-' of co-creativity), this signifies the prioritization given throughout this book to researching '*with*' (rather than on, or for); in its broadest sense, it also acknowledges the fundamental relationality of life. A relational lens, and therefore also the notion of emergence, is similarly central to how creativity is itself conceptualized. The term creativity is understood here as embodying a generative way of thinking and being, seeing and doing, arising from relational forms of knowledge-practice. An extended discussion of this conceptualization is provided below.

In fusing together collaboration and creativity into the concept of co-creativity, what I and all other contributors to this edited collection are especially interested in is socially inclusive research. Whilst at its base, all scientific scholarship is arguably creative, the degree to which individual research studies purposively aim to achieve greater social inclusivity through their chosen methodologies varies widely. By placing emphasis on social inclusivity in our conceptualization of co-creativity, this in turn highlights the political nature of a question that frames the entirety of this collection: how can co-creative research practice, as a generative process, best support the emergence of alternative—potentially even transformative—ways of being in the world?

Thus, methods and approaches are understood here as co-creative when they stimulate *alternative understandings of why and how things are, and how they could be*. Notably, though, such stimulation of alternative understandings needs to be a shared one, experienced (albeit in different ways and to different extents) by multiple persons within a research process, including both the 'researchers' and the 'researched' alike. To engage in co-creative research therefore calls for a retained sensitivity to the importance of researching 'with' *throughout* the process of doing research. Furthermore, in order to realize transformative sustainability agendas through the stimulation of alternative understandings of why and how things are, and how they could be, such alternative framings are simultaneously critical in tone. More particularly, by offering up co-creative ways of 'mobiliz[ing] the critique through the alternative—by showing that another way is possible' such methods and approaches help 'call in to question' not only how things are, but also the way in which we, or others, commonly respond to them (Hannah & Jeremijenko, 2017, p. 214). This in turn accounts for why co-creativity is directly associated through this collection with engaged scholarship.

Approached as a form of praxis, engaged scholarship is accordingly understood here as being driven not simply by a desire to interpret and understand the world, but also to change it (Cowley, 2013). Materializing such a desire requires a close and continuous alignment of thought and action: that is, it requires 'the synthesis of theory and practice and the reciprocal relationship between them' (Cowley, 2013, p. 1). All the chapters in this collection are authored by scholars committed to such a praxis-orientated form of engaged scholarship. At the same time, nevertheless, the relative umbrella coverage of the term engaged scholarship is also visible throughout (Boyer, 1990, 1996; Shultz & Kajner, 2013). Indeed, the array of cases presented in this book are variously informed by (but not limited to) principles of Participatory Action Research (see for examples Chapters 4, 8, 12, 13, 16), Militant Scholarship (see especially Chapter 3), Appreciative Enquiry (see especially Chapter 5), Care-full Scholarship (see Chapters 5 and 16) and Transformative Research (see for example, Chapters 6, 12 and 14). In their own way, each of these serve as exemplars of engaged scholarship. Collectively, in turn, they also provide a good indication of the diverse ways in which co-creativity, in association with engaged scholarship, is able to shape and enrich not only the sustainability sciences, but also the value that comes from taking the time to critically consider and reflect upon the complex nature of its contribution.

In parallel to noting the relative diversity of research approaches featured within this collection, it is nevertheless important to also acknowledge that the individual research methods are themselves all drawn from the *social* sustainability sciences. That is, their primary orientation is towards deepening our understanding and advancement of sustainability by questioning its social and cultural origins, meanings, practices, values, interpretations, structures, systems and relationships. This shared characteristic can be situated within, on the one hand, an appreciation by all the contributing authors for the vast array of epistemologies and ontologies circulating within the sustainability sciences, but also, on the other hand, a critical awareness of the ongoing dominance of more traditional and often very binary forms of problem framing. In response, whilst this edited collection is not intended as an attack on more traditional forms of research scholarship in and of themselves, it is aimed at further drawing attention to what Sandercock and Attili (2012, p. 140) eloquently summarize as:

> the many other ways of knowing that exist: experiential, intuitive and somatic knowledges; local knowledges; knowledges based on the practices of talking and listening, seeing, contemplating and sharing; and knowledges expressed in visual, symbolic, ritual and other artistic ways.

Moreover, it is particularly aimed at explaining *how* these many other ways of knowing might care-fully (Moriggi et al., 2020) and respectfully be enrolled, through co-creative forms of engaged, transdisciplinary scholarship.

To delineate this collection one step further, the type of *social* sustainability science with which it can be affiliated (as akin to the aforementioned definition of engaged scholarship as a form of praxis) is one that regularly 'places power at the centre of analysis' (Eubanks, 2012, p. 229). Indeed, whilst an ambition to advance transformational research agendas is widely shared by sustainability scientists (Kates, 2011), reflective of both a traditional orientation towards (and ongoing dominance of) the natural and technical sciences, all too often scholarship within the sustainability sciences fails to engage with the political core of sustainability. In failing to pose the question of what a more just sustainable world would look like in a far more inclusive, dialogical and expansive manner, or, to acknowledge the political underpinnings of 'sustainable development'—and indeed why both *just* sustainability and *sustainable* development remain so elusive—this in turn not only brings into doubt the appropriateness of solutions being proposed, it also impedes diagnosis of the very problems that need to be addressed (Carpenter et al., 2020; Miller et al., 2014). In short, the emergence of just and sustainable forms of transformative change are all too regularly severely curtailed (Vallance et al., 2011). It is this curtailment that the co-creative and engaged forms of scholarship endorsed within this collection seek to overcome. It is also worth emphasizing, however, that what follows is not an uncritical collection of celebratory accounts. Rather, a unifying characteristic of this entire collection is the belief that the intentional nurturing of co-creativity through research practice, holds as much potential to be mutually rewarding for all involved as it does to be highly problematic. It is in recognition of this problematic that the need for extended critical reflection by a community of engaged scholars, and in turn the aims of this book, were originally derived.

Having offered some introductory orientation to what this book is about, as well as the beliefs and assumptions on which it is based, the remainder of this chapter is structured as follows. In the next section I briefly explain more about the original motivation behind this book. This serves to further set the scene for the interest of this entire collection with sustainability as requiring just and transformative change. I then look in greater depth at the meaning of creativity, and in turn, the nurturing of co-creative methods and approaches within the sustainability sciences. I end by offering an introductory overview to each of the chapters that make up the remainder of this collection.

'Making a Difference': Transformative Research Agendas

Throughout my career I have observed that when doctoral research candidates are asked during recruitment why they are interested in undertaking a doctoral degree within the sustainability sciences, many provide an answer that is primarily centred around an inherent desire to make the world—or at least a small part of it—a more just and environmentally sustainable place. Whilst a doctoral research project is perceived (either in the short or longer term) as a means of achieving this goal, the desire to contribute new knowledge commonly remains secondary, occasionally even requiring a prompt in order for it to be mentioned at all during the initial interview. Such aspirations for making a 'real world' difference through one's doctoral research come in many different shapes and forms. Sometimes, for instance, they are heavily prioritized around a single issue, sometimes around multiple issues; sometimes they engage with the social, cultural, environmental and economic pillars of sustainability in intentional concert; in other cases, they are orientated almost exclusively towards one pillar alone. Similarly, sometimes they are seemingly driven by an ambition, or a need, to bring about change at a global scale; other times the motivation—no less powerful—is about achieving change locally in a particular place, or with a particular group of individuals. Another commonly reported ambition, both personally and professionally, is that a scientifically robust connection be established between the research and the change that the individual is seeking to bring about. In parallel, much greater recognition is increasingly being given to the importance of the research process itself, as well as—or in some cases even instead of—aspiring to any sort of pre-definable direct and immediate end result.

It has been my sustained encounter with this deeply felt need by doctoral candidates for 'making a difference'—for engaged scholarship in all its various forms—and alongside, the questions and responsibilities that my colleagues and I are then necessarily confronted with as doctoral supervisors, mentors and coordinators of sustainability science research projects, which ignited my own motivation for editing this book. Such questions include: actually, how synergistic are the institutional requirements of contributing new knowledge, and the personal desires for making a 'real world' difference, within the regulated space of a doctoral degree? To what extent, regardless of the relative breadth or narrowness of the above stated ambitions, can a difference be made (outwith academia) through doctoral or early career stage research? Or perhaps even more to the point in this context, *how* to go about trying to achieve this whilst at the same time meeting the scholarly and institutional demands of the associated academic research without detriment to the wellbeing of the researcher (including their work-life balance)? And, relatedly, what impact will it have on the morale of the researcher should they decide, part way through, that they are unable to make sufficient difference within the lifespan of the PhD? Furthermore, on the part of a supervisor or mentor, what are the best ways to support such ambition, including during the occasions when the scholarly, institutional and 'real world' demands are tending more towards being in conflict rather than synergy with one another?

In seeking through this edited collection to provide some answers to these questions, the approach taken is one that forefronts the first-hand accounts of the contributing authors' own direct experience of co-creative research practice. Notably also, the majority of the chapters are centred around research undertaken during a Doctorate. Such experiential accounts serve to richly evidence and support the critical reflections of the authors on what is involved and what is encountered when it comes to propagating co-creativity in the pursuit of transformative research. Moreover, it is hoped that the centrality given to first-hand accounts will assist the reader in relating to and subsequently building upon the learning shared by the authors in their own future research practice. In this respect also, the book has been written with a primary target audience in mind of doctoral and early career researchers affiliated to academic institutions, but also researchers working out-with academia across a range of different institutions and community settings. At the same time, though, it is hoped that the relatively wide range of disciplinary backgrounds, plus the international profile and experience of contributors, will make the collection insightful and relevant to a much broader cohort of co-creative sustainability and transdisciplinary practitioners and research mentors.

When compared to the project based short-termism that characterizes much academic research, the relative freedom of a social science doctoral research contract, plus the opportunity it commonly presents for at least three years of focused study, is in many ways more conducive to facilitating societal change. Moreover, a driving ambition to bring about change can actually be one of the best reasons for an individual wanting to embark on a PhD in the sustainability sciences. This view is built on observing (and personally relating to) the importance, the essentialness, for research encounters to have meaning not only for the research participants, but also for the researcher. That the actions and beliefs of people and communities might otherwise merely constitute or represent 'data' to be collected or 'extracted', is not enough—verging even, for some researchers, on being unethical. For it is the meaningfulness of research encounters, particularly those of a deep and/or sustained nature, that sparks the potentiality for change and gives energy to the need for action.

Furthermore, given the reality that only a small minority of those we engage with through our research practices are likely to come into direct contact with the research findings, it is the contact—the collaboration—that we have with them during and through the research process itself that is sometimes *the* chance to stimulate a change in mindsets and in practice. It is also the encounter itself that creates an immediate opportunity to acknowledge and in turn celebrate the work of many research participants, through shared recognition of their commitment to making a difference, or simply their daily struggle to overcome the micro, meso and macro level injustices of our unsustainable systems and institutions. This is particularly so for those individuals whose voices otherwise remain silenced or marginalized; but equally also for those who are driven by the ambition to change the thinking and ways of working amongst people in positions of power whose voices already get heard.

Therefore, it is my—our—contention through this collection, that dedicated time must be set aside to dwell, to reflect, on the transformational potential that comes from the process of 'doing' and 'being' engaged *through* one's research and *with* all those contributing to the research (for further discussion on this see for example Chapters 3, 5, 12, 13 and 16). It is also my contention that conceptualizing engaged scholarship as a form of co-creative praxis and, in turn, exploring co-creativity from a political starting point of engagement and inclusivity, can be of mutual reward to the practicing of both. With this in mind, as a further introduction to the chapters contained in this collection, but also more immediately as an orientation for the remainder of this chapter, three guiding questions are hereby proposed:

- What does it mean to think of research as a co-creative practice and of researchers as co-creative practitioners?
- How can engaged forms of co-creative research practice be nurtured?

o Why does attending to the (inherent) co-creativity of research matter within the sustainability sciences?

Responding to these questions and building also on the opening discussion, in the next section I offer a brief overview of the trajectory of academic debate concerning the meaning of creativity. In turn, I then further explain how co-creativity is conceptualized for the purposes of this collection.

Conceptualizing (Co-)creativity as Relational and Emergent Praxis

As a subject of study creativity, scholarship dates back to at least the nineteenth century. It is only since the late twentieth century, though, that it has really burgeoned as a field of scientific interest (Kaufman & Glăveanu, 2019). In the vast majority of the material published prior to the twenty-first century, a few overarching observations can be made; observations that, in turn, explain why it is only actually a relatively narrow stream of creativity scholarship, published during the last two decades or so, that directly informs the aims and research questions guiding this book. The first of these observations relates to how creativity has traditionally been defined; the second relates to the prevailing unit of study.

The standard definition of creativity, which has dominated its study from within the field of psychology since at least the 1950s (Runco & Jaeger, 2012), points towards the bringing into being of something new and useful. Under this definition, creativity can take the form of a thought, an action or an object, with the utility component generally depicted as representing at least some degree of 'social value' (see, for example, Helfand et al., 2016). Somewhat surprisingly, much of the wider body of creativity scholarship merely accepts and runs with this basic definition. Where substantive critique does exist, it is rightly centred around the connection that this dual emphasis on novelty and utility establishes with capitalist market pressures for the production of new and valued commodities (Liep, 2001). As Rehn and De Cock assert, for example, 'emphasis on novelty is needed to ideologically position creativity as part of an economic movement and to connect it to the modernist ideology of progress' (2008, p. 225).

In terms of the implications of such a definition for the future of creativity itself (or indeed, its relevance to the transformative agenda of this edited collection), the risks include that creativity comes to be nurtured within educational and institutional contexts merely with the intention of preparing individuals for securing 'competitive advantage over others in a world dominated by the need to achieve and accumulate' (Literat & Glăveanu, 2016, p. 330). Furthermore, as Chan (2016, p. 649) explains, in an institutional environment in which novelty is held in increasingly high regard, including attracting tangible rewards, it is likely that over time 'novelty will permeate the assumptions, perceptions, attitudes, values and methods of workers in this field'. The point being made here is that rather than serving to increase creativity, such emphasis on novelty or innovation instead merely creates 'a discourse of novelty' and 'a rhetoric of experimentation' (Chan, 2016, p. 649). Alongside this perceived erosion of creative substance, Rehn and De Cock raise further critique on the grounds that:

> We cannot allow the concept of creativity to be always-already defined by novelty, nor to fall under the ideological framework of progress and modernism [… …] The notion of novelty as defining creativity is […] not only analytically problematic, it is also uncreative as it discounts other possibilities. (Rehn & De Cock, 2008, p. 225).

In the context of academia particularly, such critique rightly raises the need for considerable scepticism and caution when it comes to responding to the seemingly ever increasing institutional endorsement for heightened evidence of creativity in scientific inquiry; an endorsement that commonly serves only to create the above stated limitations of an over emphasis on novelty. This is a point that Leitheiser et al. discuss in much greater depth in Chapter 2 of this collection (including from an early career perspective). More broadly, being an implicit point of concern for all contributors to this collection, it is also the reason why this book should not be read as an unqualified celebration of the presence and furthering of co-creativity within research practice. At the same time, though, as I return to later in this chapter, neither is it a reason for one to ignore nor dismiss the idea that attending to the (inherent) co-creativity of research practice matters within sustainability science research.

Running in parallel to the traditionally widespread acceptance of novelty and utility as being the defining components of creativity, common also to much of the existing creativity literature, has been an overriding concern with the individual as a unit of analysis; or more specifically still, the individual mind. Such an extended and prolonged degree of concentration on the individual has been attributed by some to a modernist preoccupation with an ideology of individualism. Reflective and reinforcing of this trend, is the fact that the vast majority of creativity scholarship derives from the field of cognitive psychology. Accordingly, discerning, modelling and defining the characteristics of a creative individual has tended to attract by far the most attention, including in particular investigation of the types of thinking and corresponding personality traits most supportive of generating creative ideas.[1]

Despite a scholarly interest in the working of the human mind continuing to dominate creativity research, there exists a growing cohort of researchers who regard the prevalence of this focal point as problematic:

> If there are some unifying features for the psychology of creativity that cut across the whole domain they unfortunately group around the more or less implicit belief that it is the individual mind doing the creating. (Glăveanu, 2014, p. 7)

Beyond the perceived over-emphasis placed on the individual, also contributing this critique, is a growing dissatisfaction with the tendency for much of the early scholarship to actively project and uphold an image of the creative individual as being that of 'the [predominantly male] genius, of eminent creators who almost singlehandedly revolutionize society and culture' (Glăveanu & Sierra, 2015, p. 345). Such a conceptualization, as Glăveanu and Sierra argue at length, 'is ultimately

[1] Attributed with stimulating much of this line of scholarship is Guilford (1950) and his work on divergent thinking. For a review of numerous models of creativity as being the property of the individual, see Kaufman and Glăveanu (2019).

used as a political tool to silence the claim to creativity and agency of the marginalised or oppressed' (p. 345).

Whilst more recent scholarship has largely rejected the idea that 'true' creativity is the property only of eminent individuals, its replacement with the idea that creativity resides 'within' all of us is not itself entirely unproblematic either (Glăveanu & Sierra, 2015). As Literat and Glăveanu (2016, p. 330) explain, for example, the danger here is that for those of a neo-liberalist persuasion especially, 'creativity becomes not only an individual trait, but an individual responsibility—everyone is required to cultivate his or her own creativity'. Upon this reading, by being conceived as something that we 'should' all be meeting and living up to, this in turn implies a failing of ourselves as individuals should it be discerned that we are not constantly striving to attain (ever more) exceptional standards of practice (see Chapter 2 for an extended discussion of this point). That this book does not inadvertently add to this pressure is something that has been afforded considerable thought during its compilation. Crucially, as I discuss further below, the call for the co-creative potential of research practice to be more actively nurtured is one that goes hand-in-hand with the importance of ongoing critical reflection and of an ethics of mutual care (Hartz-Karp & Stocker, 2013).

As a consequence of the prolonged interest in the role of the human mind, two further characteristics discernible within much creativity scholarship are a general propagation of the Cartesian mind–body divide, and a discounting (or even denial) of the fundamentally relational nature of our lives. Indeed, it was not until the latter stages of the twentieth century that the social dimension of creativity began to receive greater attention, albeit with the research foci initially limited merely to investigating the effect of environmental conditions on the creativity of an individual (see e.g., Amabile, 1996; Nickerson, 1998). As Glăveanu (2010) recounts, the limitation with much of this initial discussion of what came to be termed 'social creativity' (see also Purser & Montuori, 2000), is that by 'portraying the social as an *external environment*, a set of stimulations that facilitate or constrain the creative act', this in turn obscures the 'social roots' and 'social dynamics' of creativity (p. 83, original emphasis).

It is only in the last two decades that, in some areas of the creativity research field at least, a growing momentum can be observed, away from solely individualistic understandings of creativity and beyond binary notions of 'external' environmental influences. The alternative conceptualization of creativity being proposed is a relational and emergent one, understanding creativity as residing in collaborative forms of knowledge-practice. It is this conceptualization (and the largely complimentary variants thereof) that has proved influential in refining how creativity, and in turn also co-creativity, is understood for the purposes of framing this edited collection. Particularly noteworthy here are the contributions of Tanggaard (2012, 2015) on 'sociomaterial creativity', and the considerable body of work offered up by Vlad Glăveanu and colleagues on 'distributed creativity' (see also Sawyer & DeZutter, 2009; Sawyer, 2006, 2018; Ingold & Hallam, 2007).

Tanggaard (2012, p. 20) posits sociomaterial creativity as a means of promoting 'awareness of how different environments in everyday life do not merely [in]form creativity and create conditions for it, but also themselves represent a substantial component of creativity'. As she explains, 'creativity thus occurs when we develop our practices—not via isolated thought processes but as part of life itself' (Tanggaard, 2012, p. 22). Meanwhile, in considerably extending and deepening his own earlier conceptualization of creativity as simultaneously 'I/he/we' (Glăveanu, 2010),[2] Glăveanu (2014), puts forward the notion of distributed creativity. The term 'distributed' signifies a conceptualization of creativity '*not as a "thing" but as action in and on the world*' (p. 9 [original emphasis]), and accordingly, as being located not in the essence of an individual, but rather 'in-between people and objects' (2014, p. 9). As such, '...creative action is distributed between multiple actors, creations, places and times' (p. 2). Moreover, as Glăveanu (2014) goes on to explain, creativity is 'never simply distributed as an end state but always in the process of being distributed' (p. 9).

In accordance with both Tanggaard (2012, 2015) and Glăveanu (2014) (see also Glăveanu, 2015, Glăveanu et al., 2019; Sawyer, 2006,

---

[2] Furthering also the work of such as Negus and Pickering (2004); Potts et al. (2008); Sawyer and DeZutter (2009); Vygotsky (2004)

2018; Sawyer & DeZutter, 2009; Ingold & Hallam, 2007), creativity is understood in this book as a relational and emergent activity that is anchored in social practice. Notably, though, as Glăveanu himself is careful to point out, such an understanding is meant as 'an argument against individualism, *not* the individual' (Glăveanu, 2014, p. 9 (emphasis added)). Rather, the intention here is to move the discussion 'from the individual to the collective'—the latter including both human and non-human form—without 'losing the individual component from sight' in the process (Glăveanu, 2015, p. 191). It is by moving the discussion away from creativity as an essence, as derived purely from thought, or as a fixed state of achievement, that it becomes possible to understand it as a collaborative practice, a relational state of doing, being and becoming. Such an understanding is encapsulated well in the concept of distributed creativity.

Therefore, it is the concept of distributed creativity that directly informs the meaning attributed to co-creative research practice here (and similarly also it's potential for nurturing transformative change). At the same time, however, by conceptualizing co-creative research as a socially inclusive (and embodied) form of praxis, founded on the notion of researching *with*, and bringing co-creativity into contact with the notion of engaged scholarship, this serves to reinforce the political edge of co-creativity; a key dimension that is otherwise not always so apparent within existing scholarship on distributed creativity. The installation of a political lens, in turn, also helps to guard against any ongoing potential for propagating the misuse, or 'dark side' of creativity within the sustainability sciences—a point I return to below.

Research as Co-Creative Praxis

In this section, I argue that the value of a co-creative lens rests not in its use for categorizing individual research methods as either co-creative or not, but rather in understanding why it is possible to locate or install a potential for co-creativity in a multitude of research methods and approaches across the sustainability sciences. In doing so I not only account for the current absence of the term 'creativity' from much scientific discourse on research methods, but also why it matters that we bring it back to the fore. Indeed, notwithstanding a long and widely acknowledged relationship between scientific discovery and creative thinking (albeit propagated in part by the sustained emphasis on the eminent creative individual), the presence and role of creative practice in 'fieldwork' has tended to receive far less attention (although for some notable exceptions see e.g., Carpenter et al., 2020; Kara, 2015; Pauwels & Mannay, 2019).

The relative silence as to the role of creativity in such primary research settings in part reflects a history of science dominated by calls for objectivity and replicability. It also reflects a tendency towards conservativeness when it comes to the construction of what constitutes 'rigorous research' within any one discipline. Parallels can be drawn here with Lubart's (1998) broader reflection that 'culture encourages creativity in some situations and for some topics but discourages it for others' (p. 342). That creativity is either seldom referred to by name or explicitly promoted within standard research methods texts across most of the sciences is, perhaps, far more of a reflection of the culture of mainstream academic research practice, than it is an indicator of its actual presence and influence throughout the research process. As Kaufman and Glăveanu (2019, p. 3) put it: 'creativity is everywhere and nowhere in academia'.

Given the still very much dominant belief that scientific enquiry, as an exclusive professional domain, is ultimately dependent on the expertise of the professional scientist alone, it is perhaps predictable that even less attention has in turn been given to the idea of research as *co*-creative practice; that is, as being simultaneously creative and *collaborative* in nature. Despite some clear exceptions (e.g., PAR, transdisciplinary science), all too commonly within academia, 'the results of creativity are celebrated as more or less individual achievements' (Tanggaard, 2012, p. 21). The expertise, rigour and diligence with which an individual researcher—or even a whole team of researchers—plans, collects and analyses their data, is obviously central to the eventual standard of a research project. However, the value and significance of the findings are as much related to the contributions of the research participants, as to the researcher themselves.[3] As such, whilst the researcher commonly plays a key role in analysing, interpreting and recording findings, the eventual outputs are in many senses theirs alone in name only.[4]

If, however, in accordance with the above stated conceptualization of distributed creativity, research practice is instead approached as a relational process of 'engaging in shared creation', often with the result of conceiving something, in thought or in deed, that a researcher would not otherwise arrive at on their own (Lubart & Thornhill-Miller, 2019, p. 286), this begins to give it a very different emphasis. When research practice is perceived relationally in this manner we are all simultaneously researchers and research participants (see e.g., Chapters 3 and 16 this book). As Glăveanu et al. (2019, p. 2) summarize: 'even when working in solitude, we implicitly build on and respond to the views, knowledge, and expectations of other people'. It is therefore the need to encourage further critical reflection on the shared and collaborative nature of creativity within research practice, towards which this chapter (and this entire collection) seeks to contribute through its conceptualization of 'co-creativity'.

A rich and diverse literature already exists on the societal and scientific gains that can simultaneously be achieved through the adoption of appreciative, participatory, decolonial and action research orientated principles of engaged research (see e.g., Chapter 4, this book). Nevertheless, there remains much more to be understood about the relationship between creativity and collaboration in the furthering of transformative sustainability agendas more broadly throughout the sustainability sciences. The same argument applies equally when it comes to conceiving what constitutes a *creative* research method. Attention needs to be directed not to the properties of a method in isolation, but rather to its use in a way that encourages participants to think openly and differently. Accordingly, co-creative methods are ***not*** understood here as necessarily limited to those that are overtly recognizable as either collaborative (e.g., PAR,

---

[3] For an example of where this has been creatively acknowledged see Kinpaisby-Hill Mrs C. (2008). Taking stock of participatory geographies: Envisioning the communiversity. *Transactions of the Institute of British Geographers, 33*(3), 292–299.

[4] This detail being, of course, highly significant, as the attribution by name creates a sense of ownership and reinforces the privileging of the individual scholarly voice.

transdisciplinary science) or creative (e.g., art-based methods of enquiry) by design. Building on this opening clarification, and responding also to the second and third of the guiding research questions set out above, the next section therefore proceeds to discuss how co-creative research encounters might best be nurtured in a way that offers those involved the time and space to think differently; that is, to help generate alternative understandings of why and how things are and how they could be.

Nurturing Co-Creative Research Encounters

Kara (2015, p. 1) asserts that 'doing research is an inherently creative activity at all stages of the process'. As much as this accords with how I conceptualize creativity here, it is nevertheless important to also acknowledge that, in practice, considerable variation occurs in the ways in which creativity features, or is invoked, during periods of data collection. We can likely all, for example, recall moments in the process of data collection that, regardless of how standard the research method, have produced intensely inspiring sessions of creative and visionary thinking; similarly, we have also all likely experienced occasions of exposure to creative techniques that have failed to produce within us, or within other participants around us, anything remarkable at all (for further discussion of this see especially Chapters 12 and 13, this book).

As Axinte reflects upon (Chapter 12, this book), just because a research activity is designed to induce a creative encounter, or might widely be thought of as an overtly creative technique, this does not mean that it is experienced as such by the participants. Why is this? Understanding how the introduction of more overtly creative methods might be received and responded to by research participants forms a crucial part of preparing an approach that is capable of opening-up rather than closing-down the potential for co-creativity. What are the effects, for example, of confronting particular sets of citizens with a more overtly creative method or approach? Is there a danger that too innovative, alternative or very artful forms of creativity might alienate or act as a barrier to the participation of some in the research? How can those who wrongly perceive themselves as 'not creative' best be supported in engaging with more overtly creative research activities in a more rewarding way (for further discussion in a policy context, see also Chapter 16, this book)?

In unpacking the reasons for and effects of this variation in terms of how either more or less overtly creative methods might be received and responded to by participants, my interest here is in the possibilities that can arise from making researchers more conscious of their (potential) role as enablers of co-creativity. However, as explained above, in encouraging an increased awareness of the role and presence of co-creativity within the research process, this is not meant to imply a subsequent (nor universal) need for making its presence more explicit by way, for instance, of utilizing *only* more overtly creative research methods in sustainability science research. Creative methods can (and do) include the use of a rich array of overtly creative techniques, tools and other prompts designed to enable and encourage people to think and act differently from how they perhaps otherwise normally would in their everyday lives (see e.g., Chapters 14 and 15, this book). At the same time, though, there is also much potential for co-creative research practice to be nurtured through the use of more traditional social science research methods.

I am reminded here of a piece of advice given by the author Rebecca Giggs during a recent writing workshop. She explained that the purpose of structure is to enable you to be creative with the content. When creating an argument through a piece of written work, the role of structure is to support, not to overpower (Giggs, 2020). Beyond its value in guiding the process of academic writing, wider lessons can be drawn from this example when it comes to thinking about how best to nurture the co-creativity of others in a research setting. This includes, for instance, the context of encouraging community members to participate in research activities they might otherwise find challenging (see Chapter 5, this book for further discussion on the parallel role of care-full scholarship). As noted by Davies et al. (2013, p. 85) 'the provision of "safe" structure' enables people 'to take risks, to think creatively and critically, and to question'. It supports the establishment of a research environment in which the participants—or co-researchers—'feel that their contributions are valued' (Kligler-Vilenchik & Literat, 2018, p. 77).

Demonstrating to research participants that they are being listened to and heard on their own terms can, however, be more or less attainable with some methods as compared to others. In conceptualizing and encouraging the nurturing of co-creative research practice, prioritization is given in this collection to offering examples of whole approaches, but also individual methods, tools and techniques that embrace a high degree of openness in the shaping of how a participant is able to engage and respond (Sawyer & DeZutter, 2009; Tanggaard & Juelsbo, 2016). Another characteristic shared by all such examples is that they are centred around a desire to achieve meaningfulness and inclusivity, rather than innovativeness or originality through research encounters (see e.g., Chapters 3, 4, 5, 11, 12, 13 and 14). Prioritizing a meaningful encounter for the participant does not, of course, negate the importance of an original contribution to knowledge as remaining at the heart of academic research. Rather, it is by attending to the encounter in such a way that it creates and retains meaning and integrity for the participants, as well as for the researcher, which in turn gives rise to originality.[5]

To return once more to existing theories of creativity, in putting forward the dual ideas of distributed creativity and collaborative emergence, Sawyer and DeZutter (2009) suggest a group of collaborating individuals to be an ideal setting from which creativity can emerge. Within this setting, the most fertile conditions for its emergence are said to be where there is sufficient openness retained in the process to allow for at least a degree of 'unpredictability' and 'improvisation', as well as 'moment-to-moment contingency' (p. 82). Parallels can be drawn here with more open and unstructured forms of traditional social sciences research methods, as well as more visually and materially creative research methods (many of which are derived from the arts and humanities). Similarly, Tanggaard's (2012) call for much greater attention to be given to the sociomaterial dimensions of creativity accords well with the range of methods documented in this collection. This is true with regards to both the participatory and material components around which they are variously aligned. In the case of the latter, for example, as Tanggaard

[5] Originality, conceived in this way, in turn avoids an over reliance on its acknowledgement and legitimation as such being solely at the determination of academia.

asserts: 'a material can be regarded creative in its confrontation with people, who respond to the object's hardness, its softness, or whatever the object can do for them' (2012, p. 24). As a direct illustration of this, the creativity that emerges from areas such as deep mapping (see Chapters 11 and 12, this book), or community theatre (see Chapter 13), is no more located in the cognitive domain of a singular 'creative individual' than it is solely attributable to the creativity of humans alone—'the creativity of our imaginative reflections is inseparable from our performative engagements with the materials that surround us' (Ingold & Hallam, 2007, p. 3, cited in Tanggaard, 2012, p. 24).

In all of the above such cases, the interaction between the researcher and the researched is not adversely overly scripted from the outset, but is rather left very much open, allowed to find its own natural rhythm and thread. Along similar lines (albeit outside of a research setting), Sawyer (2018) uses the example of improv theatre to illustrate the productiveness of ambiguity in a collective encounter:

> As a result of unpredictability and ambiguity, even a performer doesn't know what his own creative action means. Only when the interaction continues does the meaning of a single action become clear. Performers trust the collective creativity of the group to determine their own action's meaning. (p. 284)

Not entirely dissimilar to the above example of improv theatre, the fact that academic research commonly requires advanced planning and careful forethought, does not need to foreclose the possibility of spontaneity (see e.g., Chapters 9, 15, and 16, this book). A retained willingness on the part of the researcher to deviate from the script, to seek ways of enlivening it (Hitchings, 2012), or on occasion to ignore the script entirely—if, in the actual moment, there is felt to be value in doing so—is an integral part of approaching primary research as an inclusive, emergent and situated form of practice (for further examples of this, see Chapters 3, 8 and 12, this book). The fact that this is rarely acknowledged, or reflected upon, in the process of securing ethical approval, says far more about the institutionalized nature of the ethical approval than the enactment of research itself.

From this perspective, then, knowing how to investigate an issue in situ is as much about the *way* in which a researcher approaches, presents and guides the performance of a particular method during a research encounter, as it is about the appropriate selection of the individual method itself. At the same time, exploring and remaining open to the possibilities of collaboration requires a deep acceptance of, but also the active making-of-room for the inherent unknown potential and richness of social interaction. Similarly, on the part of a research supervisor too, equipping the researcher with the ability (both mentally and emotionally) to accept as necessary the occasional occurrence of 'mess and stumbling' in the situated unfolding of research as an emergent social practice is crucial (Tanggaard & Juelsbo, 2016, p. 86). It is as important for their ability to 'become with' their research participants, as it is for a supervisor to 'become with' their student (Haraway, 2007).

Both the planned and the actual emergent meaning(s) and experience of a research encounter, can be equally as significant in shaping the researcher and the research participants' potential for co-creativity. This point is well evidenced by Moriggi (Chapter 5, this book) in her framing and practicing of arts-based creative methods through Appreciative Inquiry and an ethics of care. Her account helps to further illustrate, for instance, why transdisciplinary forms of scientific enquiry, when practiced as a form of care-full scholarship (Moriggi et al., 2020), offer such a strong foundation for nurturing co-creative research practice in and of itself. At the same time, it also evidences why 'slower' and more open forms of research enquiry can be particularly conducive to co-creative thinking, in which the time and space is made available for the researcher to understand from the perspective of the participant; and for the participant—if not to understand from the perspective of the researcher—to at least become more consciously self-aware of their own perspective (see also e.g., Chapters 6 and 16, this book).

Much is known about the potential negative effects of participating in scientific research and correspondingly the ethical steps that need to be adhered to in order to prevent any adverse effects. In contrast, little attention is commonly given to the positive impacts of a research encounter upon individual participants. More overtly co-creative techniques, such as photo-voice (see Chapters 9 and 13, this book) and guerrilla narrative (see Chapter 3, this book), for example, can also be understood as having the potential for much more dispersed and distributed forms of meaning making (see also Chapter 4, this book). As spaces for care-full reflection (Moriggi et al., 2020), they can enable insight into practice that in turn allows the participants—researchers and researched alike—to envision how to further enhance or otherwise change their approach and activity at a wider scale or in alternative settings. At the same time, as sites of embodied meaning making, more needs to be understood about how the visceral and material experiences of research participation themselves support deeper reflection, insight and self-awareness during, but also beyond, individual research encounters (see e.g., Chapters 5, 6 and 16, this book). The importance of co-creative research practice is in this sense by no means limited to the findings that are subsequently generated and shared by the researcher; it also retains the potential to contribute to the wider transformative goals and aspirations of sustainability science. In short, (co-)creative moments tend to be very memorable; in turn, as Pearson discusses in much greater depth (Chapter 6, this book) memorable moments possess an on-going potential to be birth places for transformative change.

Creating the time and space for research participants to reflect on their own existing practice, to bring into conscious thought that which is already known, including as a means of supporting their own self-evaluation of their actions and achievements to date, can play a direct role in shaping future actions. This is another sense in which the doing of research practice, in and of itself, retains the potential to be co-creative. Indeed, sometimes evidence of co-creative thinking and action becomes immediately apparent within research practice, whilst on other occasions its emergence is much slower and far less linear. Also relevant here is the connection drawn by Tanggaard (2012; see also Vygotsky, 2004, Wegener & Wegener, 2016) between continuity and renewal: '"ways of doing" already in the world are taken as starting points for new creations' (p. 20). As she points out, bringing about change in practice is not necessarily always a pre-mediated or even conscious act. Rather, realization and critical reflection around the fact that an action has brought about a change in conditions can often follow behind. On occasion, for example, it is only by momentarily stepping out of the daily routine to participate in, for instance, a qualitative research interview, that an individual is able to dwell on what they have been 'doing' and acknowledge for themselves what has (or has not) thus far been achieved (see, e.g., Chapters 8, 13 and 16, this book, for further discussion of this point in connection with specific individual methods).

Somewhat ironically, the above avocation for research methods that favour relatively low levels of structure and high levels of openness, stands in marked contrast to the ways in which the subject matter of creativity has itself predominantly been investigated. For many scholars of creativity, as a reflection of its strong disciplinary base in psychology, rigorous research is largely framed by the need for 'control and representativeness' (Mayer, 1998, p. 456). The result has been a dominance of psychometric and experimental research approaches, together with a preference for quantitative forms of research and analysis. Presumably, however, as and when a distributed understanding of creativity comes to be accepted more widely within this field, so too might the value and robustness of more qualitative, ethnographic and transdisciplinary forms of research practice.

To summarize then, nurturing creativity within research can involve both tangible and intangible elements of creative design. Similarly, creativity can be implicit or explicit in what is asked of participants, and research exercises can range from activities that are relatively mundane to those that might be thought of as far more extraordinaire. As such, co-creativity can either be an integral component of a traditional research method, or it can require that such a method be turned on its head and turned inside out. Ultimately, though, whilst some methods may lend themselves more easily to advancing the co-creative potential of research practice, the realization of this potential resides as much in the process of *doing* research and *being* engaged, as in the actual components of a particular method. When practiced reflexively and sensitively, a vast array of research methods can (and do) invoke co-creative thought and action. As noted above, but worthy of repeating: it is because of this framing of co-creativity that we do not limit ourselves in this collection to restricting the categorization of methods as creative to those that are more overtly recognizable as such by design. Nor do we attempt to construct a list of which methods are, or are not, co-creative in their composition. Doing so would be misguided at best.[6] Rather what we are most interested in here is nurturing and enabling the realization of co-creativity through the very potential of the research encounter.

Co-creativity and the Transformative Potential of Sustainability Science

> If engaged scholars are to transform the social world for the purposes of equality, they need to be examining questions and concerns that are directly relevant to the everyday lived experiences of excluded individuals and communities, questions which emerge from their own ontological understandings of what it means to be in the world. This is the basis from which changes in the structures, systems, and relations that underpin social exclusion can emerge. Thus, the focus is not just on what is known and what scholars and communities can know together, but also who they can become together. (Kajner, 2013, p. 16)

Akin to Kajner's (2013) above call with respect to engaged scholarship, if an ambition of the sustainability sciences is to make the world a better place, socially, economically and environmentally, then opportunities for increasing the contribution of academic research need to be sought, including through the pursuance of adaptive and transformative change. An underlying aim of this book is to advance understanding of the role and potential of co-creative research practice in furthering such agendas. What has been referred to elsewhere as the contemporary obsession with creativity can, however, on occasion, create an environment in which individuals feel under considerable pressure to act creatively (Weiner, 2000). This, and a wider context referred to by Rehn and De Cock (2008) as an era in which 'creativity has been corralled into the service of both big business and the nation state' (p. 229) (see above; see also Chapter 2, this book), raises the question of how to go about nurturing

[6] Including, also, for the fact that what might be received as highly creative in some disciplines may be viewed as entirely standard or even mundane in others.

and attending to co-creativity in the sustainability sciences without inadvertently suppressing the very voices (human and non-human) most in need of being heard.

Alongside the challenge of retaining integrity of research practice in the face of adverse external pressures to 'be' creative, there is, of course, also a 'dark side' of creativity that also needs to be acknowledged. Much has already been written about this in relation to both the 'creative individual' and the disastrous societal, environmental and economic impact of some creative ideas and actions (see, e.g., Cropley et al., 2010; Glăveanu et al., 2019; Chapter 2, this book). There is no doubt far more to be understood and discussed in this context with regards to the misappropriation also of co-creative research methods. Whilst the analysis of specific such examples and cases falls without the scope of this chapter, in this penultimate section I nevertheless take the opportunity to make some concluding comments about why it matters that the presence and potential of co-creativity is critically reflected upon within sustainability research especially. In short, I am driven here by the conviction that, despite its risk of being (further) co-opted by elite interests, this does not and cannot negate the potential contribution of co-creativity to achieving more just and empowering forms of transformative change (Carpenter et al., 2020). For this potential to be realized, however, it has to be practiced in a mutually ethical and integral way (Moran et al., 2014), or as Moriggi argues (Chapter 5, this book), it needs care-full scholarship.

Creativity, when understood as distributed, requires of us that we 'continuously construct ways of connecting with others and understanding them, including as a means of understanding ourselves' (Lebuda & Glăveanu, 2019, p. 4). Participatory and transdisciplinary forms of action research inquiry can be particularly effective in facilitating such mutuality of co-creative thinking between self and others through the very process of doing research. Seemingly pivotal to their effectiveness is the importance they attach to dialogue (Giri, 2002; see also Chapter 3, this book). In direct accordance with a conceptualization of creativity as distributed, a greater emphasis on dialogue, as a means of better understanding the epistemologies and ontologies of one another and of others, would surely be of benefit to all forms of sustainability science research (Lin, 2011).

It is through a shared understanding and appreciation for one another's differences in expertise, in perception and in life experience, that we are in turn able to advance our understanding about what matters and for whom, about how things are and for whom, about the reasons for why this may be the case, and about how they might be changed for the better (and of course, better for whom). Enabling others, and ourselves, to become better at thinking differently, thinking passionately about *how things could be*, rather than merely how they are and to what effect, is arguably crucial to making a positive difference in the world. Indeed, much of the importance attached in this collection to furthering our understanding of the co-creative potential of engaged scholarship, is motivated by the need to bring alternative ways of knowing and practicing into being. Towards this very goal and given also that all academic researchers have 'a social responsibility', 'using the concept of *creativity critically and reflectively* is [thus] crucial' (Glăveanu et al., 2019, p. 4 (emphasis added)). Moreover, this pairing of creativity and critical thinking ensures not only that we remain critical in our thinking about creativity, but also that we can advance our critical thinking *through* (co-)creativity. As Nickerson (1998), for example, makes clear, despite the widespread, erroneous tendency for creative and critical thinking to be contrasted as opposites to one another, they are in fact 'two sides of the same coin. Good thinking requires both and requires that there be a balance between their contributions' (p. 399); so too does engaged scholarship (MacKinnon, 2010).

Therefore, not only does this book aim to enable researchers to find co-creative ways of better understanding and interpreting what is going on around them and why this may be so, it is simultaneously also crafted towards assisting researchers to become more advanced in prompting others to think more critically, more creatively and more reflexively about their relationship with those around them (be they near or far; human or non-human). In follow-on, it asks about what actions can be taken to enhance or change that relationship in a forward looking manner. Indeed, when transformative change is framed in such a way, the dependence on this pairing of creative and critical thinking becomes even more obvious. As both Vygotsky (2004) and Vadeboncoeur et al. (2016) explain, whilst 'it is precisely human creative activity that makes the human being a creature oriented toward the future, creating the future and thus altering his own present' (Vygotsky, 2004, p. 9); simultaneously, 'it is creativity, and the human ability to see and act "other than" or "as if", to challenge and to question, that assists the creation of new practices along with the values that support them enabling the dialectic between continuity and change to become cultural transformation' (Vadeboncoeur et al., 2016, p. 300; see also e.g., Chapter 6, this book).

To conclude then, the ambition of 'making a difference' through co-creative research endeavour requires simultaneously both a clear strategy and a retained openness for the unknown and unforeseen. This in turn calls for an emphasis on iteration rather than linearity in the research process, a prioritization of dialogue, a sustained pursuit of self-reflexivity, an embracing of 'emotional, embodied and intuitive forms of knowing' (Shrivastava & Ivanaj, 2011, p. 84), and simultaneously, an altogether greater recognition of the relational nature of research practice. The latter not only attests to the importance of caring for and *with* others, but also of self-care (Tronto, 1993). Indeed, to close this introduction by indulging in what Wegener and Wegener (2016) might term an act of creative mirroring (i.e., creativity as building on the richness of what already exists), I find it stimulating to end here by drawing once more from a feminist ethics of care perspective that implicitly underpins much of this discussion and combine it with an equally powerful lesson drawn from agroecology—a field of sustainability science that arguably leads the way in achieving transformative change. What they lead me to conclude is that, in realizing its potential to contribute to making the world a better place, co-creative research practice needs to be nurtured in a way that is 'political in its perspective and dialogical in its method' (Donovan, 2006, p. 324; see also Kajner, 2013; Hartz-Karp & Stocker, 2013). By doing so, the (critical) pursuit of co-creative research in turn provides further momentum towards establishing sustainability science itself as simultaneously a science, a movement and a practice (Wezel et al., 2009). It is by adapting and pursuing sustainability science as such, that it stands the greatest chance of making a difference. I hand over, in a moment, to the contributing authors to propose and further explain some of the ways in which such guiding principles and ambitions might be taken up.

Introducing Co-Creative Research in Practice

The remainder of this book comprises of predominantly first-hand accounts of research approaches, tools and techniques that are centred around openly (and actively) collaborative and creative forms of research inquiry. In some chapters, this comes with more of an emphasis on widening collaborative practice; in others, greater emphasis is placed on overtly stimulating creative practice. In all cases, though, whether involving explicitly or implicitly collaborative and creative techniques, the approaches and methods presented are understood as nurturing co-creative research practice, due to the inherently open, respectful and relational nature of their form.

In accordance with the aims and motivation guiding this book, all lead authors of the chapters that follow are early career (doctoral or post-doctoral) researchers. More specifically they are all advocates for —and actively engaging with—transdisciplinary and participatory methodologies, and are all working within the (social) sustainability sciences. Serving to further unite them as a community of practice at the time of writing is their involvement (either directly or indirectly) with a four-year H2020 *Marie*-Skłodowska *Curie* Innovative Training Network (MSCA-ITN) entitled RECOMS. Founded on principles of transdisciplinary science, the RECOMS consortium of early career research fellows, academic mentors and expert practitioners has a shared goal of advancing 'resourceful and resilient community environmental practice'.

A cross-cutting pedagogical theme of RECOMS is visual and creative research methods. That this project ever came to be formulated around this theme is testament to a wish to enable and encourage early career social sustainability scientists to experiment with more overtly creative research techniques. Creating the time, space and resources for such experimentation in a research setting that is centred around an international network of fifteen doctoral research projects (incorporating a relatively diverse range of empirical research settings and research questions) provides a major opportunity for critical reflection on the relationship between co-creativity, research practice and transformative change. This book is one of the outputs derived from this opportunity. I end this opening chapter with a brief introduction on each of the chapters that follow, and a note of explanation on how they connect to and interrelate with one another.

Chapter 2 serves to further introduce and provide a critical contextualization for the collection. Leitheiser et al. approach the uptake of creative methods from a starting point of neo-liberalization and the on-going trend for the corporate managerialism of university research. Beginning with an example from the dark side of creativity, they discuss how the creativity of scientific endeavour is simultaneously shaped and mediated by both individuals and institutions. In the case of individuals working in managerialist universities, however, they argue that the possibility of pursuing co-creative research praxis is commonly highly constrained and at constant risk of co-optation. Supported by examples, they explain why creative methods, as a collaborative form of research practice, have potential to reinforce or to subvert the relegation of universities to mere 'factories of knowledge production'. In offering a very critical analysis, and warning against 'forced creativity', they call upon academics to collectively reflect, with eyes wide open, on 'the possibilities for action' in order that creative methods might actually support, not merely further undermine, just transformative change.

Chapters 3, 4 and 5 are united in their emphasis on the significance of positionality in research and the need for conducting research in a way which directly challenges, rather than reinforces, imbalances in power. In **Chapter** 3, Ruiz Cayuela and Armiero consider their own positionality as *militant researchers* and the foundation that 'all knowledge production is partisan'. Having explained why it is essential that academic research challenges the homogenizing discourses of the elite, they then proceed to focus on the power of narrative and its potential for supporting more inclusive, transformative, and counter-hegemonic practices of research and knowledge exchange. They dedicate the remainder of the chapter to introducing and reviewing the use of *guerrilla narrative* as a mechanism for undermining the grip of capitalism by spreading 'commoning subjectivities' within marginalized communities. In doing so they offer a range of illustrations from their work with Co-operation Birmingham, a mutual aid organization located in the west midlands of the UK. Like Ruiz Cayuela and Amiero themselves, Co-operation Birmingham is committed to igniting social transformation through the disruptive practicing of co-operation, solidarity, horizontality and care, from the ground up.

In **Chapter** 4, De la Rosa Solano et al. explore the presence and uptake of *decolonial participative approaches* in the environmental humanities. Their analysis reaffirms the validity of people's knowledge in the construction of historical narratives. It also illustrates the just and transformative potential of environmental history research when practiced more inclusively, in conformity with the principles of *decolonialism* and *Participatory Action Research*. Drawing on a series of examples from the literature of Latin America, De la Rosa Solano et al. discuss at length the value of applying a decolonial lens to the environmental humanities. Such an approach, they argue, supports not only a better understanding of our relationship with the non-human, but also one's own positionality within the research process. Taking as a case in point the role of memory as an historical resource, De la Rosa Solano et al. explain why the centring of a decolonial lens achieves a more fair and inclusive process of knowledge generation. This, in turn, they conclude, further strengthens the contribution of environmental history to achieving societal transformation.

In **Chapter** 5 the geographical setting of just research practices moves from Latin America to northern Europe. Here, Moriggi explores the transformative potential of co-creative research methods by approaching them from a starting point of *Appreciative Inquiry* and an *ethics of care*. Taking the case of Green Care, she offers an extended reflection of her own doctoral research, undertaken in Finland, on nature-based activities with a social-innovation purpose. The Participatory Action Research approach employed by Moriggi enabled her to collaborate with three different communities of green care practitioners in a manner that paid equal attention to care-full research and to 'creativity, innovation and imagination as forms of knowledge production'. In offering a first-hand critical account of her use of five different kinds of creative and arts-based methods with members of these communities over an extended period of time, Moriggi illustrates how an 'ethos of appreciation' can be 'embodied and applied in practice' and to what effect.

Chapters 6, 7 and 8 all draw our attention to the connection between co-creative methods and arts-based research. In **Chapter** 6, beginning with the role of worldviews, interests, values and ideologies in shaping human behaviour, Pearson draws our attention to the potential contribution of artistic processes as a mechanism for triggering transformative change. Focusing in on these 'inner dimensions' of sustainability, she explores the transformative capacity of arts-based creative methods when it comes to supporting '*imaginative leadership*' and '*transformative imagination*' in the arena of sustainability. That is, how they might be used to 'provoke and strengthen' more environmentally conscientious '*transformative mindsets*' through collaborative experimentation. The discussion is supported by a detailed account of two cases in which Pearson took a lead role in co-designing and implementing creative methods workshops with a range of different stakeholders. Orientated towards the more-than-human, the aim of the workshops was to stimulate deep self-reflection, as a means of opening up 'new spaces of possibility for action and perception'.

In **Chapter** 7, Van der Vaart begins with the question of 'how communities can be prepared—or prepare themselves—for a more sustainable future?'. In response, she reviews the opportunities that are created by bringing together science, arts and society, as part of a place-based transdisciplinary approach to enacting change. The chapter takes as its empirical focus a *community arts* project in the Netherlands, Grutte Pier, which was initiated by a social enterprise (PeerGrouP) specializing in the use of arts based participatory methods. In contrast to many of the other chapters in this collection, here co-creative methods therefore become the 'object' of the research enquiry. Drawing on data from in-depth interviews with project participants, but also with local residents who chose not to engage with the project, Van der Vaart analyses the impact that such arts-based initiatives can have in promoting resourceful and resilient community environmental practice. Woven throughout this analysis, however, is a critical awareness of the inherent tensions and risks associated with measuring the societal value of arts-based practice.

In **Chapter** 8, in contrast to the approach taken by Van Der Vaart, Davis et al.'s discussion is centred around a first-hand reflective account of working *with* a community as a creative practitioner. With the 'coming together' of academic researchers and art-based practitioners increasingly encouraged, this chapter is motivated by a perceived need for further critical methodological reflection on this pairing from a starting point of Participatory Action Research and the '*art of invitation*'. Drawing on a case study from the north of the Netherlands, Davies et al. collectively reflect on the challenges of working within the conditions of an 'invited space'. Notably, this includes problematizing how to engage with, understand, and respond to the needs of a community on their own terms. They discuss the risk of instrumentalizing creative methods in accordance with external interests and the challenge that this presents, including with respect to their own personal ethics and integrity. Despite the inherent difficulties of coming in as an 'outsider', the chapters of Van Der Vaart and Davies et al. both nevertheless also attest to the opportunities that this status can sometimes bring. As Davies et al. explain, realizing this potential is ultimately dependent on the sensitivity and reflexivity with which arts-based co-creativity is practiced.

Whilst arts-based methods continue to feature in **Chapters** 9 and 10, here the focus is more firmly on *visual methods*. In **Chapter** 9, Leung begins by explaining her motivation to use the technique of photo-elicitation in order to understand meaning making 'beyond word-based cognitive reflexivity'. In reflecting on her experience, she first focuses on using *photo-elicitation* as a means of softening the representational challenges of working in a cross-cultural setting within rural Japan, with a translator; and second, in response to a situation whereby her respondents—Japanese rice farmers—were neither well versed nor comfortable with providing extended amounts of verbal reflection. Notably, the photos in question are of rural artworks emplaced within the local farming landscape. Integral to Leung's account is the acknowledgement she gives to the importance, but also the difficulty, of interpreting how her respondents de-coded photographs in a way capable of encompassing 'the marked differences between Asian and Western cultures in the relations made between elements in an image'. In extending her exploration of such representational and cross-cultural challenges a step further, she also experiments with commissioning a local artist to create visual illustrations of oral quotations as an alternative form of translation. Leung's chapter therefore offers an openly critical account of the strengths and weaknesses of using creative techniques as a means of enabling cross-cultural research practice. In parallel, it also encompasses the affective dimension of *doing* cross-cultural research.

In **Chapter** 10, the problematic of visual representation is also the focal point for Baimukhamedova, this time engaged in respect to media depictions of wild animals. Drawing on the example of Eurasian Lynx, Baimukhamedova traces the historical development of human–wildlife relationships within the Bavarian Forest region of Germany. In doing so, she offers a rich reflective account of applying *visual analysis* techniques to twentieth-century media publications as a means of understanding the affective quality of wildlife images. Beyond considering the overall importance of visual imagery when it comes to understanding societal relationships with the more-than-human, Baimukhamedova's account also attests to the need for visual material to be more closely attended to within social sustainability research more broadly. As she notes, whilst the applicability of visual analysis depends on the kind of research question one wishes to ask—some answers might well be found in the visual. Moreover, in learning to practice visual analysis, one is also propelled to reflect more intently on the positionality of their own gaze.

In **Chapters** 11–14, co-creative research practice is explored in the context of *critical cartography*. In **Chapter** 11, presenting creative map-making as 'transdisciplinary and conceptually boundless', Reitz explains how they enable the coming into being of alternative forms of 'sensing, representing and relating to space'. She evidences these assertions by offering a detailed account of two creative mapping methods: *deep mapping* and *social cartography*. In reviewing these two methods, Reitz discusses how they attempt to represent the complexity and open endedness of space in a way that is not possible with traditional cartographic methods. Notably, as she highlights, they bring to the fore a need to continuously ask of ourselves questions such as 'which ideas, senses, and values are included or excluded in the mapping process? Who is heard and who is silenced? What purpose does the map serve and which transformations can it unravel?' Beyond reminding us of the 'partial, subjective nature of map making', such questions demonstrate the revealing power of co-creative methods; a power that also extends to the role of methods such as deep mapping and social cartography in furthering the inclusiveness of social sustainability science research.

Deep mapping features again in **Chapter** 12, this time as part of a first-hand critical account of its use within an ex-industrial area of the Glasgow docklands (Scotland). Authored by Humphris et al., this chapter uses deep mapping to instigate an evocative conversation about the legitimacy of marginalized uses of 'discarded' places. Making optimal use of the 'thick description' that deep mapping affords, they weave together the rich layering of informal users and uses of their case study site—a site that, within traditional urban cartography, would otherwise be depicted as a vacant and derelict space. In doing so, Humphris et al. evidence and account for why co-creative arts-based practices, such as deep mapping, constitute valuable investigative tools. Through both their written and visual analysis, they demonstrate how deep maps can bring attention to place-making to better understand spaces, reshape relationships and support communities. However, in emphasizing the political nature of deep mapping, Humphris et al. simultaneously draw attention to the *process* of deep mapping as requiring a collaborative and reflexive cycle of research, dialogue, learning and action. It is through such a co-creative approach that the possibility of giving voice to 'marginalized micro-narratives' can best be realized. As Humphris' own personal critical reflection also shows, though, such a process must on no account be approached as unproblematic in and of itself.

In **Chapter** 13, whilst creative mapping remains at the heart of the research approach, here Anxite considers its potential as a digital method, in combination with *photo-voice*, for engaging young people in the planning of city-regions. Her chapter draws on doctoral research undertaken with young people residing in south-east Wales (the 'Cardiff city-region'). Notably, this chapter demonstrates how co-creative methods can used as a means for stimulating dialogue around issues and concepts with which research participants may not otherwise be familiar, or appreciative of the relevance that they hold to their everyday (and future) lives and sense of place. Moreover, in electing to work with web-mapping and photo-voice, Anxite demonstrates how to stimulate such dialogue in a way that enables young people's views and aspirations for the future to be presented in a necessarily disruptive manner. At the same time, however,

Anxite's account gives full acknowledgement to the limitations of what can be achieved—even with co-creative methods—when working in contexts of entrenched relationships of power and top-down decision-making. She also offers constructive reflections on the challenges of using visual and creative methods in group settings when working as a lone academic researcher.

In **Chapter** 14 Ramirez Aranda and Vezzoni further extend the discussion of digital participatory mapping that was initiated by Anxite in **Chapter** 13. In exploring the possibilities that digital tools offer for the achievement of 'more democratic and inclusive participation processes', they specifically consider the use of participatory mapping web apps as a means for facilitating decision-making around the planning, public use and protection of greenspace both locally and from afar. By way of illustration, they offer a technical account of the co-creation of two innovative online participatory GIS platforms ('My Green Place' (Belgium) and 'Greenmapper' (the Netherlands)). Whilst the examples given clearly demonstrate the potential value of digital methods in widening participation and challenging existing forms of accountable decision-making (both locally and at a distance), Ramirez Aranda and Vezzoni end with a series of critical reflections drawn from their first-hand involvement with these two cases. Notably, this includes guarding against the risk of digital participatory methods being taken up as a means of 'managing discontent through "artwashing"'. Whilst capable of making a strong contribution to transdisciplinary research, Ramirez Aranda and Vezzoni firmly categorize participatory web apps as a 'complementary tool', not a 'cure all'.

In **Chapter** 15 the importance of promoting co-creation as a basis for urban planning and policy decision-making is further considered in the context of grand challenges and 'wicked' environmental problems. Here, however, the focus for Rădulescu et al. is on Living Labs as a creative and collaborative planning method. Rădulescu et al. begin by offering a typology of *Living Labs,* based on an extensive review of the various ways in which they are interpreted and characterized within the current literature. Drawing also then on their own first-hand experience of involvement, they critically consider their potential for supporting the advancement of participatory practices within the domain of water infrastructure and spatial planning. Establishing planning as a fundamentally collaborative and inclusive form of practice, they argue, is essential to the future safeguarding of local environmental resources. Supported by a series of practice-orientated examples from three (past and present) water infrastructure projects in the Netherlands, the authors end by offering a series of recommendations for optimizing their use as a participatory planning tool.

In seeking to try and 'make a difference' through their research, commonly social sustainability scientists seek out the involvement of community groups, social movements, NGOs and/or marginalized or disempowered individuals as research partners. Less common is the active selection of private businesses or governmental institutions. As this collection evidences, though, there is sometimes as much to be gained from working with those who are otherwise depicted as being part of 'the problem'. In **Chapter** 16, Giambartolomei et al. offer an extended reflection on the dual challenges and rewards of engaging in transdisciplinary science where it involves a government institution, and where it constitutes the central pivot of a doctoral degree. At the same time, however, they argue for the potentially profound value of such a coupling when it comes to making a difference with one's research. Key here is the opportunity that such 'spaces in-between' present for outing the emotional and embodied dimensions of collaborative 'doing' *transdisciplinary research*. Drawing on Giambartolomei's first-hand experience of working with the Welsh Government, and supported further by a conceptual lens of care, the authors explain how more meaningful relationships between academics and policy-makers might be established and nurtured. In particular, they discuss the opportunities that transdisciplinary science creates for reinvigorating reflexive forms of governance and, in turn (in their research case at least), a willingness by governmental institutions to trial more (co-)creative and care-full forms of natural resources management.

The relationship between research and policy decision-making continues to be the main point of orientation in **Chapter** 17. In this final chapter of the collection, Zolyomi addresses the issue of how to make policy-makers take notice of, engage with, and act upon, the results of sustainability science research. Taking as a principal illustration the case of biodiversity loss, she pays particular attention to the contribution of creative methods when it comes to achieving *policy impact*. The discussion is supported by a review of existing literature and first-hand experience on how to go about communicating research findings at the European Union level. The experience is derived from working with a conservation and advocacy NGO. Zolyomi's account remains sensitive to the fact that very often a lack of financial resources further heightens the challenges that have to be confronted by researchers in order to be heard by policy-makers. Accordingly, alongside the role of creative methods she also pays close attention to the importance of message framing and to the channels of delivery. Zolyomi's work is driven by the conviction that understanding how best to communicate research to policy-makers is 'pivotal for a more sustainable future'. It is with this conviction and her accompanying reflections on how best to craft impactful messages that this edited collection is drawn to a close.

Acknowledgements This research received funding from the European Union's Horizon 2020 research and innovation programme under the Marie Sklowdoska-Curie Grant Agreement No. 765389.

References

Amabile, T. M. (1996). *Creativity in context*. Westview.

Boyer, E. (1996). The scholarship of engagement. *Journal of Public Service and Outreach, 1*(1), 11–20.

Boyer, E. (1990). *Scholarship reconsidered: Priorities of the professoriate*. Carnegie Foundation for the Advancement of Teaching

Carpenter, J., Horvath, C., & Spencer, B. (2020). Co-creation as an agonistic practice in the favela of Santa Marta, Rio de Janeiro. *Urban Studies*, pp. 1–18

Chan, J. (2016). Creativity and culture: A sociological perspective, chapter 31. In V. P. Glăveanu (Ed.), Th*e Palgrave handbook of creativity and culture research*, Palgrave studies in creativity and culture (pp. 639–660).

Cowley, N. (2013). What is praxis? Discussed in relation to Hegel, Marx, Nietzsche and Sartre. *Graduate and Post-Graduate E-Journal Volume 4*, pp. 1–8 (accessed April 28th 2021). https://www.waikato.ac.nz/__data/assets/pdf_file/0005/149261/NatalieCowley.pdf

Cropley, D. H., Cropley, A. J., Kaufman, J. C., & Runco, M. A. (Eds.). (2010). *The dark side of creativity*. Cambridge University Press.

Davies, D., Jindal-Snape, D., Collier, C., Digby, R., Hay, P., & Howe, A. (2013). Creative learning environments in education—A systematic literature review. *Thinking Skills and Creativity, 8*, 80–91

Donovan, J. (2006). Feminism and the treatment of animals: From care to dialogue. *Signs, 31*(2), (Winter), 305–329.

Eubanks, V. (2012). Feminist phronesis and technologiesof citizenship. Chapter 11 In B. Flyvbierg, T. Landman, & S. Schram (Eds.), *Real social science: Applied phronesis* (pp. 228–245). Cambridge University Press.

Giggs, R (2020, July 29). *RECOMS training event 5—Writing workshop* (Online training event).

Giri, A. K. (2002). The calling of a creative transdisicplinarity. *Futures, 32*, 103–115.

Glăveanu, V. P. (2010). Paradigms in the study of creativity: Introducing the perspective of cultural psychology. *New Ideas in Psychology, 28*(1), 79–93.

Glăveanu, V. P. (2014). *Distributed creativity: Thinking outside the box of the creative individual*. Springer.

Glăveanu, V. P. (2015). Creativity as a social act. *Journal of Creative Behavior, 49*, 182–197.

Glăveanu, V. P., Hanson, M. H., Baer, J., Barbot, B., Clapp, E. P., Corazza, G. E., Hennessey, B., Kaufman, J. C., Lebuda, I., Lubart, T., Monuori, A., Ness, I. J., Plucker, J., Reiter-Palmon, R., Sierra, Z., Simonton, D. K., Neves-Pereira, M. S., & Sternberg, R. J. (2019). Advancing creativity theory and research: A socio-cultural manifesto. *The Journal of Creative Behavior, 0*(0), 1–5

Glăveanu, V. P., & Sierra, Z. (2015). Creativity and epistemologies of the South. *Culture & Psychology, 21*(3), 340–358

Guilford, J. P. (1950). Creativity. *American Psychologist, 5*, 444–454.

Hannah, D., & Jeremijenko, N. (2017). Natalie Jeremijenko's new experimentalism. In R. Gruisin (Ed.), *Anthropocene feminism,* chapter 9 (pp. 197–219). University of Minnesota Press.

Haraway, D. J. (2007). *When species meet. Post humanities series* (Vol. 3). University of Minnesota Press.

Helfand, M., Kaufman, J. C., & Beghetto, R. A. (2016). The four-C model of creativity: Culture and context, chapter 2. In V. P. Glăveanu (Ed.),

The *Palgrave handbook of creativity and culture research,* Palgrave studies in creativity and culture (pp. 15–36).
Hitchings, R (2012). People can talk about their practices. *Area,* 44(1), pp. 61–67.
Ingold, T., & Hallam, E. (2007, red.). Creativity and cultural improvisation: An introduction. In E. Hallam & T. Ingold (Eds.), *Creativity and cultural improvisation* (pp. 1–24). Berg.
Hartz-Karp, J., & Stocker, L. (2013). Deliberative democracy, a collaborative action oriented learning process for a more sustainable future, chapter 9. In L. Shultz & T. Kajner (Eds.), *Engaged scholarship: The politics of engagement and disengagement* (pp. 121–138). Sense Publishers.
Kajner. (2013). Beyond the binary Chapter 2. In L. Shultz & T. Kajner (Eds.), *Engaged scholarship: The politics of engagement and disengagement* (pp. 9–20). Sense Publishers.
Kara, H. (2015). *Creative research methods in the social sciences: A practical guide* Policy Press.
Kates, R. W. (2011). What kind of a science is sustainability science? *Commentary PNAS, 108*(9), 19449–19450.
Kaufman, J. C. & Glăveanu, V. P. (2019). A review of creativity theories: What questions are we trying to answer? In J. C. Kaufman & R. J. Sternberg (Eds.), *Cambridge handbook of creativity* (2nd ed.) (pp. 27–43). Cambridge University Press. Accessed via https://www.researchgate.net/publication/334749545_A_review_of_creativity_theories_What_questions_are_we_trying_to_answer
Kligler-Vilenchik, N., & Literat, I. (2018). Distributed creativity as political expression: Youth responses to the 2016 U.S Presidential election in online affinity networks. *Journal of Communication, 68*, 75–97.
Kinpaisby-Hill, M. C. (2008). Taking stock of participatory geographies: Envisioning the communiversity. *Transactions of the Institute of British Geographers, 33*(3), 292–299.
Lebuda, I., & Glăveanu, V. P. (2019). Re/searching the social in creativity, past, present and future: An introduction to the Palgrave handbook of social creativity research, chapter 1. In I. Lebuda & V. P. Glăveanu (Eds.), *Palgrave handbook of social creativity research*, Palgrave studies in creativity and culture (pp. 1–9). Palgrave.
Liep, J. (2001). Introduction. In J. Liep (Ed.), *Locating cultural creativity* (pp. 1–13). Pluto Press.
Lin, Y.-S. (2011). Fostering creativity through education—A conceptual framework of creative pedagogy. *Creative Education, 2*(3), 149–155.

Literat, I., & Glăveanu, V. P. (2016) Same but different? Distributed creativity in the internet age. *Creativity. Theories—Research—Applications, 3*(2), 330–342

Lubart, T. I. (1998). Creativity across cultures, chapter 17. In R. J. Sternberg (Ed.), *Handbook of creativity* (pp. 339–350). Cambridge University Press.

Lubart, T., & Thornhill-Miller, B. (2019). Creativity: An overview of the 7C's of creative thought, chapter 15. In R. Sternberg & J. Funke (Eds.), *The psychology of human thought: An introduction* (pp. 277–305). Heidelberg University Publishing.

MacKinnon, C. (2010). Engaged scholarship as method and vocation Catharine. *Yale Journal of Law and Feminism, 22*(2), 193–205. https://digitalcommons.law.yale.edu/yjlf/vol22/iss2/2

Mayer, R. E. (1998). Fifty years of creativity research, chapter 22. In R. J. Sternberg, (Ed.), *Handbook of creativity* (pp. 449–460). Cambridge University Press.

Miller, T. R., Wiek, A., Sarewitz, D., Robinson, J., Olsson, L., Kriebel, D., & Loorbach, D. (2014). The future of sustainability science: A solutions-oriented research agenda. *Sustainability Science, 9*, 239–246.

Moran, S., Cropley, D., & Kaufman, J. (Eds.). (2014). *The ethics of creativity*. Palgrave.

Moriggi, A., Soini, K., Franklin, A., & Dirk, R. (2020). A care-based approach to transformative change: Ethically-informed practices, relational response-ability, and emotional awareness. *Ethics, Policy and Environment, 23*(3), 281–298.

Negus, K., & Pickering, M. (2004). *Creativity, communication and cultural value*. London: Sage Publications

Nickerson, R. S. (1998). Enhancing creativity, chapter 20. In R. J. Sternberg (Ed.), *Handbook of creativity* (pp. 392–430). Cambridge University Press.

Pauwels, L., & Mannay, D. (Eds.). (2019). *The SAGE handbook of visual research methods* (2nd ed.). Sage.

Potts, J., Hartley, J., Banks, J., Burgess. J., Cobcroft, R., Cunningham, S., & Montgomery, L. (2008). Consumer co-creation and situated creativity. *Industry and Innovation, 5*(5), 459–474.

Purser, R., & Montuori, A. (2000). *In search of creativity: Beyond individualism and collectivism*. Paper presented at the Western Academy of Management Conference, Kona, Hawaii.

Rehn, A., & De Cock, C. (2008). Deconstructing creativity, chapter 18. In T. Rickards, M. A. Runco & S. Moger (Eds.),*The Routledge companion to creativity*. Routledge.

Runco, M. A., & Jaeger, G. J. (2012). The standard definition of creativity. *Creativity Research Journal, 24*(1), 92–96

Sandercock, L., & Atilli, G. (2012). (Un)settling a settler society: Film, phronesis and collaborative planning in small-town Canada, chapter 8. In B. Flyvbierg, T. Landman, & S. Schram (Eds.), *Real social science: Applied phronesis* (pp. 137–166). Cambridge University Press.

Sawyer, R. K. (2018). An interdisciplinary study of group creativity, chapter 18. In R. J. Sternberg (Ed.), *Handbook of creativity* (pp. 280–290). Cambridge University Press: Cambridge

Sawyer, R. K. (2006). *Explaining creativity: The science of human innovation.* Oxford University Press.

Sawyer, R. K., & DeZutter, S. (2009). Distributed creativity: How collective creations emerge from collaboration. *Psychology of Aesthetics, Creativity, and the Arts, 3*(2), 81–92.

Shrivastava, P., & Ivanaj, S. (2011). Transdisciplinary art, technology, and management for sustainable enterprise . *Transdisciplinary Journal of Engineering & Science, 2*, 81–92.

Shultz, L., & Kajner, T. (2013). *Engaged scholarship: The politics of engagement and disengagement*. Sense Publishers.

Tanggaard, L. (2015). Pathways, chapter 12. In V. Glăveanu, L. Tanggaard & C. Wegener (Eds.), *Creativity: A new vocabulary* (pp. 96–103). Palgrave Macmillian.

Tanggaard, L. (2012). The sociomateriality of creativity in everyday life. *Culture & Psychology, 19*(1), 20–32.

Tanggaard, L., & Juelsbo, T. (2016). Mess, chapter 10. In V. Glăveanu, L. Tanggaard & C. Wegener (Eds.), *Creativity: A new vocabulary* (pp. 78–86). Palgrave Macmillian.

Tronto, J. C. (1993). *Moral boundaries.* Routledge.

Vadeboncoeur, J. A., Perone, A., & Panina-Beard, N. (2016). Creativity as a practice of freedom: Imaginative play, moral imagination, and the production of culture, chapter 16. In V. P. Glăveanu (Ed.), Th*e Palgrave handbook of creativity and culture research,* Palgrave studies in creativity and culture (pp. 285–305).

Vallance, S., Perkins, H. C., & Dixon, J. E. (2011). What is social sustainability? A clarification of concepts. *Geoforum, 42*(3), 342–348.

Vygotsky, L. S. (2004). Imagination and creativity in childhood. *Journal of Russian and East European Psychology, 42*, 7–97.

Weiner, R. P. (2000). *Creativity and beyond: Cultures, values, and change*. State University of New York Press.

Wegener, C., & Wegener, G. (2016). Mirroring, chapter 11. In V. Glăveanu, L. Tanggaard & C. Wegener (Eds.), *Creativity: A new vocabulary* (pp. 87–95). Palgrave Macmillian.

Wezel, A., Bellon, S., Doré, T., Francis, C., Vallod, D., & David, C. (2009). Agreocology as a science, a movement and a practice: A review. *Agronomy for Sustainable Development, 29*, 503–515.

2

Painting Outside the Lines: Transgressing the Managerial University, Avoiding Forced Creativity

Stephen Leitheiser, Rubén Vezzoni, and Viola Hakkarainen

Introduction

"The act of creation is, I have said, the same in science as in art. It is a natural, human, living act" (Bronowski, 1968). Jacob Bronowski arrived in Nagasaki in 1945 as a mathematician who had worked to develop efficient British bombing strategies during World War II. After being sent to document the destruction following the dropping of the atomic bomb with a team of fellow scientists, he left Nagasaki as a humanist philosopher who would go on to devote his remaining career to foregrounding the importance of human-created values in science, and the fundamental connections between imagination, science, and the arts (Bronowski, 1956; Bronowski et al., 1964). Seeing the wreckage to which his field

S. Leitheiser (✉)
University of Groningen, Groningen, The Netherlands
e-mail: srleitheiser@protonmail.com

R. Vezzoni · V. Hakkarainen
University of Helsinki, Helsinki, Finland

A. Franklin (ed.), *Co-Creativity and Engaged Scholarship*,
https://doi.org/10.1007/978-3-030-84248-2_2

of scientific work had contributed, Bronowski was faced with the reality that science is not a purely mechanical, neutral, or indifferent collection of observed facts. Instead, he would come to understand science as a creative and imaginative system of knowledge, underpinned by human values, with a blurred, rather than clear-cut line between production and use. Deeply affected by his experience in Nagasaki, Bronowski experienced first-hand the danger of disconnecting science from human values and judgement. He would come to argue that, in achieving its greatest discoveries and usefulness, science had always been humanistic. For Bronowski (1985), science at its best was (1) anti-authoritarian, (2) rooted in human experience, and (3) interconnected with and immersed in nature. It was only when scientists, and more importantly the institutional wholes of which they were part, lost touch with these value-based roots that science could become a "bag of tricks" deployed in the service of a callous bureaucracy intent on preserving its *status quo* (*ibid*., p. 264).

Bronowski's story illustrates the complex interaction of lived human experience with subjectivity and understanding of science. It shows the dynamic interplay of experience, values, and worldviews, and in turn, how this shapes approaches to scientific inquiry. This interplay marks the difference between whether one views science as a mechanical set of indifferent facts, or something that is creative, informed by values and context, and conditioned by its use. Just as our human values influence the ways in which we seek to understand the world through science, reason and the things we learn about the world through the scientific method also influence our values and particular normative valuations and prescriptions (Sayer, 2011).

The main theme of this collective book, creative methods (CMs), represents an attempt to contribute to critical discussions about how the process and pursuit of research may be more conducive to (1) making people question established ways of thinking and acting, and (2) building a more inclusive approach to research in which unheard voices are empowered (see Franklin, this book). However, as an approach to conducting research and exchanging knowledge, CMs, just like any other methods, are embedded into human value systems that influence how they are used to produce knowledge, and how that knowledge will be applied (or not applied) in practice (Harré, 1981; Longino, 1990).

Here we wish to highlight the reflexive character of scientific investigation, which is particularly prominent in the case of humanistic and social sciences. The personal beliefs, motivations and expectations of the researchers play a role in determining *what gets discovered* and *for what purpose*, e.g., what kinds of questions are asked and what kind of evidence counts as valid to confirm a hypothesis, across disciplines. Value assumptions, whether epistemic, moral, or political, shape the content of science and its application. Yet this normative shaping of scientific inquiry does not end with the individual researcher, but is mediated through the wider social environment (Sayer, 2011). In particular, we contend that the content and application of science in society is influenced by the institutions that employ researchers and the funders that provide the basis for their material existence. Therefore, a critical discussion of CMs would be incomplete without a structural analysis of the values embedded into the wider contextual environment in which CMs emerge: university systems that are increasingly managerial (Deem et al., 2007; Leišyte, 2015; Shepherd, 2018).

Managerialism is an ideology that is predicated on the universalized application of private sector values and practices, and namely corporate-style management, into all spheres of society (Chauvière & Mick, 2013; Deem, 2001). Management becomes "*hyper*-management" in which "management, as a form and as a process, becomes an end in itself, a self-serving entity" (Barberis, 2012, p. 327). Applied in a university setting, managerialism "colonizes" (Klikauer, 2015) the values traditionally associated with higher education (e.g., truth, autonomy, democracy, or the public good) (Giroux, 2010). Below, we outline what we identify as the main values of the managerial university that are sustained through four major driving forces: an environment of funding scarcity; a logic of competition to secure funding; the implementation of accountability metrics to rank competitors; and the creation of incentives for obedience. First, however, a few disclaimers. We acknowledge that managerialism is not a uniform blueprint, but is rather a pattern in which a more general organizational approach has been applied in various local contexts across the globe (Deem, 2001; Pusser et al., 2011). Nor is the managerial ideology ubiquitous among all academic staff (see, e.g., Connell, 2019). Nevertheless, its system of ideas, ideals, manners, and thoughts has been applied by many sitting in influential positions, and mediates academic work, even for those individual academics who may work to resist (Anderson, 2008; Evans, 2020). Finally, we have nothing against management as such, just *hyper*-management, where its application is counter-productive or inappropriate.

The aim of our critique is to point to how the values and practices of managerialism embedded into universities place inherent constraints on those wishing to bring creativity—as in originality and imagination—into the academic process, regardless of whether it is in the research process itself or in its application. Moreover, as many researchers who are interested in CMs may also be motivated to achieving transformative real-world outcomes (to fostering, e.g., sustainability, resilience, etc.), we wish to highlight the tremendous uphill battle that they may face within the confines of managerial universities. We do not do this to promote feelings of hopelessness, but rather to shed illusions, and warn about the ever-present danger of co-option. This applies even in the case of novel research approaches that may, on the surface, appear to be different and subversive, of which CMs are an illustrative example. Bringing in creativity may just as easily be used to affirm the *status quo* as to challenge it (Mould, 2018). By highlighting the managerial university's constraints, we wish to motivate strategic thinking for political action and coalition building outside of and beyond academic work. This chapter provides the analytical grounds from which collective practices can derive strength and cohesion. However, our reflections do not fall in the binary trap "don't act, just think". Beyond provocative statements *à la* Slavoj Žižek, our approach is more that of Noam Chomsky, who in a recent interview suggested to "look around, analyse the problems, ask yourself what you can do and set out on the work!".[1] Our task in this chapter is to organize our analysis of the problems, and motivate ourselves and others to reflect on possibilities for action.

If one views individual researchers who use CMs as painters who sketch out original and imaginative ways of approaching and disseminating research, our chapter provides an analysis of the canvas on which

[1] Interview by Zeit Campus, 14 June 2011.

they are painting: university institutions. We argue that the managerial university is not a blank canvas on which creativity can be painted. Rather, the canvas of managerialism is defined by a particular set of values that discourages painting outside the lines. In other words, managerialism produces structural impediments to CMs (Bullen et al., 2004; Connell, 2019). Failure to recognize that CMs are simply tools, free to be co-opted and stabilized into this restrictive context, may amplify the latent risk of CMs being reduced to a nice-sounding *bag of tricks*, that reinforces the *status quo* under a new coat of paint. We call this latent risk the danger of "**forced creativity**",[2] of which we provide two illustrative examples: "artwashing" and "funding tricks". We do not wish to suggest that these dangers are *particularly* unique to CMs as such. Similar challenges are certainly faced by a variety of heterodox thinkers who wish to generate new approaches in science and higher education, or even to hold onto old traditions (Brown, 2010). Our contribution also aims to contribute to this wider discussion, and debates on the role of the university in society. However, in this context of this book, we wish to use the example of CMs to illustrate that even those approaches that may appear to be novel and radical are not inherently resistant to the managerial university's distorting influence. We contend that CMs will only be truly *creative* to the extent that they are able to resist the trends of the managerial university, which aim to co-opt and appropriate their heterodoxy or stabilize their transformative character. If researchers truly wish to address power relations in the research process, give a voice to the voiceless, and break free from the shackles of the strictly positivist paradigm in social sciences, their efforts must be extended beyond the processes of data collection and dissemination of research, and towards building wider coalitions to intervene in technocratic and managerial takeover.

The chapter begins with an historical context that has contributed to the institutionalization of these managerial values in universities across the globe, after which we outline the values of the managerial university (accountability, competition, and obedience) that find fertile ground

[2] The inspiration for this concept comes from Graeber (2018) who uses the term in his book *Bullshit Jobs*.

in an environment of funding scarcity. Next we move on to define our understanding of forced creativity and give two illustrative examples. We suggest that those interested in CMs also bring creativity into the transgression of established academic protocols that bring about the risks of forced creativity.

The Managerial University and the De-Politicization of the Public Sphere

The Western university as an institution has its roots in the millenary field of political philosophy (Arendt, 1958; Wolin, 1960). In this traditional understanding, *the political* is an open public realm in which a plurality of possible visions for organizing social and ecological relationships are formed; contested and debated, and contingently agreed upon and institutionalized (Swyngedouw, 2018). The pluralistic character of healthy democratic politics is nurtured by the diversity of interpretations of social existence. Forms of governance can be said to be democratic to the extent that they allow this diversity to thrive. Although dissent may always be challenged, it is never suppressed. The space where this process of political deliberation occurs has traditionally been referred to as the "public sphere" by democratic theorists (e.g., Habermas, 1990): a domain of civic thought and normative discussion on matters of general interest that is separate from both the state and the market (Holmwood, 2017). The public sphere as such is a political space that cannot be reduced to *purely* rational, technical, or scientific calculations (Arendt, 1958; Mouffe, 2005). It is the space in which meaning, social difference, normative thought, and science enter into a collective process of making sense of the world, evaluation, and prescription. Surely, this ideal has not always existed in practice throughout history, and relationships of power have always determined the conditions and constraints of debate in the public sphere and in the academy (Mouffe, 1992; Tierney & Lechuga, 2005). That being said, the university has, in different times and places throughout history, been a stronghold in the process of democratizing society (Deem & Eggins, 2017; Giroux, 2009; Tierney & Lechuga, 2005). Throughout the twentieth century, however, several philosophers began to warn that such public spaces for distinctly political thought—e.g., plurality, dissent, and open debate on matters of general interest—were becoming increasingly narrowed in society at large (Arendt, 1958; Bronowski, 1956; Wolin, 1960). As autonomous and free inquiry are a central component of an open democratic society, universities were a major part of this closure (Giroux, 2009; Holmwood, 2017).

Managerialism draws many similarities with Taylorism, Fordism, technocracy and other traditions of techno-scientific standardization and social homogenization, which of course did not begin in university institutions (see, e.g., Scott, 1998 for a broader, more detailed historical perspective). Here, we trace the modern origins of the managerial ideology into institutions of higher education back to the early twentieth century in the United States, where in the 1910s, an intense debate was waged over educational reform. David Snedden's social efficiency approach (rooted in vocationalism and the production of obedient workers for the capitalist economy) was here pitted against John Dewey's liberal approach (rooted in a desire to create free citizens empowered by independent thought) (Labaree, 2010). The social efficiency approach saw education as a form of training—"something like filling a vessel with water", or imparting pre-existing knowledge on passive subjects (Chomsky, 2012, p. 56). The liberal approach, on the other hand, saw education more like the nurturing of a tree, or "providing the circumstances in which the normal creative patterns will flourish" (*ibid.*). While Dewey's name and approach may be more recognizable in the present day, Snedden's social efficiency approach would go on to be much more influential in shaping education policy for the remainder of the twentieth century (Labaree, 2010). In the context of a rising working class consciousness at the end of the nineteenth century and early twentieth century (Chomsky, 2012; Goodwyn, 1978; Ware, 1929), Snedden's approach resonated strongly with the powerful coalition of state and capitalist elites. Not only did they view social criticism and moral and political philosophy of the liberal approach as a threat to the *status quo*, they were also keen to have the burden of training obedient workers covered by public subsidy (Labaree, 2010). This also helps to explain the trend of prioritizing STEM (science, technology, engineering, and mathematics) subjects vis-à-vis humanities (namely moral and political philosophy) in many university institutions (Palumbo & Scott, 2018).

In the post-World War II era, the social efficiency model for the university was increasingly globalized. Managerialism moved beyond the United States, as universities became increasingly viewed as tools for driving global economic development in the broader context of the Cold War and European reconstruction (Adler et al., 2007; Palumbo & Scott, 2018; Schrum, 2012). During this time, the Rockefeller, Carnegie, and Ford Foundations partnered with the American state to steer the evolution of universities in the *so-called* developing world[3]; this effectively established a relationship of dependency and instituted a global system of training (rather than education) in line with metropolitan development (Connell, 2019). Business schools rose to power in universities around the world, and an administrative and behavioural approach to social science became more and more hegemonic (Pettigrew et al., 2014; Schrum, 2012). Approaches that foregrounded human values and judgement, creativity and imagination—e.g., liberal arts, or moral and political philosophy—were therefore displaced by a quest to systematically model human behaviour with unified general theories, based on (a perversion of) physical sciences (Klikauer, 2015; Schrum, 2012; Wolin, 1960). According to Wolin (1960), the expansion and fragmentation of social science into disparate disciplines throughout the nineteenth and twentieth centuries was predicated on an eclipse of moral and political philosophy: "While one flourishes, the other flounders in uncertainty of what, if anything, constitutes its subject-matter" (p. 288). As social sciences became increasingly separated and siloed, i.e., into categories of sociology, economics, psychology, etc., and detached from philosophy they also began to distance themselves from the normative critiques that had been attached to positive description in social sciences during (and before) the Enlightenment (Sayer, 2011). Moreover, evaluation and judgement became largely taboo for (social) scientists in

[3] We emphasize "so-called" here in order to distance ourselves from the normative view that depicts most of the world's countries as lacking "development" and celebrates the progress achieved by relatively few countries that has come at the expense of negative environmental and social externalities that are mostly experienced by those living in other parts of the world (see also Gibson-Graham et al., 2013).

general (*ibid.*). This included the evacuation of meaningful critiques of dominant (corporatist) ideologies (Chomsky, 2000), and debates over the fundamental role that universities should play in society (Deem & Eggins, 2017; Pusser et al., 2011). Consequently, foundational dissent has to a great extent been marginalized in wider public debates, with the academy positioned as a central node in the military–industrial–academic complex (Giroux, 2015). In light of this history, we understand the managerial university as a prime contributor to de-politicizing the public sphere and stabilizing the *status quo*.

As anticipated in the introduction, knowledge creation is not a value-free process and the society-science relationship is not linear (Turnhout, 2018). Since science does not exist above and outside of society, science *qua* institution can also not be said to be completely politically neutral. As in a jury trial, the role of scientific expertise is to *augment* and *sharpen* democratic, ethical and political discourse through technical fact-finding and bias mitigation; it is not to give a fixed decision-making blueprint to a passive population (Follett, 1930; Hansson, 2004). On the contrary, in political debates science can (and should) inspire a democratic discussion of what constitutes the most desirable direction (Sayer, 2011). The democratic character of discourse is amplified by the extent to which different interests are taken into account, including that of those generations yet to be born or of non-human species. Scientific findings can elucidate this multiplicity of positions and augment our collective intelligence.

The unwillingness to recognize the political aspects that influence the institution of science paradoxically expose it to appropriation by those in relative positions of power in society. Following Bronowski's definition of science laid out above, higher education and research (and the university by extension) cannot be truly scientific without being anti-authoritarian; that is, maintaining an environment of radically open and critical thought, using different lenses and approaches to investigate truth in understanding society and its relationship with the natural world, and exploring the possibilities for alternative ways of thinking about and organizing socio-natural relationships (cf. Deem & Eggins, 2017). In this sense, science can be seen as a "deeply democratic principle, since it rejects all claims to absolute certainty and insists on open, undominated dialogue as the basis for correcting errors and advancing knowledge" (Wright, 2006, p. 94). The difference between these formulations of science and dogmatic *scientism* is the capacity to critically engage with the ideologies and values of the wider social environment that guide and mediate scientific inquiry (Chomsky, 2008; Popper, 1979). In the case that institutions of knowledge production fail to embody open democratic principles, they run the risk of propagating a mythical science (i.e., anti-scientific scientism), which—as the sole and infallible arbiter of truth that can objectively conduct human decision-making from the outside—can be contorted to suit the whims of powerful agendas: e.g., *we* [the people] have no choice but to do *X*, because Science has told *us* [the techno-managerial elite] *Y*. In short, science can also become the "bag of tricks" Bronowski (1985) has warned us about above: deployed in the service of a callous bureaucracy in order to narrow the spectrum of valid thought and debate in the public sphere.

A failure to recognize these risks becomes more problematic as researchers are increasingly called to engage in real world processes to facilitate the application of scientific knowledge and address the wicked and complex problems that humanity is currently facing (e.g., climate change, the destruction of the biosphere, peak oil and peak soil, global inequality) (Blythe et al., 2018). In this socio-environmental context, there is an increasing need for researchers to develop the philosophical underpinnings of action-orientated knowledge production while being able to produce actionable knowledge (Nagatsu et al., 2020). This necessitates that researchers challenge their own biases and assumptions related to global changes. It also includes insulating science per se from perversion by the same forces that may be driving such problems (Shrivastava et al., 2020).These issues have been brought into the spotlight in several fields of research, including, e.g., sustainability science (see e.g., Clark, 2016; Kates et al., 2001; Miller, 2013), in which the position of researchers as detached and objective observers of facts is already well established as false (Wittmayer & Schäpke, 2014). In this context, many researchers are increasingly reflexive to the implications of their own positionality and the normative stances they imply in the pursuit of actionable knowledge (Hölscher et al., 2017; Wittmayer & Schäpke, 2014). However, even in fields of research that would ideally carry this reflexive awareness, attempts to escape the de-politicized landscape of the public sphere have been unsuccessful (Nagatsu et al., 2020). For example, Fazey et al. (2018) point out that although the need for transformative research is recognized, the majority of resources and attention are directed towards more conventional approaches. Radical approaches or innovations are often seen to be co-opted into old patterns, and realign with, rather than challenge, existing trajectories and power dynamics (Blythe et al., 2018; Fazey et al., 2018; Kläy et al., 2015).

Although change and innovation are encouraged as a central part of managerialism in universities (Barberis, 2012), the trajectory and boundaries of change are determined by unaccountable forces (e.g., philanthropic foundations, market forces, the European Commission). Ultimately this points us back to the managerialist framework (the canvas) that encourages the production of checklists and simplified results that can be easily operationalized, and discourages the nuanced complexities of political realities and social differentiation (Blythe et al., 2018; Scoones, 2009). Therefore, if originality, a willingness to deviate from norms, and explorative thinking that deviates from traditional paths are the essence of CMs (Kara, 2015; Richards, 2010) the approach of individual researchers is not the only thing that needs to become more creative. It is more importantly the creative approach to the organization and funding of knowledge production in universities, and the processes through which their roles in society are negotiated, that must be challenged. Since we have identified de-politicization (i.e., control and sterilization of the public debate) as the crucial barrier to creativity and CM, the next section spells out in more detail the particular value characteristics of control that we believe should be confronted.

The Values of Managerial Knowledge Production

Don't Bite the Hand that Funds You

As Bavington (2002) has shown, the roots of the word "management" stem back to the Italian word *maneggiare*, which in the sixteenth century originally referred to the rearing of wild horses. Likewise, we have defined the managerial university as rooted in control—keeping research and higher education on a leash, and ultimately marginalizing meaningful dissent in the public sphere. The process of bringing managerial practices and values into universities has been sustained by four major driving forces: an environment of funding scarcity; a logic of competition to secure funding; the implementation of accountability metrics to rank competitors; and the creation of incentives for obedience. In looking closer at these forces, this section will explain more specifically how control is maintained. We do this by deconstructing the narratives used to justify New Public Management (NPM) reform, and providing evidence to suggest that these reforms are, in essence, about keeping academia on a leash.

Managerialism has relied on a marketization approach to funding public institutions that is consistent with general NPM reforms (Irzik, 2007; Palumbo & Scott, 2018). In this approach, universities act as corporations in a competitive market, instead of functional parts of a whole (Connell, 2019). This has made competing to maximize —or at least maintain—access to resources the driving organizational force of university governance (Palumbo & Scott, 2018). The market-based approach has fostered a culture of ruthless competition for academics among and against each other in order to fund their work (and consequently, their economic survival). In order to rank the competitors and determine who would receive funding, managers from states, supranational institutions, and university administrations have standardized the measurement of individual academic and university performance (Lynch, 2015). Muller (2018) refers to the resulting system as a "tyranny of metrics": a faith in objective and quantifiable measurability as a replacement for subjective and qualitative human judgement. The tyranny of metrics can also be described with what Deem et al. (2007) have described as an "institutionalized distrust", rooted in a pervasive suspicion that seeks to strictly monitor staff, curtail their room for autonomy and improvisation, and have them constantly justify their work and activities (Adler & Borys, 1996; Graeber, 2018). As we have stated above, the claim that such "objective" measurements are devoid of social values is illusory (cf. Sayer, 2011).

These metrics are, in fact, defined from above in line with the values and material interests of states, supranational institutions, corporations, and private philanthropic foundations who have the capacity to allocate grants and other resources to fund research. While funding institutions may claim impartiality, a closer look shows that many—private foundations and corporations in particular—are not as politically neutral as they may purport (Lynch, 2015). Apart from governments, none of these institutions are subjected to democratic control, and may be driven by private interests that are indifferent or even antithetical to public or common interests (Irzik, 2007). The lack of oversight for these funding institutions is particularly concerning in the case of private philanthropic foundations. Private foundations have been described as "black boxes", immune from public oversight, with largely unrestricted "hyper-agency"—i.e., "the ability to shape socio-political frameworks and matrices in which networked governance occurs" (Jung & Harrow, 2015, p. 49). It is argued that many of these private foundations are driven by "philanthro-capitalism" (Garcia-Arias, 2019; Mediavilla & Garcia-Arias, 2019; Silver, 1998). In this sense, the hyper-agency of philanthropists can be used as a sort of masked lobbying.[4] That is, facilitating the production of knowledge that is ostensibly in the public interest, but is primarily driven by private agendas. The opaque nature of foundations allows them to bypass society's democratic structures and advance an "economic model of investment and political model of control" under the guise of generosity (Shiva & Shiva, 2018, p. 120). One example is the Bill and Melinda Gates Foundation, which funds research and development programmes for public health and agriculture around the world to the tune of billions of dollars per year (Biovision Foundation for Ecological Development and IPES-Food, 2020; McCoy et al., 2009; Shiva, 2016). The foundation's approach to funding research leads to the promotion of certain paradigms (e.g., centralized industrial agriculture, privatization of medical systems, etc.) at the expense of others (see also Vanloqueren & Baret, 2009). In fact, all funding institutions exert some measure of control over science policy and research

[4] See for example the arguments of the Reese Committee investigation of tax-exempt foundations in the United States in the 1950s (Gideonse, 1954).

content (Gläser & Laudel, 2016). Funders effectively hold the *reins* of control to direct society's production of knowledge and are the factual *managers* of the managerial university (Lynch, 2015).

In line with corporate governance, which aims to give shareholders more control over management, funders exert their control through top-down monitoring and assessment. This has a disciplinary effect on dissent for academic staff, in line with the old adage, "don't bite the hand that feeds you". It creates barriers to levying independent foundational critiques of funding institutions (Pusser et al., 2011), and incentivizes and selects for the perpetuation of dominant ideological paradigms—e.g., those in power within the current system self-select for those who share their worldview and values (Mitchell & Fazi, 2017). This perpetuation is further reinforced by a growing "reserve army" of precarious academic labourers (e.g., PhDs and post-docs) who work on short-term contracts without job security (cf. Ginsberg, 2011). While the numbers of these precarious labourers continue to grow by many estimates (e.g., the number of doctoral graduates in OECD countries grew by *40%* from 2000–2009), the secure and tenured positions do not (Worms & Boman, 2017). Academic labour, in turn, becomes devalued and easily replaceable, creating a further incentive for staff to *toe the line* in a positional competition game, or to simply take their skills to the (corporate) private sector where earning potentials are higher, or at least more secure.

Three main rationalizations are used to justify NPM reforms promoting accountability and competition. First, they are said to foster societal engagement and innovation by eliminating freeriding privileges for "ivory tower" academics with tenure. In this narrative, NPM reforms are carried out in order to provide more value for the taxpayer who assumes the role of shareholder. Value is understood in economic terms as a return on investment (Halffman & Radder, 2015), and science is evaluated based on its ability to facilitate the creation of wealth or jobs (Jasanoff, 2005). However, instead of eliminating privileges for "freeloaders", such collaborations have created a new set of privileges for (corporate) actors in the military–industrial–academic complex. These private actors—unbeholden to any notion of the public good or wider social responsibilities—have in turn been able to leverage their funding capacities and therefore outsource the risk of research and development

(R&D) to publicly subsidized universities, while privatizing the benefits (Mazzucato, 2011; Palumbo & Scott, 2018; Schugurensky, 2006). A prime example of this is the US Bayh-Dohl Act (1980) that allowed inventions discovered with public funds to be patented for private gain (Irzik, 2007). In effect, this has created a system of technology transfer from the public to the private sector (*ibid.*; cf. Mazzucato, 2011) under the guise of ostensible "societal engagement".

Another argument for the reforms is that they control for quality in research. In reality, there is evidence to suggest that the opposite is true. Competition has actually led to a race to the bottom: spreading the work of academics increasingly thin, subjecting them to higher levels of stress and anxiety, and therefore negatively impacting the quality of their work (Berg, 2015). While *any* person may win the competition of being the best scholar, not *every* person can win this competition. The zero-sum logic of competition (again not only for prestige, but also for institutional funding, and economic survival of individual researchers) requires the acceptance that while some may win the competition, others will lose. Obviously, this is not a new dynamic, as any attempt to make use of limited resources (e.g., funding, job positions) involves a certain degree of competition. What we are highlighting are the dangers of placing competition as a *core principle* of academic life. For example, a survey of more than four thousand UK academics conducted by the Wellcome Trust showed that only 32% of respondents agreed that "*healthy* competition" was encouraged in their working environment, while 78% agreed that competition had created unkind and aggressive research conditions (Wellcome Trust, 2020).

Managerialism has accelerated the trend of relentless competition, the influence of which has been multiplied by the increasing hordes of a "reserve army" of junior academic staff. With the introduction of these high stakes, academics are incentivized to produce scientific knowledge at an increasingly rapid pace in order to stand out from their competitors. Today, the average academic publishes approximately six times as many papers as if they were working a century ago (Larsen & von Ins, 2010). This increased production has come with a detriment to the substance of scientific output. The phenomenon has been referred to as "scientific salami slicing" (Ding et al., 2019); this describes how academics separate research articles into the "minimum publishable unit" with the goal of maximizing the number of publications they can achieve from the same study (cf. Halffman & Radder, 2015). Moreover, duplicate publications are also common practice. A recent study shows that up to 20% of new publications in certain fields of research have reported the same results as in previous publications (Lai et al., 2020). Rather than sharpening quality, managerial reforms have engendered a situation where "knowledge" is overproduced. Is the latest article motivated by a piercing new insight? A novel contribution to knowledge? Or is it intended to pad the author's *curriculum vitae* due the coming expiration of their temporary contract? Ultimately, this overproduction devalues quality work by leaving researchers to search for the needle of quality in a seemingly infinite haystack of overproduced publications. A prime example (but by no means the only example) is the journal *Sustainability*. At the time of writing, the journal has planned over 150 special issues for 2021 alone, and in 2018 had more than 200.[5] Standard issues, which in 2019 were bi-monthly, often include more than 500 articles. Additionally, at the time of writing the "Article Processing Charge" for *Sustainability* was more than €1,700 per paper. This fantastic amount of papers could hardly be thoroughly digested by even the most astute of readers who is interested in keeping up on all of the latest debates in sustainability and sustainable development.

One is left to wonder, *who* and *what* are all these papers for? What is actually motivating their authorship and publication?

According to a blog post from Arjen Wals (2019), a sustainability researcher in the Netherlands, contemporary academia's publish or perish culture has led to a troubling paradox, in which "everybody is writing while nobody seems to be reading, really, which means that everybody is writing for nobody". This, in our view, is a race to the bottom: the illogical result of a university system based on managerial values and practices that incentivize quantity of publications over quality, and facilitate the commodification of publicly funded knowledge for private accumulation of profit.

[5] See: https://www.mdpi.com/journal/sustainability/special_issues?page_count=100&page_no=31&search=&section_id=0&sort=deadline&view=open (Accessed 26 Feburary, 2021).

Finally, reforms have been justified on the basis that they produce efficiency. However, there is (even more) evidence to suggest that many of the outcomes have been quite inefficient. Academic staff are subjected to a great number of "box-ticking" rituals in which they must constantly assess and justify their own work (Graeber, 2018). This can result in a paradox in which *more* university time and resources are allocated to monitoring and applying for further funding, than in doing actual research and education. For example, one study in the Netherlands estimated that approximately *one quarter* of the research budget for a federal subsidy programme for Dutch universities is spent on "overheads of writing, reviewing, and allocating" applications for the budget itself (Halffman & Radder, 2015, p. 169). "Ironically", under the tyranny of metrics, Muller (2018, p. 75) emphasizes, "in the name of controlling costs, expenditures wax". Following a recent study, European universities spend approximately *€1.4 billion* every year to fund failed grant applications.[6] If efficiency is the goal, would it not actually be more expedient to make resources available to responsible and autonomous academics in the first place?

To conclude, the expansion of competition has found fertile ground in shrinking public budgets and precarious funding conditions for students and employees. These dynamics have been reinforced by a standardized accountability system that rewards obedience and filters out dissent through groupthink and fear of being replaced. The managerial university relies on a simplification of parameters to quantify research output, which ends up prioritizing quantity over quality. Time consuming and thorough investigations of complex issues are devalued, as scholars are encouraged to "publish or perish". For academics, we argue that these values encourage what Graeber (2018) has called "forced creativity". The danger of forced creativity is that, while CMs may be different on the surface, they fail to break from the chains of the institutional context within which they operate, leaving their creativity forced, and substantively hollow. Although research activities have been extended to involve actors outside of academia, the pursuit of unspecified impact can come

[6] https://www.timeshighereducation.com/news/billions-lost-in-bids-to-secure-european-union-research-funding (Accessed 26, Feburary, 2021).

with undesirable consequences, which have hardly been recognized and studied as a result of more action-oriented research projects (Louder et al., 2021). The problem starts from the profoundly diverse (and often not explicitly stated) epistemic assumptions and what counts as impact (*ibid*). However, under pressure to avoid failure (Davies et al., 2021), and achieve the maximum impact of measurable output, academics may be led to blindly chase any kind of creativity that sets them apart, without asking important questions such as "who benefits and loses [...] and how this can be justified" (Turnhout, 2018, p. 368). In other words, as the individual researcher strives to survive in an increasingly demanding and competitive "industry", the risk is that their mobilization of creative methods exacerbates the problems they intend to address, rather than providing a solution.

Now that we have sketched out our critical analysis of the managerial university and its role in suppressing open democratic politics, we will look closer at the prospect of bringing creativity into the research process in the confines of the managerial university.

The Danger of Forced Creativity

Latent Risks in Creative and Arts-Based Methods

Michel Foucault famously explained that the point of his critique was not that "everything is bad", but rather that "everything is dangerous" (cited in Galliers et al., 2011, p. 177). Likewise, our message here is not that CMs are inherently *bad*. It is rather that, especially in the context of the managerial university we outlined above, CMs are dangerous: their use can be co-opted into a coercive maintenance of the *status quo*, as much as they can to liberating empowerment. In order to raise awareness about the dangers, we provide two practical examples of forced creativity. In doing so, we intend to outline the major risks of which researchers thinking of using CMs should be aware.

Artwashing

Much like the "greenwashing" tactics employed by corporate polluters (Athanasiou, 1996), artwashing is a de-politicizing strategy that may be used by powerful actors to manage and placate discontent by giving a "cool" and "artsy" appearance to elite agendas (Novak, 2019). One example is Florida's (2005) "creative class" concept, in which arts and creativity were used to generate a positive vibe for elite development projects that drive gentrification, privatization and marketization of urban space (Ruck, 2020). Similarly, the work of researchers who use CMs may (unintentionally) be appropriated to give a favourable "spin" to potentially unpopular messages of management (Barberis, 2012, p. 330).

Artwashing can also be used to give symbolic recognition to communities, while obscuring deeper foundations of their disempowerment. Mirroring Nancy Fraser's notion of "progressive neoliberalism", CMs may contribute to superficial *recognition* (surface reallocations of respect), while masking inequalities of *distribution* (share of material resources), and *representation* (share of decision-making and political equality) that also contribute to disempowerment (Fraser, 2005, 2016). As researchers use CMs to bring recognition to the struggles of communities, they run the risk of subjecting communities to a form of "tokenism" (Arnstein, 1969), which gives an illusory appearance of participation and inclusion. Instead of drawing attention to the root causes of community disempowerment, under the pretence of using CMs researchers may nominate themselves as a spokesperson for communicating community needs and desires (Kouritzin & Nakagawa, 2018). This hierarchical positioning can put the researched community into a subordinate level of representation, with the risk that their demands are interpreted in line with the researcher's positionality. Namely, this can create an extractive relationship between the researcher and the community, in which the experiences and actions of the community are used as a resource to advance the career of the researcher, while the goal of amplifying voices becomes secondary.

This "empowerment without power" is a direct consequence of de-politicization, as critical questioning of the social and ecological relationships that determine undemocratic representation and detrimental material conditions are left out of the picture. An a-critical use of CMs runs the risk of aestheticizing community members in their present state, leaving them embalmed rather than providing them with the means to emancipate themselves in the future. In an environment where the underlying goal of research that uses CMs is likely to be focused on quantitative personal academic career development (e.g., citations), such extractive tendencies are an ever present risk.

Funding Tricks

External funding conditions in the larger political economy described above often come with an imperative of constant innovation, novelty, and adjustment (Palumbo & Scott, 2018). Funding tricks, in which creativity is performed for the sake of helping one stand out in comparison to other funding applicants, are an inevitable danger of responding to these incentives. Moreover, as funding calls mostly require that applicants fit within a format that is predetermined by funders who expect certain outcomes, funding tricks are particularly prone to propagating forced creativity.

The research funding strategy of the European Commission, through the European Research Council (ERC), is a good illustration of this. Access to funding is granted according to a specific jargon and sophisticated rules. This creates incentives for the formation of a specialized body of experts whose aim is not to write meaningful research applications, but to work as intermediaries between the source of the funding and the researcher. This incentivizes the production of "nice-sounding" proposals that embellish and "dress up" business as usual responses to funding calls (cf. Cornwall & Brock, 2005). If one must "sell" their research proposal in order to avail in competition, they are incentivized to employ deceptive tactics used by marketers or public relations firms (Frankfurt, 2009). This practice is common enough that it has found expression in at least two European languages. A German word, *Förderantragsjargon* (funding application jargon), describes the practice of creating token participation in response to the EU's Smart City funding calls (the EU requires participation, so applicants include it in the proposal without the intent of actually incorporating it) (Follmann et al., 2020). An Italian term, *euro-progettazione*, in use since the late 1990s, describes a specific discipline that literally translates into "European project-making".

Several higher education courses are nowadays available for those who want to master the litanies of EU funding applications. When the content of the research has a similar or even secondary relevance compared to the jargon used to present it, forced creativity may grow in the guise of CMs. It is the stratified governance of highly bureaucratic organizations such as the EU that is inevitably entailing a certain degree of resistance to innovation (Banchoff, 2002). This institutional inertia creates niches of privilege. This is what Hoenig (2017) defines the "new scientific elite", which emerges according to centre-periphery-structures due to historical path-dependency and accumulation of knowledge in certain geopolitical locations.

Conclusions

"[…] if 'the revolution will not be televised', it certainly won't be peer-reviewed". (Davies et al., 2021, p. 5). So far, we have offered a critical review of the managerial university, its embedded values, and the dangers of employing creative methods (CMs) in such an environment. As we noted at the beginning of this chapter, CMs have been framed as an individual responsibility of a researcher, to fulfil the new multi-faceted role of knowledge producer, knowledge translator, communicator, co-designer, and implementers of action (Freeth et al., 2019; Horlings et al., 2020; Wittmayer & Schäpke, 2014). Although individual reflexivity and understanding one's own normative position as a moral and political agent in a changing world may be crucial, we ask for understanding CM in the light of the greater structures of academia. We wish to stress that the picture we have highlighted is one of a collective problem that cannot be addressed through a purely individual struggle. An understanding of the wider context—which we referred to earlier as the "canvas" on which researchers "paint"—and a willingness to creatively transgress established academic structures and protocols (Temper et al., 2019) are of utmost importance if we as academics wish to move towards truly creative academic practice while avoiding the latent risks of forced creativity. In other words, one cannot be truly creative in a transgressive manner if one does not know exactly what they are transgressing. Transgression of the managerial university that enables an environment of creativity will have to include various radical interventions into its sustaining forces—namely, *funding*, *competition*, and *obedience*.

In preparing this chapter, both in conceiving it and drafting it, we were further persuaded by several one-on-one discussions with scholars who, despite coming from very different contexts and backgrounds, described the same feeling of working in a deteriorating environment where much of their activity made little sense (see also Berg (2015) who conveys a similar experience in the long process of writing his critical article on neoliberalization of universities). For us, the awareness that we are not alone in feeling somewhat lost and hopeless in the halls of the managerial university is comforting and empowering. This more realistic, even stoic understanding of the institutional setting can give young researchers who may be looking to engage in heterodox and transformative approaches a more coherent picture of what they are up against. By no means do we wish to chastise the individuals who are not willing to engage in this struggle. But we do hope that other young academics who are disillusioned by their institutional environments dedicate themselves to further political action and coalition-building beyond the constraints of the managerial university.

Acknowledgements This research received funding from the European Union's Horizon 2020 research and innovation programme under the Marie Sklowdoska-Curie Grant Agreement No. 765389.

References

Adler, P. S., & Borys, B. (1996). Two types of bureaucracy: enabling and coercive. *Administrative Science Quarterly, 41*, 61–89.

Adler, P. S., Forbes, L. C., & Willmott, H. (2007). *Critical Management Studies*, 1–61.

Anderson, G. (2008). Mapping academic resistance in the managerial university. *Organization, 15*, 251–270. https://doi.org/10.1177/1350508407086583

Arendt, H. (1958). *The human condition* (2nd ed.). The University of Chicago Press.

Arnstein, S. R. (1969). A ladder of citizen participation. *Journal of the American Institute of Planners, 35*, 216–224. https://doi.org/10.1080/01944366908977225

Athanasiou, T. (1996). The age of greenwashing. *Capitalism Nature Socialism, 7*, 1–36.

Banchoff, T. (2002). Institutions, inertia and European union. *Research Policy, 40*, 1–21.

Barberis, P. (2012). The managerial imperative: Fifty years' change in UK public administration. *Public Policy Administration, 28*, 327–345. https://doi.org/10.1177/0952076712458789

Bavington, D. (2002). Managerial ecology and its discontents: Exploring the complexities of control, careful use and coping in resource and environmental management. *Environments, 30*, 3–22.

Berg, L. D. (2015). Rethinking the PhD in the age of neoliberalization. *GeoJournal, 80*, 219–224. https://doi.org/10.1007/s10708-014-9574-6

Biovision Foundation for Ecological Development, IPES-Food. (2020). Money flows: What is holding back investment in agroecological research for Africa? Biovision Found. Ecol. Dev. Int. Panel Expert. Sustain. Food Syst.

Blythe, J., Silver, J., Evans, L., Armitage, D., Bennett, N. J., Moore, M.-L., Morrison, T. H., & Brown, K. (2018). The dark side of transformation: Latent risks in contemporary sustainability discourse. *Antipode, 50*, 1206–1223. https://doi.org/10.1111/anti.12405

Bregman, R. (2017). *Utopia for realists: And how we can get there*. Bloomsbury.

Bronowski, J. (1985). Science as a humanistic discipline. *Leonardo, 18*, 261. https://doi.org/10.2307/1578078

Bronowski, J. (1968). The creative process. *Journal of Creative Behaviour, 2*, 63–70. https://doi.org/10.1002/j.2162-6057.1968.tb00083.x

Bronowski, J. (1956). *Science and human values*. Julian Messner.

Bronowski, J., Steele Commager, H., Allport, G., & Buck, P. (Eds.). (1964). *Imagination and the university*. University of Toronto Press.

Brown, W. (2010). Political theory is not a luxury: A response to Timothy Kaufman-Osborn's "political theory as a profession." *Political Research Quarterly, 63*, 680–685. https://doi.org/10.1177/1065912910369843

Bullen, E., Robb, S., & Kenway, J. (2004). "Creative destruction": Knowledge economy policy and the future of the arts and humanities in the academy. *Journal of Educational Policy, 19*, 3–22. https://doi.org/10.1080/0268093042000182609

Chauvière, M., & Mick, S. (2013). The French sociological critique of managerialism: Themes and frameworks. *Critical Sociology, 39*, 135–143.

Chomsky, N. (2000). *Rogue state*. South End Press.

Chomsky, N. (2008). The responsibility of intellectuals. In A. Arnove (Ed.), *The essential chomsky*. The New Press.

Chomsky, N. (2012). *Democracy and Education. Counterpoints, 422*, 55–70. https://doi.org/10.1017/CBO9781107415324.004

Clark, W. C. (2016). Crafting usable knowledge for sustainable development. *Proceedings of the National Academy of Sciences, 113*, 4570–4578.

Connell, R. (2019). *The good university: What universities actually do and why it's time for radical change*. Zed Books Ltd.

Cornwall, A., & Brock, K. (2005). What do buzzwords do for development policy? A critical look at "participation", "empowerment" and "poverty reduction." *Third World Quaterly, 26*, 1043–1060. https://doi.org/10.1080/01436590500235603

Cruikshank, B. (1999). *The will to empower*. Cornell University Press.

Davies, T., Disney, T., & Harrowell, E. (2021). Emotion, space and society reclaiming failure in geography: Academic honesty in a neoliberal world. *Emotion Space and Society, 100769*. https://doi.org/10.1016/j.emospa.2021.100769

Deem, R. (2001). Globalisation, new managerialism, academic capitalism and entrepreneurialism in universities: Is the local dimension still important? *Comparative Education, 37*, 7–20. https://doi.org/10.1080/03050060020020408

Deem, R., & Eggins, H. (Eds.). (2017). *The university as a critical institution?, Higher education research in the 21st century Series*. Sense Publishers. https://doi.org/10.1017/CBO9781107415324.004

Deem, R., Hillyard, S., & Reed, M. (2007). *Knowledge, higher education and the new managerialism: The changing management of UK universities*. Oxford Univeristy Press.

Ding, D., Nguyen, B., Gebel, K., Bauman, A., & Bero, L. (2019). Duplicate and salami publication: A prevalence study of journal policies. *International Journal of Epidemiology, 49*, 281–288. https://doi.org/10.1093/ije/dyz187

Evans, M. (2020). Navigating the neoliberal university: Reflecting on teaching practice as a teacher-researcher-trade unionist. *British Journal of Sociology of Education, 41*, 574–590. https://doi.org/10.1080/01425692.2020.1748572

Fazey, I., Schäpke, N., Caniglia, G., Patterson, J., Hultman, J., van Mierlo, B., Säwe, F., Wiek, A., Wittmayer, J., Aldunce, P., Al Waer, H., Battacharya, N., Bradbury, H., Carmen, E., Colvin, J., Cvitanovic, C., D'Souza, M., Gopel, M., Goldstein, B., ... Wyborn, C. (2018). Ten essentials for action-oriented and second order energy transitions, transformations and climate change research. *Energy Research & Social Science, 40*, 54–70. https://doi.org/10.1016/j.erss.2017.11.026

Florida, R. (2005). *Cities and the Creative Class*. Routledge.

Follett, M. P. (1930). *Creative experience*. Longmans, Green and Co.

Follmann, A., Leitheiser, S., & Kreitschmer, H. (2020). *Smart und oder partizipativ? Eine kritische Betrachtung der Smart City Cologne* (forthcoming). sub/urban.

Frankfurt, H. G. (2009). *On bullshit*. Princeton University Press.

Fraser, N. (2005). Reframing justice in a globalizing world. *New Left Review, 36*, 69–88.

Fraser, N. (2016). Progressive neoliberalism versus reactionary populism: A Choice that Feminists Should Refuse. *NORA—Nordic Journal of Feminist and Gender Research, 24*, 281–284. https://doi.org/10.1080/08038740.2016.1278263

Freeth, R., Clarke, E. A., & Fam, D. (2019). Engaging creatively with tension in collaborative research. In A. Valerie, John A. Brown & D. W.-T. Harris (Ed.), *Independent thinking in an uncertain world*. Routledge.

Galliers, R., Currie, W., Willcocks, L., & Lioliou, E. (2011). 'Everything is dangerous': Rethinking Michel Foucault and the social study of ICT. In R. D. Galliers & W. L. Currie (Eds.), *The oxford handbook of management information systems: Critical perspectives and new directions*. Oxford University Press.

Garcia-Arias, J. (2019). *A critical perspective on development economics philanthrocapitalism: How to legitimize the hegemony of the rich with a "good vibes" discourse.*

Gibson-Graham, J. K., Cameron, J., & Healy, S. (2013). *Take back the economy: An ethical guide for transforming our communities*. University of Minnesota Press.

Gideonse, H. D. (1954). A congressional committee's investigation of the foundations. *Journal of Higher Education, 25*, 457–463.

Ginsberg, B. (2011). *The fall of the faculty*. Oxford University Press.

Giroux, H. A. (2015). *University in chains: Confronting the military-industrial-academic complex*. Routledge.
Giroux, H. A. (2010). Bare pedagogy and the scourge of neoliberalism: Rethinking higher education as a democratic public sphere. *The Educational Forum, 74*, 184–196. https://doi.org/10.1080/00131725.2010.483897
Giroux, H. A. (2009). Democracy's nemesis: The rise of the corporate university. *Cultural Studies—Critical Methodologies, 9*, 669–695. https://doi.org/10.1177/1532708609341169
Gläser, J., & Laudel, G. (2016). Governing science: How science policy shapes research content. *European Journal of Sociology, 57*, 117–168. https://doi.org/10.1017/s0003975616000047
Goodwyn, L. (1978). *The populist moment: A short history of the agrarian revolt in America*. Oxford University Press.
Graeber, D. (2018). *Bullshit jobs: A theory*. Simon & Schuster.
Habermas, J. (1990). *Strukturwandel der Öffentlichkeit*. Suhrkamp Verlag.
Halffman, W., & Radder, H. (2015). The academic manifesto: From an occupied to a public university. *Minerva, 53*, 165–187. https://doi.org/10.1007/s11024-015-9270-9
Hansson, S. O. (2004). Seven Myths of Risk. *Risk Management, 7*, 7–17.
Harré, R. (1981). Philosophical aspects of the micro-macro problem. In K. Knorr-Cetina & A. V. Cicou-rel (Eds.), *Advances in social theory and methodology: Towards an integration of micro and macro sociologies*. Routledge and Kegan Paul.
Hoenig, B. (2017). *Europe's new scientific elite: Social mechanisms of science in the European research area, Europe's new scientific elite: Social mechanisms of science in the European research area*. Taylor and Francis. https://doi.org/10.4324/9781315446042
Holmwood, J. (2017). The university, democracy and the public sphere. *British Journal of Sociology of Education, 38*, 927–942. https://doi.org/10.1080/01425692.2016.1220286
Hölscher, K., Wittmayer, J. M., Avelino, F., & Giezen, M. (2017). Opening up the transition arena: An analysis of (dis)empowerment of civil society actors in transition management in cities. *Technological Forecasting and Social Change, 0–1*. https://doi.org/10.1016/j.techfore.2017.05.004
Horlings, L. G., Nieto-Romero, M., Pisters, S., & Soini, K. (2020). Operationalising transformative sustainability science through place-based research: The role of researchers. *Sustainability Science, 15*, 467–484. https://doi.org/10.1007/s11625-019-00757-x

Irzik, G. (2007). Commercialization of Science in a Neoliberal World. In A. Bugra & K. Agartan (Eds.), *Reading Karl Polanyi for the twenty-first century: Market economy as a political project* (pp. 135–153). Palgrave Macmillan.

Jasanoff, S. (2005). *Designs on nature: Science and democracy in europe and the United States*. Princeton University Press.

Jung, T., & Harrow, J. (2015). New development: Philanthropy in networked governance—Treading with care. *Public Money Management, 35*, 47–52. https://doi.org/10.1080/09540962.2015.986880

Kara, H. (2015). *Creative research methods in the social sciences: A practical guide*. Policy Press.

Kates, R., Clark, W., Corell, R., Hall, C., & Lowe, I. (2001). Sustainability science. *Science, 292*(5517), 641–642.

Kläy, A., Zimmermann, A. B., & Schneider, F. (2015). Rethinking science for sustainable development: Reflexive interaction for a paradigm transformation. *Futures, 65*, 72–85. https://doi.org/10.1016/j.futures.2014.10.012

Klikauer, T. (2015). What is managerialism? *Critical Sociology, 41*, 1103–1119. https://doi.org/10.1177/0896920513501351

Kouritzin, S., & Nakagawa, S. (2018). Toward a non-extractive research ethics for transcultural, translingual research: Perspectives from the coloniser and the colonised. *Journal of Multilingual and Multicultural Development, 39*, 675–687. https://doi.org/10.1080/01434632.2018.1427755

Labaree, D. F. (2010). How Dewey lost: The victory of David Snedden and social efficiency in the reform of American education. In D. Tröhler, T. Schlag & F. Osterwalder (Eds.), *Pragmatism and modernities* (pp. 163–188). Sense Publishers.

Lai, C., Sbidian, E., Giraudeau, B., & Le Cleach, L. (2020). Twenty percent of secondary publications of randomized controlled trials of drugs did not provide new results relative to the primary publication. *Journal of Clinical Epidemiology, 117*, 20–28. https://doi.org/10.1016/j.jclinepi.2019.09.012

Larsen, P. O., & von Ins, M. (2010). The rate of growth in scientific publication and the decline in coverage provided by science citation index. *Scientometrics, 84*, 575–603. https://doi.org/10.1007/s11192-010-0202-z

Leišyte, L. (2015). Changing academic identities in the context of a managerial university: Bridging the duality between professions and organizations. In W. Cummings & U. Teichler (Eds.), *The relevance of academic work in comparative perspective* (pp. 59–73). Springer.

Longino, H. (1990). *Science as social knowledge*. Princeton University Press.

Louder, E., Wyborn, C., Cvitanovic, C., & Bednarek, A. T. (2021). A synthesis of the frameworks available to guide evaluations of research impact

at the interface of environmental science, policy and practice. *Environmental Science & Policy, 116*, 258–265. https://doi.org/10.1016/j.envsci.2020.12.006
Lynch, K. (2015). Control by numbers: New managerialism and ranking in higher education. *Critical Studies in Education, 56*, 190–207.
Mazzucato, M. (2011). *The entrepreneurial state*. Demos.
McCoy, D., Kembhavi, G., Patel, J., & Luintel, A. (2009). The bill & Melinda Gates foundation's grant-making programme for global health. *Lancet, 373*, 1645–1653. https://doi.org/10.1016/S0140-6736(09)60571-7
Mediavilla, J., & Garcia-Arias, J. (2019). Philanthrocapitalism as a neoliberal (development agenda) artefact: Philanthropic discourse and hegemony in (financing for) international development*. *Globalizations, 16*, 857–875. https://doi.org/10.1080/14747731.2018.1560187
Miller, T. R. (2013). Constructing sustainability science: Emerging perspectives and research trajectories. *Sustainability Science, 8*, 279–293.
Mitchell, W., & Fazi, T. (2017). *Reclaiming the state: A progressive vision for sovereignty for a post-neoliberal world*. Pluto Press.
Mouffe, C. (1992). Democratic citizenship and the political community. In C. Mouffe (Ed.), *Dimensions of radical democracy: Pluralism, citizenship, community* (pp. 225–239). Verso. https://doi.org/papers2://publication/uuid/9C739CCF-6B42-483C-AA66-B35CF7D2FA6E
Mouffe, C. (2005). *On the political*. Routledge.
Mould, O. (2018). *Against creativity*. Verso.
Muller, J. Z. (2018). *The tyranny of metrics*. Princeton University Press.
Nagatsu, M., Davis, T., DesRoches, C. T., Koskinen, I., MacLeod, M., Stojanovic, M., & Thorén, H. (2020). Philosophy of science for sustainability science. *Sustainability Science, 15*, 1807–1817. https://doi.org/10.1007/s11625-020-00832-8
Novak, D. (2019). The arts of gentrification: Creativity, cultural policy, and public space in Kamagasaki. *City and Society, 31*, 94–118. https://doi.org/10.1111/ciso.12195
Palumbo, A., & Scott, A. (2018). *Remaking market society: A critique of social theory and political economy in neoliberal times*. Routledge.
Pettigrew, A. M., Cornuel, E., & Hommel, U. (Eds.). (2014). *The institutional development of business schools*. Oxford Scholarship Online.
Popper, K. R. (1979). *Objective knowledge: An evolutionary approach*. Oxford University Press.

Pusser, B., Kempner, K., Marginson, S., & Ordorika, I. (Eds.). (2011). *Universities and the public sphere: Knowledge creation and state building in the era of globalization*. Routledge.
Richards, R. (2010). Everyday creativity: Process and way of life— Four key issues. In J. C. Kaufman & R. J. Sternberg (Eds.), *The cambridge handbook of creativity* (pp. 189–215). Cambridge University Press.
Ruck, A. (2020). Artwashing Education? *International Journal of Art & Design Education, 39*, 405–417. https://doi.org/10.1111/jade.12290
Sayer, A. (2011). *Why things matter to people: Social science*. Cambridge University Press.
Schrum, E. (2012). To "administer the present": Clark Kerr and the purpose of the postwar American research university. *Social Science History, 36*, 499–523. https://doi.org/10.1215/01455532-1717154
Schugurensky, D. (2006). The political economy of higher education in the time of global markets: whither the social responsibility of the university? In R. Rhoads & C. Torres (Eds.), *The university, state, and market: The political economy of globalization in ThSe Americas*. Stanford University Press.
Scoones, I. (2009). Livelihoods perspectives and rural development. *Journal of Peasant Studies, 36*, 171–196.
Scott, J. C. (1998). *Seeing like a state: How certain schemes to improve the human condition have failed*. Yale University Press.
Shepherd, S. (2018). Managerialism: An ideal type. *Studies in Higher Education, 43*, 1668–1678. https://doi.org/10.1080/03075079.2017.1281239
Shiva, V. (2016). *Who really feeds the world?* North Atlantic Books.
Shiva, V., & Shiva, K. (2018). *Oneness vs. the 1%: Shattering illusions, seeding freedom*. Spinifex Press.
Shrivastava, P., Stafford Smith, M., O'Brien, K., & Zsolnai, L. (2020). Transforming sustainability science to generate positive social and environmental change globally. *One Earth, 2*, 329–340. https://doi.org/10.1016/j.oneear.2020.04.010
Silver, I. (1998). Buying an activist identity: Reproducing class through social movement philanthropy. *Sociological Perspectives, 41*, 303–321.
Swyngedouw, E. (2018). *Promises of the political: Insurgent cities in a post-political environment*. The MIT Press.
Temper, L., McGarry, D., & Weber, L. (2019). From academic to political rigour: Insights from the 'Tarot' of transgressive research. *Ecological Economics, 164*. https://doi.org/10.1016/j.ecolecon.2019.106379
Tierney, W. G., & Lechuga, V. M. (2005). Academic Freedom in the 21st Century. *Thought & Action*, 7–22.

Turnhout, E. (2018). The politics of environmental knowledge. *Conservation and Society, 16*, 363–371. https://doi.org/10.4103/cs.cs

Vanloqueren, G., & Baret, P. V. (2009). How agricultural research systems shape a technological regime that develops genetic engineering but locks out agroecological innovations. *Research Policy, 38*, 971–983. https://doi.org/10.4324/9781315666396

Wals, A. (2019). Publish AND perish: how the commodification of scientific publishing is undermining both science and the public good [WWW Document]. Learn. Sustain. times Accel. Chang.

Ware, N. (1929). *The labor movement in the United States, 1860–1895: A study in democracy*. Vintage Books.

Wellcome Trust. (2020). *What researchers think about the culture they work In.* https://doi.org/10.1042/bio20200032

Wittmayer, J. M., & Schäpke, N. (2014). Action, research and participation: Roles of researchers in sustainability transitions. *Sustainability Science, 9*, 483–496. https://doi.org/10.1007/s11625-014-0258-4

Wolin, S. S. (1960). *Politics and vision: Continuity and innovation Western political thought*. Little, Brown and Company.

Worms, J.-C., & Boman, J. (2017). The post-doc problem [WWW Document]. *European Science Found*. https://www.esf.org/news-media/esf-blog/post/the-post-doc-problem/. Accessed 11 June, 20.

Wright, E. O. (2006). Compass points. *New Left Review, 93–124*. https://doi.org/10.1016/b978-0-12-374722-8.00035-9

3

Cooking Commoning Subjectivities: Guerrilla Narrative in the Cooperation Birmingham Solidarity Kitchen

Sergio Ruiz Cayuela and Marco Armiero

Introduction

Capitalism is a system inherently unequal and undemocratic (Wood, 1995). In the capitalist social and productive organization, a small elite is constantly accumulating wealth by dispossessing the rest of the population of their labour (waged or unwaged) and transforming people and the environment either into resources to be exploited or socio-ecological dumps for the toxic remains of production and consumption. These facts, which remain hidden in plain sight for most of us, lead to an uncontestable conclusion: capitalism is not sustainable and very few people benefit from it. Why then would the majority of the population accept their subaltern position with all its consequences (dispossession

S. Ruiz Cayuela (✉)
Rachel Carson Center, Munich, Germany

M. Armiero
Institute for Studies on the Mediterranean, CNR, Rome, Italy

Environmental Humanities Laboratory, KTH, Stockholm, Sweden

A. Franklin (ed.), *Co-Creativity and Engaged Scholarship*,
https://doi.org/10.1007/978-3-030-84248-2_3

of their labour, denial of opportunities for future generations, oppression of minorities, toxicity and socially constructed diseases, to mention a few)? There is no simple answer to this question. Perhaps, Gramsci can help us to understand the intricacies and even contradictions of capitalist success through the concept of hegemony (Ramos, 1982). Capitalist elites exercise their power through coercion, violence, and expropriation, but also by "winning over" subaltern classes, imposing on them a hegemonic discourse that reproduces subalternity while convincing subalterns that they can change their conditions through "hard work" and competition. Discourses have always been a key tool in normalizing injustice and inequality. In fact, the origins of capitalism are closely linked to the spread of discourses of racial superiority and even dehumanization of Indigenous peoples in the colonies and women everywhere (Federici, 2004). As the consolidation of patriarchal and colonial structures have proved in the subsequent centuries, discourse formation is closely interlinked with material conditions of life.

The use of discourse is not only something from the past. In the current neoliberal era of capitalism, elites are crafting intricate narratives to legitimize austerity, precarity, environmental degradation and coloniality among other things. Naomi Klein, for example, describes how advocates of neoliberalism portray devastating catastrophes, such as natural disasters or terrorist attacks, as opportunities to implement free-market policies in what she describes as a 'shock doctrine' (2007). This requires the creation of a dehumanizing discourse in which disasters are assessed quantitatively and even the loss of human lives is evaluated in economic terms. Economistic assessments of disasters are then used to justify private investment, which is presented as the only possible way to revitalize the battered local economy. It is precisely this widespread perception of capitalism as the only possible system that Mark Fisher calls 'capitalist realism' (2009). Borrowing a quote originally attributed to Jameson and Žižek, Fisher asserts that "it is easier to imagine the end of the world than the end of capitalism" (*Ibid.*, p. 2). He goes on to unpack this idea by describing how cultural agencies, including the media and the educational system, work in ways that preclude the possibility of even imagining alternatives. An important parcel of this strategy relies on the systematic erasure of non-mainstream (hi)stories of resistance and lived alternatives; another world is not only impossible in the present, it must disappear from our stories about the past and from our imagination of the future. It might be worth mentioning here, as example, the recent decision in the UK by the Department of Education to forbid the use in schools of any material produced by anti-capitalist groups (Busby, 2020). The concept of 'capitalist realism' highlights the tight connection between discourses and material conditions of life, as the narratives that narrow the realm of possibility also constrain transformative thought and action (Fisher, 2009, p. 16). Rancière examines closer the mechanisms of legitimization of inequality and injustice in the creation of what he calls the 'distribution of the sensible': the perception and normalization of what constitutes common sense, and what is excluded from it (2004/2013). In fact, Rancière claims that history is a form of fiction: "Politics and art, like forms of knowledge, construct 'fictions', that is to say material rearrangements of signs and images, relationships between what is seen and what is said, between what is done and what can be done" (*Ibid.*, p. 39). Therefore, Rancière expands the analysis to aesthetic forms beyond the narrative in analysing discourse formation, and the way it shapes our perception of the world and the realm of political possibility. In political ecology, Stefania Barca (2014) and Armiero et al. (2019) have both argued that the imposition of environmental injustice always comes hand-in-hand with the imposition of a toxic narrative,[1] which either silences or normalizes injustice. This means that the struggles for environmental justice are always also struggles for narrative justice.

Although sophisticated, the capitalist systems of normalization described above do not go uncontested. The totalizing and homogenizing forces of capital cannot stop the constant emergence of cracks that, although usually not deep enough to threaten the system, constitute alternatives to a few people and prefigure ways of inhabiting the world that do not abide to capital (Holloway, 2010). A form of resistance that has been repeatedly used across multiple geographies and historical moments is that of commoning. Colonized indigenous peoples,

[1] Toxic narratives are those rhetoric dispositifs that silence, invisibilize or normalize injustices, often resulting in blaming the victims for their conditions (Armiero et al., 2019).

exploited factory workers, peasants in the global South, urban dwellers fighting for the right to the city, women fighting against patriarchy, in all these examples and beyond, commoning has been experienced as a radical alternative to the totalitarianism of capitalist realism that imposes individualism and private property.

We frame commoning as the socio-ecological infrastructure that (re)produces commons through care, sharing, and inclusion, therefore sabotaging the wasting relationships that produce inequalities through extraction, privatization and exclusion (Armiero, 2021). Through commoning, commoners do not only share and have access to a set of resources, they are also entitled to decide on the ways of using and sharing them, while enhancing relationships of cooperation and mutuality among them and with the environment. In this sense, we argue that commoning has the emancipatory potential to advance socioecological relationships based on cooperation, horizontality, openness and care. The beauty of commoning resides in that it is not only a form of resistance, but it performs an alternative: while capitalism sees commons as a thing to be expropriated and monetized, commoners practice commoning as a set of socio-ecological relationships that reproduce both commons and commoners. As De Angelis (2017) reminds us, only the commoning of socially reproductive activities (such as food growing, care work or energy provision) can bring about truly emancipatory commons that pose a viable alternative to capital. While capitalism frames social reproduction as a set of processes that reproduce labour power, emancipatory commoning puts the reproduction and wellbeing of the commoners at the very centre. Mainstream discourses and toxic narratives have worked hard to conceal these practices from the public eye. As Marina Sitrin and Darío Azzellini put it: "[o]fficial history … is told by the 'victors'" and "[t]hey have no interest in telling the history of people taking their lives into their own hands" (2014, p. 8).

In this chapter we aim to expand what we have called "guerrilla narrative" (Armiero et al., 2019), proposing it as a powerful tool for subaltern communities to resist marginalization and oppression. Our aim is to explore the possibilities of "guerrilla narrative" to uncover stories of commoning that challenge homogenizing discourses, toxic narratives and capitalist legitimacy. We want to explore the power of narrative strategies to expand the commons by advancing the production of commoning subjectivities. By counter-narratives we mean discursive and material strategies that re-invent the possibilities of the present while practising antagonist collective identities.

Section 3.2 deals with our positionality as militant researchers, our methodological choices, and the rationale for using Cooperation Birmingham as a case study. In Sect. 3.3, we go deeper into the concept of "guerrilla narrative", focusing on the possibilities that it offers in a context of commoning. In Sect. 3.4 we introduce the case study. First, we characterize the permanent crisis of social reproduction that is taking place in Birmingham. We then describe the foundation and basic dynamics of Cooperation Birmingham, a mutual aid organization in which commoning practices thrive. In Sect. 3.5 we examine in-depth the narratives and the co-production process of the Cooperation Birmingham newsletter, which we analyse within the guerrilla narrative framework. In Sect. 3.6 we broaden the scope of guerrilla narrative by examining how the cooking and caring at the solidarity kitchen were central in creating commoning subjectivities. We argue that these very material practices hold an inherent narrative power, and that literary forms of guerrilla narrative are enhanced by them.

Democratizing Knowledge Through Militant Research

To better grasp the scope of this chapter, it is crucial to understand our positionality. We both take seriously the need for transformative change, and are active members of several political and environmental groups. We consciously engage in commoning practices in our everyday lives and in our academic work. We recognize that knowledge and power are closely linked while denying claims of neutrality and objectivity in our own research. Instead, following the tradition of militant and other action oriented approaches to research, we take sides and produce knowledge that aims to advance specific struggles (Derickson & Routledge, 2015; Halvorsen, 2015). We recognize academia as a site of political struggle, where knowledge production can be directed either to reinforce the *status quo*, often under the pretence of scientific neutrality, or to achieve social transformations. Thereby, in producing politically loaded research, we maintain our scholarly integrity, we do not falsificate our sources, we do not conceal information for the sake of our argument, while engaging in "research that produces knowledge for social struggle and is itself a form of political intervention" (Dalton & Mason-Deese, 2012, p. 445). Following Sandra Harding's "strong objectivity" approach (1995), we maintain that producing situated knowledge does not jeopardize but rather enhances the quality of that knowledge. It is this approach that informs our intentions with this chapter. We aim to advance knowledge on guerrilla narrative and its potential to contribute to the expansion of the commons, both theoretically and in practice.

Armiero et al. (2019) have defined guerrilla narrative as the sabotage of toxic narratives, or, in other words, the occupation of that space with counter-hegemonic storytelling. They have employed guerrilla narrative mainly as a tool to uncover the toxic legacy of capitalism in the lives of subalterns. With this chapter we aim to mobilize guerrilla narrative as a creative path to nurture alternatives to capitalist realism, especially in the forms of commoning. This chapter is grounded on real life struggles. Our insights and reflections are aimed at supporting the mutual aid efforts of Cooperation Birmingham, an organization based in the city of Birmingham, United Kingdom. The first author of this chapter is an active member in several political organizations and community groups in the West Midlands of England (the region where Birmingham is located), and is one of the co-founders of Cooperation Birmingham. We hope that our insights will help advance the goals of the organization.

Cooperation Birmingham is a mutual aid organization established in March 2020 that has been active in providing relief to people living in poverty and self-isolating during the Covid-19 pandemic. However, members of the organization see the recent sanitary crisis as the tip of an iceberg that has been forming during the last decades with the dismantling of the welfare state and the harshening of the living conditions of the subalterns in the UK. Therefore, the long-term goal of the project is to bring together several local organizations (both formal and informal) in order to provide a social and material infrastructure for enhancing the empowerment and autonomy of marginalized communities (Ruiz Cayuela, 2020a). At the same time, Cooperation Birmingham works on spreading a culture of self-organization and solidarity in the city, as viable alternatives to capital and the state. Even if the concept of "guerrilla narrative" is not explicitly used by members of Cooperation Birmingham, the values represented in the organization, the fact that it emerges from the community, and the importance given to diverse narrative practices that emerge from below, make it a suitable case study for this chapter.

Data collection for this chapter is closely linked to the material co-produced by Cooperation Birmingham. Our main source is the four issues of the newsletter that the organization published between May and August 2020.[2] The newsletter was widely distributed through different channels. Printed copies were delivered with meals, made available for free at the local Warehouse Cafe, and given to participants of Cooperation Birmingham to share with whom they wished. The newsletter was also distributed online through Cooperation Birmingham's blog and social media, both as a pdf and as a podcast. It has become an open space for people to express their feelings and ideas. In order to complement and contextualize the newsletters, we use other material posted on social media and on the Cooperation Birmingham website; we also use minutes from the meetings of the organization, which are accessible to the public in an open online forum.[3] Finally, we also rely on field notes and personal experiences from the first author, who has been actively involved in the project. This connects our work with militant ethnographic scholarship and practice, which favours a qualitative approach in which the experience of the researcher is emphasized (Juris, 2007). It is important to stress that we have chosen to place the co-produced newsletters and other narrative practices at the core of the discussion and theoretical development. By doing this, and in line with the recent scholarship in "guerrilla narrative", we aim to democratize knowledge production, legitimize different formats as valid sources of knowledge,

[2] You can access all the newsletters through Cooperation Birmingham's blog: https://cooperationbirmingham.org.uk/blog/

[3] https://forum.cooperationbirmingham.org.uk/.

and to implicitly acknowledge all the contributors as co-producers of this chapter.

Guerrilla Narrative

We started to speak about guerrilla narrative in 2017, when a modest grant allowed the Environmental Humanities Laboratory[4] to launch the ToxicBios project. The aim was to gather stories of contamination as experienced and narrated by affected individuals and communities and make them available in an online, open access archive. Our inspiration was the massive EJAtlas,[5] coordinated by Joan Martinez Alier, the largest open access worldwide database on environmental conflicts. The idea was to explore environmental justice controversies from a humanities perspective, building on the assumption that every environmental justice struggle is also a struggle over narratives. Stefania Barca (2014) has spoken of narrative injustice, silencing crucial information and suppressing stories that do not fit into the mainstream celebration of economic growth. Armiero et al. (2019) have built their guerrilla narrative proposal in opposition to what they call "toxic narratives", that is, the rhetoric device operationalized to blame the victims for any kind of problems they are experiencing while naturalizing socio-ecological injustices. Guerrilla narrative works within and against the toxic narratives; while the latter constitute the narrative infrastructure supporting othering and oppression, the former sabotages that infrastructure fostering alternative memories and counter-hegemonic ways of reproducing them.

Toxic narratives are especially instrumental in maintaining the *status quo* when large environmental disasters expose the socio-environmental injustices that are underneath those exceptional events. In those cases, the toxic narrative infrastructure provides explanations of the disaster that never question its causes while promoting an anesthetized memory of it, purified from anger and outrage. Think for instance of the Vajont Dam

[4] The Environmental Humanities Laboratory is based at the KTH Royal Institute of Technology (Stockholm, Sweden) working at the intersection of environmental humanities and political ecology.

[5] https://ejatlas.org/.

Disaster that in 1963 killed almost 2,000 people in the Italian Northeast. There, the toxic narrative implied the naturalization of the event, with both scientific experts and journalists explaining it as a natural disaster, blaming the geology of the mountains rather than the negligence of the corporation or the state. Exemplary of this naturalization is what Dino Buzzati, an influential writer and journalist, wrote immediately after the disaster:

> A stone falls into a glass of water and the water is spilled on the tablecloth. That's it. But that the glass was hundreds of metres high and the stone was as big as a mountain; and below, on the tablecloth, there were thousands of human beings who could not defend themselves. It is not that the glass was intrinsically broken: therefore, we cannot call monsters those who built it, as in the case of the Gleno disaster. The glass was built perfectly ... Once again the fantasy of nature has been bigger and smarter than the fantasy of science. Although defeated in open battle, nature takes its revenge from behind.[6]

Evidently, the attempt to naturalize the event was key, not only for producing a pacified memory but also as a strategy to absolve the corporation and the public officials from their responsibilities. As any efficient toxic narrative, the one about the Vajont also led to the erasure of that story from the collective memory of the nation, and the imposition of a defused local memory where pain and mourning should be performed in tidy and pacified manners. The clash between a guerrilla narrative approach and the mainstream toxic narrative became clear in the story of the two Vajont cemeteries, brightly narrated by the Italian writer Lucia Vastano (2008). In 2000, the original cemetery, built by the survivors after the disaster and inhabited by personal memories and rage, was razed to the ground and replaced with a new cemetery, built by the authorities following the scheme of the war memorials, therefore, completely anonymized and pacified.

Or we can mention the parents who have lost their children due to rare oncological illnesses in the Neapolitan region and have been accused of

---

[6] Dino Buzzati, 'Natura crudele', Il Corriere della Sera, 11 October 1963, quoted here from Armiero (2011).

transforming a private suffering into a public fact. In this case, guerrilla narrative implied to counter-act against the toxic mainstream interpretations of the health crisis occurring in the region; this either blamed the victims (if they got sick, it was because of their lifestyle) or denied the very crisis ("there is no evidence of a correlation between contamination and health problems"). The visual project "Postcards from the Land of Fires", realized by the photographer Mauro Pagnano, was an example of guerrilla narrative, a way of telling the story of suffering and contamination from an embodied point of view.[7] The project gathered a collection of photographs depicting mothers in the rooms that were once occupied by their deceased children, each of them holding in her hands a photograph of the child. According to several commentators, this project was inappropriate because suffering should remain a private issue not something to use in the public sphere. Again, we see toxic narratives silencing injustices and defusing rage, versus guerrilla narrative, reclaiming the right to remember and to tell the stories of oppression and violence.

As these two examples help to clarify, we envision guerrilla narrative as the ensemble of practices that resist toxic narratives while proposing alternative (hi)stories and identities. In this sense guerrilla narrative is not simply the unheard story of oppression reclaimed from the memory dump; rather, guerrilla narrative is the practice of reimagining subaltern stories, storying them, and making collective identities. If it is true that the first step to crush a community is to take its history away (Klein et al., 2009), regaining control of the ways of remembering and storytelling is first and foremost an act of sabotage. This is what we can learn from Indigenous people who have been fighting against the erasure of their stories and memories for centuries, to the point of materially disappearing from the face of the earth; it is telling that the Zapatistas' covering of their faces was explained as a way of making visible those who had been invisibilized by centuries of colonial oppression (Khasnabish, 2013, pp. 12–13).

The toxic narrative infrastructure does not only conceal socio-ecological injustices, it prevents even the possibility of seeing them and imagining another world. This is why we decided to speak of guerrilla

[7] https://mauropagnanophotographer.viewbook.com/homepage/album/terra-dei-fuochi.

narrative and not simply of oral history, although oral history is an important root of guerrilla narrative. Guerrilla narrative implies recognizing that a counter-hegemonic storytelling does not occur in a political vacuum; rather, it strives to emerge under the harsh repression and authoritarianism of mainstream narratives. Given the disparities of the conflicting forces and the violence of toxic narratives, guerrilla narrative is the only realistic choice available to enhance counter-hegemonic visions. As Vitaliano Ravagli and Wu Ming (2005, pp. 148–149) have written:

> To understand something, you need to crumble the myth as it has been handed down to us and dig out from the ruins the living stories. Those that no one has told. The axes to dig up.

The idea that stories are axes to dig up, tools to sabotage the toxic narrative infrastructure that controls the systems of feeling and memories is at the core of the guerrilla narrative project.

Oral history has also aimed at recovering untold (hi)stories while including subjects who have been generally excluded into historical narratives. Guerrilla narrative is a close relative of oral history, but it has a clearer political stance and an antagonist character: subaltern stories do not add nuances to mainstream narratives, they dismantle them. Furthermore, guerrilla narrative recognizes the plurality of means beyond orality through which subaltern people build counter-hegemonic storytelling, including arts, written documents, people's schools, or interventions into the mainstream organizations of public memories. Black Lives Matter, for instance, has questioned racist and colonial monuments and other toxic narratives inscribed into the texture of our collective lives (Lai, 2020).

We envision guerrilla narrative more as a DIY practice than a method. The guerrilla narrative bricolage nature refers to both the radical rejection of the researcher/researched dichotomy and to the creative mobilization of what is already available. While challenging the professionalization of knowledge production, guerrilla narrative humbly acknowledges that counter-hegemonic storytelling has always occurred without any need to be codified by academics.

The ToxicBios project provides the largest empirical experiment in guerrilla narrative, to date gathering about 70 autobiographies, mostly in video formats, but also texts, audios and other more artistic formats (including songs and poems). Although the project has an unquestionable anthropocentric focus and an inclination towards individual narratives, in its realization, that is, in the bricolage of counter-hegemonic storytelling, it often challenges these limits. Several storytellers have included in their autobiographical accounts of contamination of non-human companions, such as for instance fish,[8] trees[9] or a river.[10] The tension between individual and collective stories almost explodes in the choral narration of Enzo's biography, which as the title clearly states, he would have never told himself.[11] Instead, six friends, all militants in the same grassroots organization, decided to narrate Enzo's story, therefore, pushing back against the borders that police individual and collective identities. Similarly, in the ToxicBios project there were other collective stories, told by groups of people rather than individuals.

We envision guerrilla narrative not as a methodology, but a DIY assemblage of existing practices that have been employed broadly beyond the use of that specific label. We have mentioned, for instance, Black Lives Matter's challenge of racist and colonial monuments and the Zapatistas fight against invisibilization, but we could also include a *No una de menos* attack on the codified symbols of patriarchy, being them a statue, the usual all-male syllabus, or the functioning of our languages.[12] Environmental justice movements have often cultivated some forms of counter-hegemonic narratives, preserving their histories and building

---

[8] Arlindo Marques and the Tejo river pollution, available at http://www.toxicbios.eu/#/stories.

[9] Angela Rosa, fighting oil and natural gas exploration, available at http://www.toxicbios.eu/#/stories.

[10] António Pinto and Rosa Maria Pratas from ADACE in Aveiro, available at http://www.toxicbios.eu/#/stories.

[11] Enzo Tosti would never tell his story, available at http://www.toxicbios.eu/#/stories.

[12] In order to give a few concrete examples of this, we can mention the repeated attack against the statue of a famous Italian journalist, Indro Montanelli, accused of raping an African teenager during the colonial war in Ethiopia. The number of initiatives sanctioning all-male syllabi in university courses has skyrocketed and it would be imposisble to list all of them. Similarly, the struggles for more inclusive languages have become crucial both in social movements and in academia.

positive identities. This is the case, for instance, of the movement generated around the ex-SNIA Viscosa factory in Rome, Italy, especially in its effort to recover the (hi)stories of resistance against the toxic regime of the factory through the recovery of workers' files, abandoned in the building, and the creation of a self-managed archive (Tola, 2019).

Narrowing down to academic and research practices, we can mention the collective Guerrilla Cartography, for instance, that seems to be inspired by a similar counter-hegemonic approach in their production, together with communities, of thematic atlases.[13] Directly inspired by Toxic Bios is the Guerrilla Digital Public History seminar created by Shawn Graham at Carleton University in Ottawa, Canada, which asks the crucial question: "What are the stories in Ottawa that require a guerrilla digital public history?"[14] In Tuzla, Bosnia Herzegovina, a group of researchers has joined forces with workers creating the Workers' University; this has produced narratives, even a graphic novel, on the present and past struggles in the city's chemical factory.[15]

All these examples demonstrate that, as we have argued above, guerrilla narrative is both the very stories produced through it and the process of producing/looking for them. Just as commons cannot be decoupled from commoning, that is from the socio-ecological practices (re)producing commons, in the same way, counter-hegemonic stories are not independent from guerrilla narrative, that is, from the narrative practices (re)producing those stories. Guerrilla narrative and commoning are bound together as performative practices that produce very material outputs (counter-hegemonic stories and commons) as well as socio-ecological subjectivities. In other words, guerrilla narrative or commoning are not "natural" products of a specific kind of community, rather in practising them new communities are co-produced.

---

[13] https://www.guerrillacartography.org/.

[14] https://shawngraham.github.io/guerrilla-dh/#.

[15] https://reclaimingdita.com/thestory.

From crises of Social Reproduction to Commoning and Mutual Aid

Birmingham is the second most populated city in the UK within municipal boundaries, with over 1 million inhabitants. It is a very ethnically diverse, working-class city with a strong presence of migrant communities. Birmingham has been in an almost permanent crisis for years, and a significant proportion of its population live in poverty. A report released in June 2020 reveals that by that time, Birmingham had a 14.5% rate of claimant unemployment, compared to a 7.8% for the whole country (Birmingham City Council, 2020). Another report published by the Office for National Statistics in 2018 indicated that Birmingham was one of the cities with most non-permanent workers in the UK, with around 8% of the active population working 'zero-hours' contracts, seasonal or casual contracts (Gouk & Rodger, 2018). These and other dire economic statistics reflect the dramatic conditions in which people in Birmingham are forced to live. However, it is by looking at the 'non-productive' activities and relationships that we can better understand the context. Birmingham hosts the two districts that top the national ranking of child poverty. In fact, one third of the children in the city live in poverty (Francis-Devine, 2020). Many people struggle to cover even their basic needs, including food. In 2017, for example, 33,500 people in Birmingham used food banks (Belcher, 2018), and the number has kept rising in recent years. Housing and hunger crises have become the norm for a considerable number of Brummies, and the situation is currently being further aggravated by the Covid-19 pandemic (Lawrence, 2020).

However, as Massimo de Angelis (2007) puts it, these are just 'horror statistics' to which we have grown accustomed. The truth is that behind the cold numbers, there are people suffering and struggling. The most obvious group are workers who have been made redundant, those who have been forcibly turned non-permanent, and even those who still keep their jobs but who are constantly burdened with still more tasks and feel that they could be the next to be fired. Families living in poverty, and especially children, have also been enduring a stressing time due to the controversial withdrawal (and almost immediate restoration) of subsidized school-meals during mid-term 2020 (Brewer, 2020).

And we should not overlook the devastating effects on mental health of economic hardship combined with a culture of individual responsibility and shaming of failure. In times of crisis, this has usually led to dramatic increases in the depression and even suicide rates (Zapata Hidalgo, 2020). This multidimensional and holistic understanding of the context leads us to an interpretation of crises as more than just falling rates of profit. Subalterns experience crises in very material ways, as a retrenchment of their level of well-being and even as a struggle to stay alive. The other side of an economic crisis, thus, is a multiplicity of 'crises of social reproduction' (Caffentzis, 1999). In fact, it is this dual character that makes crises disciplinary tools instrumental for the normal functioning of globalized capitalist markets (De Angelis, 2007). Therefore, what is at risk is not the reproduction of capital, but the reproduction of life. The crises of social reproduction skyrocketed when austerity policies were implemented after the 2008 economic crisis. Between 2010 and 2019, for example, the British government "announced more than 30 billion pounds ... in cuts to welfare payments, housing subsidies and social services" (Mueller, 2019), a further dismantling of the already diminished welfare state at the expense of the most marginalized. This trend of rampant neoliberalization can be traced back several decades and allows us to find the narrative foundations that normalize the extreme situation lived by the subalterns in the UK today: contempt for 'the other', fierce competition, and extreme individualism. This is the toxic legacy of Thatcher's foundational credo: "there is no such thing as society, there are individual men and women".

Cooperation Birmingham is an initiative ignited by a group of people involved in political organizations, community groups and workers' and housing cooperatives. Inspired by Cooperation Jackson[16] in the US and their quest for economic democracy (Akuno & Nangwaya, 2017), Cooperation Birmingham aims to become an active partnership between formal organizations committed to social transformation (e.g., cooperatives or unions) and politicized grassroots organizations. The idea is that the former can materially and logistically support the latter, thus

---

[16] In fact, when Kali Akuno (spokesperson of Cooperation Jackson) visited Birmingham in May 2019, he was invited and hosted by several members of what would become Cooperation Birmingham.

enabling an expansion of autonomous commoning practices in the city. The organization was supposed to start building a base of support, developing a participatory model, and gradually becoming active through 2020. However, when Covid-19 struck and deepened the manifold crises of social reproduction described above, members of Cooperation Birmingham felt the urge to provide crisis relief and stepped forward. In March 2020, Cooperation Birmingham started running a solidarity kitchen, a self-organized effort to deliver a daily healthy and hearty warm meal to people in need and/or self-isolation (Ruiz Cayuela, 2020b). Between March and August they delivered over 20,000 meals relying entirely on donations, infrastructural support from local co-ops, and the voluntary work of over 200 participants. The solidarity kitchen was framed as a mutual aid project. Decision-making was made in open online assemblies that all participants were encouraged to attend. An open online forum was enabled where all the minutes were made public and everyone could add items to the meetings' agendas or raise discussions. The kitchen crew and drivers were always given a meal in exchange for their work.

In addition to the solidarity kitchen, Cooperation Birmingham also produced and distributed reusable protective face masks used during the pandemic. It is also interesting to see how the solidarity kitchen has had spin-offs with a more sustainable scope, such as a food delivery workers' cooperative that is already running, and a compost production project that is still under discussion.

The values enacted and the strategies developed during this time make Cooperation Birmingham a clear example of commoning. In fact, members of Cooperation Birmingham have been inspired by commoning theories and experiences when planning a strategy of consolidation and expansion of the organization. On the one hand, this strategy aimed to expand the material autonomy and social reproduction capacity of Cooperation Birmingham; on the other hand, it intended to produce new commoning subjectivities within and beyond the borders of the organization (Ruiz Cayuela, 2021). In the next section, we will use a guerrilla narrative lens to investigate the narrative strategies used by Cooperation Birmingham that contributed to the creation of commoning subjectivities. Broadly speaking, the organization was consciously focused on dismantling the idea of charity, which is hegemonic in the UK third sector, and replacing it with discourses and practices of solidarity and mutual aid. Various forms of communication were used to convey this message, including direct conversation with occasional participants and food recipients, posts on social media, open online discussions on the forum, or website information and articles. However, the most focused and sustained effort to challenge toxic narratives and create anew was the co-production of a newsletter.

The Cooperation Birmingham Newsletter

The Cooperation Birmingham newsletter takes the form of an A3 sized triptych, with articles on one side and artwork on the other so it can be used as a poster. Around 300 physical copies of each issue were printed and distributed, but it was also posted online as a pdf and as a podcast. The newsletter was first edited in mid-May, less than two months after the solidarity kitchen started running. By that time, the solidarity kitchen was working smoothly and Cooperation Birmingham was gaining popularity. Although some of the inherent values were being practised on the ground (avoiding gatekeeping practices for example, "we ask no questions and we take no money"), the prioritization of the material emergency relief was somehow watering down the political nature of the organization. The general feeling in the group was that Cooperation Birmingham was successful in delivering meals, but not messages. Some members called this fact to attention, and proposed the creation of a newsletter. The newsletter was conceived as an open space where everyone related with Cooperation Birmingham or sister organizations could write about a variety of topics of interest. It was a co-produced effort where an open and horizontal organization was trying to show with practical examples that cooperation, solidarity and self-organization are all valuable practices for the subaltern communities.

One of the main goals of the Cooperation Birmingham newsletter has been to challenge and dismantle mainstream toxic narratives that seek to divide subaltern communities and pit them against each other. This was clearly stated in the very first issue, where individualism was tackled with a poster displaying the message "all we have is each other" (see Fig. 3.1), and an article by the same title proclaimed that "only in cooperation (and not in competition, like we have been told) we thrive". In that same issue, narratives that criminalize the poor and hold them accountable for their situation were addressed with an article about rent strikes. In it, the anonymous author went on, affirming that "housing IS healthcare" [original emphasis] and that "[e]victions in the middle of a pandemic are a health hazard", to finalize the article with stories of successful rent strikes happening at the time. In the second issue, a guest article by a member of the sister organization Cooperation Town (from Kentish Town, London) contested the framing of the pandemic as a natural disaster by highlighting the already existing crises of social reproduction that many communities were facing before the pandemic, and the dismantling of public services that has taken place over the last decade. In that same issue, members of Cooperation Birmingham wrote a statement explaining their decision to refer to people involved in the project as 'participants' instead of 'volunteers'. They associated the term 'volunteer' with the practice of charity, and explained how it hides power relationships. Rejecting hierarchical structures within organizations and between 'volunteers' and recipients, the authors wrote: "we do not work for anyone but for the people involved... We are all participants and we can all participate!".

The third and fourth issues, launched in June and July 2020 during the peak of the Black Lives Matter movement worldwide and in the UK, were especially vocal against racism. Right after the murder of George Floyd,[17] Cooperation Birmingham encouraged people to attend local anti-racist protests by acknowledging that "racial discrimination and oppression is also happening here in the UK, where it is linked to a colonialist past and present". In the fourth issue, an article titled Black Lives Matter celebrated the toppling of the statue of slave trader Edward Colston in Bristol, and responded to prime minister Boris Johnson's attempt to stop the widespread protests by stating that Britain is not a racist country. "This is a lie", the anonymous author wrote before

[17] George Floyd was a black man who was murdered by a police officer on 25 May 2020 in Minneapolis. His death sparked a global upsurge of the Black Lives Matter movement that took the form of demonstrations and riots against racialized police brutality all around the world.

It is a basic right to have access to food and support. This is unquestionable.

What is Cooperation Birmingham?
We are a cooperative of working class people living in and around Birmingham. We want to build an organisation that provides mutual aid to us and and our surrounding communities. We want to build a solidarity economy that works for us and not for the rich. We currently run a solidarity kitchen and a mask making project, but we aim to expand by including a second kitchen, a composting site and many other things!

Access to food in the U.K is grossly unequal: through Trussell Trust food banks, the largest provider, people can only access food with a voucher. Vouchers are given out only via 'care professionals' such as health visitors, school counsellors, social workers etc.The vouchers ensure that only those deemed deserving by the state are able to access food. This can often exclude homeless people, asylum seekers, and many others who might not check the tick box but will need food.

Our aim is not to give charity to the 'needy', but instead to be a platform where people can contribute according to their abilities and receive what they deserve. This means that we need help with things such as cooking, delivering, sewing masks, social media, or admin and we want to support people in doing so - even though this is a crisis we can still learn and grow together. If you would like to contribute, you can reach through email, social media or just tell one of us.

This is the first edition of our newsletter, in which we want to provide some context to what we are doing and hear the voices of fellow Brummies. If you are either actively participating in Cooperation Birmingham or receiving food from us, we would like to share your story and hear your thoughts, reach us at info@cooperationbirmingham.org.uk

Over 5,500 meals delivered

SOLIDARITY IN SEPARATION

It takes soap & a lot of to run a kitchen
The Digbeth solidarity kitchen is our main project. We deliver around 150 free meals every day to people around the city. We ask no questions and we take no money, it's solidarity without conditions.

On a normal day, there are usually 17 people working for the meals to be cooked, and distributed. Five people work at the kitchen of the Warehouse Cafe from 11,30am to 6pm cooking and packing the meals. Usually eight drivers deliver the food all around the city, following a strict health and safety protocol at every single stop (you can find our guidance in podcast form online). And four people are doing backroom work, which includes confirming and taking orders, creating routes, coordinating with the kitchen and offering remote support to drivers.
As for last week, the total hours of work by all people involved in the daily operation of the solidarity kitchen amounted to 2,400. At minimum wage rate, that would mean almost £25,000. Obviously none of us are paid for this work, so these astonishing numbers show that this would be possible without the solidarity of the people from Birmingham, and also the enormous shortfall in financial support for people and communities from the government. However, those numbers do not capture the whole magnitude of the solidarity kitchen.

In order to make the daily operations happen, there is much work that needs to be done in advance. Someone needs to contact all the people involved and make sure that they know what to do. Someone else needs to work on social media and outreach so potential food recipients know about us. Someone else needs to work on raising funds and managing finance so we have enough to keep cooking. And the same again next week!

Tech for good
We have set up Cooperation Birmingham with an open tech infrastructure, we have a forum on our website which people can use to get involved in the project and a map of all mutual aid support in the city, public finances and we make sure we keep all the important personal data safe. We are also working on a rota app to make it easier to sign up to a shift and an exciting new technology called Co-op Cycle – something like Deliveroo but run on solidarity instead.

All we have is each other
Since the Covid-19 pandemic started you have probably heard about mutual aid groups blooming all around the country. It is really inspiring to see this boost of solidarity in times of crisis! But… what exactly is mutual aid?
Mutual aid is the very simple idea that only in cooperation (and not in competition, like we have been told) we thrive. It is the idea that we help each other and we don't leave anyone behind. But above all that, it is the idea that together we are powerful.

Caring for and sharing with your family and your neighbours is a very simple and beautiful way of practising mutual aid. But in Cooperation Birmingham we want to bring the concept to a new level. We want to create a large community of solidarity able to make collective decisions and work for the common good.

Resistance is fertile
We get most of our food from other groups in the city, lots of the veg is intercepted from supermarkets landfill still perfectly ripe. Although there are some good things in the food industry there's a whole lot that's rotten too.

Food should be valued more in society from seedlings to meals and beyond. Our kitchen produces a mountain of vegetable peelings weekly. To close that loop we want to set up a composting service working in local communities across the city - if we've got this much imagine how much a city of 2 million produces, yet the council doesn't collect it, all that energy and nutrients gone up in smoke at Tyseley. If soil, worms and getting your hands dirty is something you find fun send us an email and pitch in!

Rent Strike?
The pandemic is having a huge economic impact, millions are struggling to afford to live and renters are coming together to say housing IS healthcare. Evictions in the middle of a pandemic are a health hazard all round and good housing was too expensive and hard to come by before this started anyhow.

HOUSING IS HEALTHCARE

One heartening story is of a landlord in Los Angeles, California, who sent an email demanding full rent payment to all his tenants but copied them all in by accident. Now able to contact each other, the tenants organised a rent strike across the city, withholding rent so they can pay for food! Closer to home in Lancaster students who have had to leave accomodation to be home with parents are on strike for refunds for housing they will never use.
If you have any problems with your landlord or paying your rent give ACORN Birmingham a shout birmingham@acorncommunities.org.uk

A news bulletin for extraordinary times. Submissions to info@cooperationbirmingham.org.uk 15/05/20 coopbrum

Fig. 3.1 First issue of the Cooperation Birmingham newsletter. Source: https://cooperationbirmingham.org.uk/

backing the defiance with telling numbers. These are just some examples that show how the newsletter was used by Cooperation Birmingham to directly confront narratives normalizing individualism, competition, the criminalization of poverty, classism and racism. The first step towards building commoning subjectivities is uncovering and unlearning deeply held toxic narratives that translate into isolation, division and discrimination; and therefore threaten the emancipatory character of commoning practices.

However, Cooperation Birmingham's newsletter was not limited to the denunciation of mainstream narratives and their noxious effects. The construction of alternative narratives was at least as important. In fact, the sabotaging of a toxic narrative and the building of alternatives happen simultaneously and are inherently connected. Let us emphasize, 'alternative narratives', in plural, because against the imposition of a single homogenizing story, members of Cooperation Birmingham consciously sought to include a diverse array of perspectives in their newsletter. The first issue of the newsletter included a brief description of the organization that described Cooperation Birmingham as a mutual aid network. In the same issue, an anonymous author sought to explain the idea of mutual aid in simple terms, and finished by stating: "We want to create a large community of solidarity able to make collective decisions and work for the common good". The second issue gave concrete shape to those ideas through the contributions of two members of Cooperation Birmingham. Bea Hughes, a kitchen participant, described her experience as enjoyable and empowering. Her relaxed tone helped to tear down the psychological barrier between 'the masses' and 'the vanguard', making of 'joyful militancy' an easily relatable feeling. Shamima Akhtar, a food recipient, reported the huge value that solidarity and care have for the subalterns' bodies and minds: "This is the type of unquestioned support 'vulnerable' people like me need, rather than charity-based support that puts pity at its centre". In that same issue, the newsletter pointed towards the formation of a political subject beyond the locality by including an article written by a member of London based Cooperation Town. Only by creating an autonomous wider network of solidarity "we can resist going back to the harmful 'normality' … and plot our way towards a better future" they asserted.

The third issue included an anonymous proposal for addressing food sovereignty issues within Cooperation Birmingham, and framed the creation of a decommodified food system as a way towards autonomy and socio-environmental justice: "If the people who grow, prepare, distribute and eat food can be freed from needing to spend and earn money, a fairer food and farming system is possible". Still reinforcing the idea of solidarity networks and the forging of political subjects on a wider scale, the third issue also included an article by a member of the local red gym, who called for more comradely and enjoyable collective spaces where people from all genders and ethnicities can feel comfortable to exercise their bodies. In the same number, one of the chefs of the solidarity kitchen told the epic story of how the radically democratic structure of the anti-imperialist movement in Poland allowed him to get highly valuable cheffing skills. In the fourth issue, the article 'Common People' called for "reclaim[ing] a new commons as a way to provide for ourselves". The anonymous author pointed towards three main pillars that should inform all commoning practices, and thus all the activities of Cooperation Birmingham: solidarity, self-organization and direct democracy.

All these examples show how the Cooperation Birmingham newsletter has been building autonomous narratives that portray 'joyful militancy' within solidarity networks not only as a real possibility, but as materially desirable for people in need. Grounded on direct relatable examples, Cooperation Birmingham has been trying to build new subjectivities based on cooperation, autonomy, diversity, mutual aid, radical democracy and self-organization. In other words, the newsletter has aimed to build a political subject ready to reclaim and inhabit the emancipatory commons.

We argue that the diverse practices of co-production that were (and are still being) distilled into the Cooperation Birmingham newsletter constitute a form of storytelling that can be framed within the guerrilla narrative strategy. As we have seen, the articles and artwork included in the newsletter specifically tackle mainstream toxic narratives, taking them out of their apolitical vacuum, and dismantling them through first-hand experiences, practices and knowledge. At the same time, those articles are drawing from the concrete experiences of Cooperation Birmingham, its members, and sister organizations to build a diversity of autonomous narratives through which the subaltern can reclaim agency and lead the expansion of the commons.

Another feature that highlights the guerrilla character of the Cooperation Birmingham newsletter is the focus on the process. For it is not only the final publication, but the different steps in the co-production process, including the articulation of ideas and the challenging of pre-established roles, that produces commoning subjectivities. Participants of Cooperation Birmingham did not have previous experience in editing or publishing, so being able to create the newsletter felt like a huge success and reinforced comradeship and dignity among the people involved. These types of achievements constitute 'small victories' that can help to forge collective identities.

Written or recorded stories can be powerful tools for conveying ideas, values and even worldviews; the reader will surely relate to experiencing small (or big!) epiphanies while reading a book or listening to a song. In line with that, contributors to the newsletter are hopeful that the articles published and the pieces of artwork have helped to disseminate commoning values among the hundreds of people who received each issue on paper, and those who read or listened to them online. However, the collective process of exploration and co-production plays an even more important role in the course of building commoning subjectivities. Several contributors to the Cooperation Birmingham newsletter are people living in poverty, from diverse backgrounds and ethnicities. Most of them had been forced, through toxic narratives and disciplinary measures, to see themselves as passive recipients of information, to think that they did not have anything important to say. One of the biggest achievements of the newsletter has therefore been the sparking of a collective process of empowerment in which all, but especially contributors from marginalized backgrounds, have broken boundaries previously imposed by the hegemonic distribution of the sensible. They have rebelled against the unidirectionality and homogeneity of toxic narratives and have become active storytellers, builders of alternative identities and stories. This was especially visible to the editors, whose labour of guidance and support with highly insecure potential contributors was crucial to the co-production of the newsletter. In fact, not everyone who expressed willingness to participate in the project was able to overcome those barriers. Therefore, the very process of creating the Cooperation Birmingham newsletter had a deep effect in politicizing the people who contributed to its realization. The process of co-production of the newsletter did not only produce publications, but also a wider community of commoners, a group of particular socio-ecological subjectivities.

Finally, the production of the Cooperation Birmingham newsletter did not follow a structured and planned process that could be labelled as 'methodology'. It assembled a multiplicity of DIY existing narrative and artistic practices of different forms through a collective process that created a common ground while valuing their diversity. All these features not only frame the production of the Cooperation Birmingham newsletter as a guerrilla narrative practice, but provide a starting point for broadening the scope of the concept. This is a crucial step that can open new dimensions to be examined through the lens of guerrilla narrative. In fact, we have a direct example in the daily activities of the solidarity kitchen. Could we consider the work, care, affections and solidarity that took place among participants of Cooperation Birmingham embodied performances of guerrilla narrative?

Embodying Guerrilla Narrative in the solidarity kitchen

Silvia Federici and Nicole Cox have argued that traditionally the Left has been quite blind towards what occurs in the kitchen (and in the bedroom). Reproductive work and gender oppression have not received enough attention in the Left strategy for emancipation. As Federici and Cox write, "the struggle which the Left offers to the wageless, the "underdeveloped," is not a struggle against capital, but a struggle for capital, in a more rationalized, developed, and productive form" (1975/2012, pp. 29–30). However, the kitchen is not only a space of gender oppression and unpaid care work. In *Re-enchanting the world*, Silvia Federici (2018) reflects specifically on the collective kitchens organized by activist women engaged in various struggles. Federici mentions women's activities at the Standing Rock Camp, in North Dakota, which supported more than seven thousand people providing food, supplies and child care (Federici, 2018, p. 4). Quoting Raúl Zibechi, Federici reminds us of the 15,000 grassroots organizations that in the 1990s were providing food for children and neighbourhoods in Lima. As Federici argues, kitchens in social movements are relevant because they remind us of "the need for a politics that refuses to separate the time of political organizing from that of reproduction" (*Ibid*., p. 7).

One might say that the problem is not in working in the kitchen per se, rather in the kind of social relationships in which that work and space are embedded. Instead of the heart of the home or of a deeply gendered space, the common kitchen is a queer space where politics and aromas mixed with friendship and humour. Placing the kitchen at the centre of social mobilization implies a shift in the ways in which activists think of politics and engage with the communities around them. Caring becomes more relevant than leading, listening to the needs of people a more useful skill than mastering the arts of public speaking. Learning how to run a collective kitchen exercises the capacities to work together towards a common aim. A revolution built around the kitchen does not sever body and mind, collective dreams and individual needs, the discussion about the structures and the small gestures through which another world gleams in the capitalist desert.

We argue that, when analysing the production of the Cooperation Birmingham newsletter as a guerrilla narrative strategy, we should not sever the editorial activities from what happens in the kitchen. We have analysed the texts published in their newsletters but we should not forget that those A3 flyers were mostly delivered with a hot meal. Is it actually correct to disentangle those words from the tastes of the food, the comforting presence of someone bringing it to the front door, the laugh and the sweat shared in preparing it, the joy and the stress enacted? The centrality of the words, whether spoken in the assemblies or written in a flyer, poses a contradiction that risks leading us back to the same old politics, one where the kitchen is a private space, caring a gendered task and the revolution a business for disembodied militants. With their solidarity kitchen, Cooperation Birmingham has practised guerrilla narrative with vegetables, pasta and their own bodies; that kitchen has told thousands of people a story of solidarity and resistance, of empowerment beyond charity, of rage beyond frustration and individual failure. It has not been a kitchen created by commoners, but the other way around: the commoning practices that have taken place in the very acts of mutual aid have forged collective identities based on emancipatory value practices.

When we started to reflect about Cooperation Birmingham and their commoning experience, we almost immediately focused our attention on the newsletter. It was in the written texts that a counter-hegemonic narrative had to be found. Almost as an involuntary reflex, we were ready to reproduce the usual fracture dividing the guts and the politics, the kitchen and the assembly hall. We should not be too harsh with ourselves; after all, as we have reported above, the participants in Cooperation Birmingham also thought that the distribution of meals was taking over the political content of their work. This is why they started the newsletter—to convey their politics and to be explicit about their aims. However, at the end of our reflection, we ask ourselves whether it was actually the newsletter conveying Cooperation Birmingham politics or the food delivered with it. Ours is perhaps only a provocation; cooking and writing, delivering meals and managing a website were all part of the same commoning experience that sabotaged the toxic narratives of individualism and emergency while prefiguring another way of being together. Nonetheless, remarking that cooking together in a collective kitchen is an exercise of guerrilla narrative is crucial because, too easily, we tend to end up with a word-centred politics where the space of caring and commoning is reduced to a symbolic instance. Instead, with Silvia Federici we argue for the centrality of the "reproductive side of political work—the dinners together, the songs that strengthen our sense of being a collective subject, the affective relations we develop among each other" (Federici, 2020, p. 126). Cooking and writing, running a newsletter and a collective kitchen are two sides of the same commoning practice; we consider them as two languages that together deliver new counter-hegemonic narratives. In this case guerrilla narrative looks like a warm soup made of vegetables, stories and ties. And it tastes like a joyful revolution.

Conclusions

In this chapter, we have addressed the need for strategies that challenge the sophisticated discourses used by elites to socially legitimize and normalize oppression and exploitation (to varying degrees) of a majority. We have briefly commented on the work of several authors who highlight diverse features of these totalizing devices that they characterize as 'shock doctrine', 'capitalism realism', 'distribution of the sensible' and 'toxic narratives'. However, they all reach a similar conclusion: narrative violence is being used to impose ways of living that lock-in and enhance privilege, inequality and environmental degradation. Based on our personal experience as militant researchers, and particularly our work on the Toxic Bios project, we have analysed the potential of the guerrilla narrative praxis to contest capitalism realism and create emancipatory alternatives. In particular, we have examined the narrative strategies used by Cooperation Birmingham that contribute to a material expansion of commoning, a set of relationships based on cooperation, solidarity, horizontality and care.

The case of the Cooperation Birmingham newsletter has contributed to the still scarce literature on guerrilla narrative by providing a detailed case study that confirms some of its defining traits. The articles published in the newsletter conform a diverse mix of topics and perspectives that challenge mainstream narratives associated with capitalist values while simultaneously normalizing commoning practices. However, confirming the process-oriented character of guerrilla narrative, we have found out that the published outcome is just the tip of the iceberg, and that all the invisible activities associated with the publication hold a great potential for building subjectivities based on cooperation and solidarity. In fact, the co-production process of the Cooperation Birmingham newsletter brought about the collective empowerment of many contributors, who were able to switch from passive objects to active subjects in history, subverting the dominant distribution of the sensible.

Following this line of analysis, in which we avoid reducing guerrilla narrative practices to mere outcomes, has allowed us to reach what is probably the main contribution of this article: the narrative power of material practices of care and solidarity. As we saw in the everyday activities of the Cooperation Birmingham solidarity kitchen, it is the labour, the interaction, and the multiple relationships forged among the participants of the kitchen that created the emergence of commoning subjectivities among them. Melissa García-Lamarca (2017) reached a similar conclusion when she examined the process of political subjectivation that took place at the Spanish anti-eviction popular movement (PAH). As she asserts, through experiencing "equal, non-commodified, and solidaristic relations" (429), a process of subjectivation took place that contested the common feeling of individual failure, and normalized collective and autonomous action when struggling against evictions. What these examples have in common is that material practices of social reproduction (like food or housing) are also narrative practices with the potential of subverting deeply embedded notions of capitalism realism and building commoning subjectivities. When we speak of guerrilla narrative, therefore, we need to start thinking beyond ink, paper, and even film reel. We need to mobilize expansive strategies that acknowledge and take advantage of the materiality of guerrilla narrative.

Acknowledgements This research received funding from the European Union's Horizon 2020 research and innovation programme under the Marie Sklowdoska-Curie Grant Agreement No. 765389.

References

Akuno, K., & Nangwaya, A. (Eds.). (2017). *Jackson rising: The struggle for economic democracy and black self-determination in Jackson*. Daraja Press.

Armiero, M. (2011). *A rugged nation*. White Horse Press.

Armiero, M. (2021). *Wasteocene. Stories from the global dump*. Cambridge University Press.

Armiero, M., Andritsos, T., Barca, S., Brás, R., Ruiz Cauyela, S., Dedeoğlu, Ç., Di Pierri, M., Fernandes, L. D., Gravagno, F., Greco, L., & Greyl, L. (2019). Toxic bios: Toxic autobiographies—A public environmental humanities project. *Environmental Justice, 12*(1), 7–11.

Barca, S. (2014). Telling the right story: Environmental violence and liberation narratives. *Environment and History, 20*(4), 535–546.

Belcher, A. (2018, April 25). *Foodbanks in Birmingham—Here's your nearest foodbank and how to access them*. Birmingham Live. https://www.birminghammail.co.uk/news/midlands-news/foodbanks-birmingham-heres-your-nearest-14500125

Birmingham City Council. (2020). *Birmingham labour market update*, Q2 2020. Inclusive Growth Directorate.

Brewer, F. (2020, October 30). *The tories could feed poor children—They just don't want to*. Novara Media. https://novaramedia.com/2020/10/30/the-tories-could-feed-poor-children-they-just-dont-want-to/

Busby, M. (2020, September 27). Schools in England told not to use material from anti-capitalist groups. *The Guardian*. https://www.theguardian.com/education/2020/sep/27/uk-schools-told-not-to-use-anti-capitalist-material-in-teaching

Caffentzis, G. (1999). On the notion of a crisis of social reproduction: A theoretical review. In M. Dallacosta & G. F. Dallacosta (Eds.), *Women, development and labor of reproduction: Struggles and movements* (pp. 153–187).

Dalton, C., & Mason-Deese, L. (2012). Counter (Mapping) actions: Mapping as militant research. *ACME: An International E-Journal for Critical Geographies, 11*(3), 439–466.

De Angelis, M. (2007). *The beginning of history: Value struggles and global capital*. Pluto Press.

De Angelis, M. (2017). *Omnia sunt communia: On the commons and the transformation to postcapitalism*. Zed Books.

Derickson, K. D., & Routledge, P. (2015). Resourcing scholar-activism: Collaboration, transformation, and the production of knowledge. *The Professional Geographer, 67*(1), 1–7.

Federici, S. (2004). *Caliban and the witch*. Autonomedia.

Federici, S. (2020). *Beyond the periphery of the skin: Rethinking, remaking, and reclaiming the body in contemporary capitalism*. PM Press.

Federici, S., & Cox, N. (1975/2012). Counter-planning from the kitchen: wages for housework, a perspective on capital and the left. In S. Federici (Ed.), *Revolution at point zero: Housework, reproduction and feminist struggle* (pp. 28–40). PM Press.

Federici, S. (2018). *Re-enchanting the world: Feminism and the politics of the commons*. PM Press.

Fisher, M. (2009). *Capitalist realism: Is there no alternative?*. John Hunt Publishing.

Francis-Devine, B. (2020, April 29). *Poverty in the UK: statistics*. House of Commons Library Number 7096. House of Commons.

García-Lamarca, M. (2017). Creating political subjects: Collective knowledge and action to enact housing rights in Spain. *Community Development Journal, 52*(3), 421–435.
Gouk, A., & Rodger, J. (2018, May 10). *Inside the Birmingham gig economy where shocking number of workers are on temporary contracts.* Birmingham Live. https://www.birminghammail.co.uk/news/midlands-news/inside-birmingham-gig-economy-shocking-14642036
Halvorsen, S. (2015). Militant research against-and-beyond itself: Critical perspectives from the university and occupy London. *Area, 47*(4), 466–472.
Harding, S. (1995). "Strong objectivity": A response to the new objectivity question. *Synthese, 104*(3), 331–349.
Holloway, J. (2010). *Crack capitalism.* Pluto Press.
Juris, J. (2007). Practicing militant ethnography. In D. David Graeber, S. Shukaitis, A. Negri & E. Biddle (Eds.), *Constituent imagination: Militant investigations, collective theorization* (pp. 164–176). AK Press.
Khasnabish, A. (2013). *Zapatistas: Rebellion from the grassroots to the global.* Zed Books.
Klein, N. (2007). *The shock doctrine: The rise of disaster capitalism.* Macmillan.
Klein N., Whitecross, M., & Winterbottom, M. (2009). *The shock doctrine*, documentary. Dogwoof.
Lai, T.-H. (2020). Political vandalism as counter-speech: A defense of defacing and destroying tainted monuments. *European Journal of Philosophy, 28*(3), 602–616.
Lawrence, F. (2020, April 11). UK hunger crisis: 1.5m people go whole day without food. *The Guardian.* https://www.theguardian.com/society/2020/apr/11/uk-hunger-crisis-15m-people-go-whole-day-without-food
Mueller, B. (2019, February 24). What is austerity and how has it affected British society? *The New York Times.* https://www.nytimes.com/2019/02/24/world/europe/britain-austerity-may-budget.html
Ramos, V. Jr. (1982). The concepts of ideology, hegemony, and organic intellectuals in Gramsci's Marxism. *Theoretical Review*, 27(3–8), 34. https://www.marxists.org/history/erol/periodicals/theoretical-review/1982301.htm
Rancière, J. (2004/2013). *The politics of aesthetics. The distribution of the sensible.* Bloomsbury Publishing.
Ravagli, V., & Wu, M. (2005). *Asce di guerra.* Tropea.
Ruiz Cayuela, S. (2020a). An alternative economy for Birmingham. In Solidarity not charity: Mutual aid in Europe. *Green European Journal, 20*, 100.

Ruiz Cayuela, S. (2020b). Organising a solidarity kitchen: Reflections from Cooperation Birmingham. *Interface: A journal for and about social movements, 21*(1), 304–309.

Ruiz Cayuela, S. (2021). Bridging materiality and subjectivity: Expanding the commons in Cooperation Birmingham. *Antipode, 53*(5), 1546–1570.

Sitrin, M., & Azzellini, D. (2014). *They can't represent us!: Reinventing democracy from Greece to Occupy*. Verso Books.

Tola, M. (2019). The archive and the lake: Labor, toxicity, and the making of cosmopolitical commons in Rome, Italy. *Environmental Humanities, 11*(1), 194–215.

Vastano, L. (2008). *Vajont, l'onda lunga: quarantacinque anni di truffe e soprusi contro chi sopravvisse alla notte piu crudele della Repubblica*. Ponte alle Grazie.

Wood, E. M. (1995). *Democracy against capitalism: Renewing historical materialism*. Cambridge University Press.

Zapata Hidalgo, M. (2020). Depressió i recuperació en mig de la voràgine. *Catarsi Magazin, 3*, 74–81.

4

Participative and Decolonial Approaches in Environmental History

Sofía De la Rosa Solano , Alex Franklin, and Luke Owen

Introduction

Environmental history is at the forefront of transdisciplinary methodological innovation. Understood as the field in charge of researching the mutual relationships between humans and non-humans through time,

Some of the literary quotations included in this chapter have been translated from Spanish to English by the first author.

S. De la Rosa Solano (✉)
Groningen University, Groningen, The Netherlands
e-mail: s.de.la.rosa.solano@rug.nl

A. Franklin · L. Owen
Centre for Agroecology, Water and Resilience (CAWR), Coventry University, Coventry, UK
e-mail: alex.franklin@coventry.ac.uk

L. Owen
e-mail: ab2943@coventry.ac.uk

A. Franklin (ed.), *Co-Creativity and Engaged Scholarship*,
https://doi.org/10.1007/978-3-030-84248-2_4

environmental historians argue that a broad set of skills and methodologies can be used for this type of research.[1] Indeed, some authors use carbon data, pest analysis, animal biology or Geographical Information Systems (GIS) to enrich narratives about the past that had previously been based mainly on written archival sources. Most of these methods aim to include biological and physical data about non-humans. In parallel, methods from social sciences and other humanities have also been included in environmental history research, such as discourse analysis, landscape reconstructions based on artistic representations, or participative methods. The inclusion of qualitative methods aims at better understanding the notions and imaginaries that determine society's relationships with non-humans.[2]

Focusing on social research methods, this chapter seeks to explore the relationship and use of decolonial participative approaches in environmental history. By doing so, we aim to contribute to the discussion on the methodological challenge of doing environmental history. Through four sections, a three-element proposal is outlined involving: first, the need to use methods from social sciences to recollect historic data through interacting with people and places. Second, to challenge the notions of legitimacy inherited from historic research methods, by validating people's knowledge as fundamental sources in the construction of environmental history narratives. And third, to think of environmental history research as an exercise that can be infused with transformative power for environmental justice. The main argument is that decolonial and participative methods are useful tools to build environmental histories that are more inclusive and communicate better with today's society. We also argue that using participative and decolonial approaches contribute to environmental awareness and political action, making environmental history a powerful discipline in contributing to a decolonial

[1] This definition of environmental history is based on McNeill (2003).

[2] To illustrate this point, one example can be found in river histories. One environmental history that is based on bio-physical data and methods can be found in Kraikovski and Lajus (2017). On the other hand, authors such as Vladimir Sánchez have used press analysis as sources to understand the evolution of society-river relationships (Sánchez-Calderón, 2017).

environmental justice.[3] This is in no way a proposal to dispose of other ways of researching environmental history. This suggestion also acknowledges that there are limitations to what participative and decolonial methods can contribute to environmental history and to environmental justice movements more broadly.

Even though there are many examples of environmental history studies with complex and multi-disciplinary methodologies, we argue there is still work to be done in building an inclusive and decolonial approach to this discipline.[4] To understand what this approach would entail, in this chapter we mostly focus on literature from Latin America on historical research, participation and environmental history. As will be shown, in this region there has been a long-term influence of the decolonial turn, a long tradition of participative methods, and a strong and consolidated environmental history body of literature. For scholars outside of Latin America, this discussion becomes relevant as it deals with methodological approaches to overcome exclusionary and institutionalized narratives of the past. It enriches the dialogues between different traditions of historical research by translating the Latin American discussions into English. It also enters the debate in Anglophone literature of interpreting the themes and focus of the Latin American environmental historiography as *tragic*. According to Mark Carey, there is a prevalence in this region of a "pervasive declensionist narrative, which is to say, stories of imperialist extraction and environmental degradation except when conservationists could successfully prevent destruction" (Carey, 2009, p. 222). The contribution of this chapter to this discussion will be made by contextualizing the construction of this *declensionist* narrative. Carey's warnings of the limitations of this approach are valid. However, the thinking behind this narrative can be useful to strengthen a scholarly tradition that builds critical and inclusive research, which engages with

[3] I adhere to the Álvarez and Coolsaet approach to environmental justice from a decolonial perspective. This proposal entails the acknowledgement of the colonial difference from a subaltern perspective. It is place-based and admits "capitalist destruction of nature as operating through heterogeneous mechanisms that are typically more brutal in places marked by colonialism and constructed as the periphery of the world-system" (Álvarez & Coolsaet, 2020, p. 15).

[4] For a global state of the art in English of environmental history, see Hughes (2016).

the larger discussions of equality. Lastly, this is an approach that hopes to bring attention to the communication of historical work to a larger audience, as "history can be characterized by its communicative nature, since it does not exist if it is not told" (Gallini et al., 2015, p. 9).

Since this chapter discusses the use of participative methods for environmental history, these are broadly defined as those involving people's knowledge for scientific research. They have a long trajectory in social sciences, while their use in humanities is increasing. The relevance of participative approaches for environmental history is that it helps historians to get involved with communities and landscapes that are not reachable through other methods. By involving researchers in the specific places they are researching, participative methods are producing situated knowledge. Situated knowledge has been broadly explored within decolonial studies as a way of overcoming Universalist assumptions that expand further inequalities. They are an integral part of the decolonial turn; a social, intellectual and cultural movement that aims to give a "new understanding of the global and local relations…as contestation of the Western eurocentric modernity, global capitalism and colonialism, which are an inseparable trilogy" (Curiel, 2014, p. 49).

The first section of this chapter reviews how historic research practiced inclusivity in general, since for historians, participative methods are less common than in other disciplines, such as anthropology and sociology. The second section discusses how participative research as a movement developed at the same time that historians were searching for more inclusive methods. In the third section, we illustrate how the characteristics of participative approaches, such as inclusivity and critical thinking, can be used for environmental history through the concept of memory. Lastly, we offer some final concluding thoughts.

"To Include" in Historical Research

This section explains how the discipline of history has worked the issue of inclusivity; this lays the ground for the subsequent use of participative approaches. This is important in thinking about participative methods for environmental history because it highlights the difficulties, inherited from the main historic discipline in hearing subaltern voices. It also sheds light on the specificities of the work of the historian; it therefore helps explain the ways and challenges of using people's knowledge to build legitimate narratives of the past.

It is first important to understand that historical research is tasked with examining the past while wearing the spectacles of the present, or that "writing history is an optimistic exercise towards the future motivated by concerns from the present" (Gallini et al., 2015, p. 12). As Eric Hobsbawm states, the past is "a permanent dimension of the human consciousness, an inevitable component of the institutions, values and other patterns of human society" (Hobsbawm, 2011, p. 54). Moreover, the discipline of history is the way in which society interprets the past and the realities and structures we inherit from it, from a scientific point of view. These interpretations, that are partial and only cover specific aspects of the past, are known as historiography, meaning, how history has been written. Charles Bergquist describes historiography as explanations of "how in the past observers and schools of thought—each influenced by historical processes, national loyalties, ethnics and of class, and intellectual currents and cultural perspectives—interpreted a determined historic reality" (Bergquist, 1989, p. 212).

In broad terms, history understood as a scientific discipline comes from the discussions of the eighteenth and nineteenth centuries (mostly in Europe) that looked to legitimize the project of modernity, where absolute truths could be achieved through the scientific method.[5] Those notions are challenged today, and history has long since moved from the pretensions of creating absolute narratives of the past. Nonetheless, most of the methods of history reflect the interests and discussions of the eighteenth and nineteenth centuries, and have retained the goal of finding *reliable* ways to understand the past. The consequence is that historic methods have focused on finding ways to guarantee that reliability. Furthermore, in broad terms, most historic research is done by following the material traces (known as sources) of the past in the present, in the form of archives, buildings, images, objects, landscapes

[5] On the debates around these ideas, see Wallerstein (1994; Lander (2000).

and others. Moreover, there is a strict process of challenging the legitimacy of the source and the information it provides, known as *source criticism.*

This process relies heavily on the availability of sources and the fact that the researcher deems the contents of these sources to be legitimate. Prioritizing written documents is based on a situated idea of scientific knowledge that corresponds to the interests of imperial powers, categorized in decolonial literature as the Global North.[6] As such, the discipline of history in its origins responded to particular interests and focused on those aspects of what was available and deemed legitimate: mainly studies of aristocracy and political events. Soon, intellectuals noticed that this approach was exclusionary of large segments of the population, and that findable and legitimate sources only gave a particular type of information, since "different types of knowledge are expressed in different ways, which are doubtlessly difficult to decode" (Gallini et al., 2015, p. 21). Therefore, historians started looking for ways to be more inclusive in their analysis of the past.

This search started a long tradition of including popular voices in history, forming what historians like Ranahit Guha and Mauricio Archila have called the quest for "the voices silenced by History" (Archila Neira, 2005; Guha, 1982). Some authors trace this interest for more popular and less aristocratic voices within the disciplinary praxis of history to Marx (Burke, 1984). Another precedent is what in the eighteenth century was called popular history, when intellectuals were developing arguments in the construction of nation states (Burke, 1984). However, the most clear and recognized reference to include working classes in historic research is the work of Edward Palmer Thompson in "*The making of the English working class*" in 1963 (Thompson, 1963). In the late twentieth century more movements joined, such as the *Annales* in France, the British Cultural Marxism, the Indian subaltern studies and others. Together, they consolidated what came to be known as "history from below". This subsequently became an integral part of fields of history such as social history, cultural history, labour history and others.

---

[6] In opposition of the Global South, a category that contains the spaces of oppression or affectation by colonialism, imperialism and capitalism. On the Global South and Global North categories, see Fernandez et al. (2014).

Practitioners of the history from below face the challenge of scarce availability of sources. This is because, unlike other approaches, the archives of the working classes are less preserved or less available than others. As Mauricio Archila points out, it is no coincidence that, in parallel with the growing interest in popular classes, oral history was also in a process of consolidation within history, as both "methodological appendix or epistemological alternative" (Archila Neira, 2005, p. 297). Oral history is a complex and rich approach to the studies of the past. It has been described by Alessandro Portelli as a work of relationships, mainly between past and present in "an effort to establish, through memory and narrative, what the past means to the present" (Portelli, 2009, p. 21). As a legitimate data collection approach, it only gained acceptance from academic historians in the Global North after World War II (Archila Neira, 2005). Combined with methods from other disciplines such as anthropology and sociology, massive oral archives were created to analyse the horrors of the war that did not leave a paper trace, giving birth to many epistemological and methodological discussions on how to collect and analyse this data within the professional practice of history.

In the Global South, the high rates of analphabetic population and scarcity of written documentation made historical research particularly difficult. In the case of Latin American, bottom-up approaches were rapidly increasing in the second half of the twentieth century. This is because the scarcity of written sources was met with the strong influence of Marxist thinking within intellectuals, making oral testimonies a more appealing source for the understanding of the class struggle. With time, the oral testimony in history evolved from being a complement for written sources to being the core of an epistemological approach that aimed to hear the voices silenced by historiography. In this context, the oral traditions that already existed independently from academia "formed the memory in which historic methods were supported…becoming another source to get a better understanding of the past" (Archila Neira, 2005, p. 300). However, historians were not the only ones in Latin America who felt more willing to work with these sources for researching the past. On the contrary, sociologists and anthropologists were leading the discussion. The next section explores how participation has been constructed in the social sciences, in parallel to the discussions on its inclusion in history.

Participation, Decoloniality and History

There is not one unique way in which to incorporate public participation in research. Also, its use does not necessarily align with the decolonial turn; rather it depends on how, why and the extent to which it is used. For example, since the nineteenth-century anthropologists have used ethnography to understand the dynamics within communities, sometimes under a colonial gaze (Tax, 1992). In contrast, decolonial participative methods have a strong relationship to the methodological approach known as "action research". Because of this, it aims not only at understanding realities, but rather at transforming them. There are many versions of action research; it takes "many forms depending on the particular context and issues involved" (Kindon et al., 2007, p. 1), and the differences "can be political, practical and epistemological" (Kemmis et al., 2014, p. 4). For that reason, we limit my focus here to its development in Latin America, where the approach evolved hand-in-hand with decolonial thinking.

Action research was first mentioned as a methodology of the social sciences, to achieve changes in society, in a publication in 1946 by Kurt Lewin in the United States of America (Lewin, 1946). Lewin challenged the separation between the production and use of knowledge, and thought of ways in which that distance could be reduced. However, he worked with a focus on "social improvement" that, even though it seemed to consider knowledge production from communities themselves, was still largely based on authoritarian views of change towards specific forms of development. Based on this approach, action research flourished during the last century in Anglophone literature, hand-in-hand with organizational science, aiming to improve problem solving and social engineering (Rahman & Fals Borda, 1992).

By 1980 action research was already an established methodology; with multilingual literature written about it, it was being applied on all continents, including by various globally recognized organizations such as the Food and Agricultural Organization (FAO) and the United Nations Educational, Scientific and Cultural Organization (UNESCO) (Rahman & Fals Borda, 1992). Since then, many other studies have been published, and its effectiveness and relevance have been widely discussed among academics and practitioners (Kemmis et al., 2014). Unlike the Anglophone context, in other parts of the world action research occurred and developed mainly outside academia, but that "occurrence" outside academia is part of its identity (Salazar, 1992). In Latin America, action research emerged as a critique of the separation between communities and researchers, a distance that seemed to prevent scientific work from actually improving the living conditions of those communities (Salazar, 1992).

A vivid debate among social science practitioners working in Latin America took place in the decades of 1960s–1980s, at the core of which was known as the "New Social Sciences" (Zamosc, 1992). These debates showed a critical attitude to imported models of analysis, since they were found to be too neoliberal and foreign for the Latin-American realities. These "New Social Sciences" were influenced by new approaches to Marxism, the pedagogy of the oppressed, and the liberation theology. Central to these was the assumption that deep engagement of the researcher was necessary if the goal of academic work is to achieve change in society (Salazar, 1992). One of the ways in which this movement developed was through direct work with communities. Moreover, it was critical with the notion of objectivity that was embedded in the Anglophone version of the participant observation method. Central in these debates were anthropologists and sociologists working with indigenous and peasant communities, looking to de-construct the colonial and imperial background of social sciences, particularly in the Global South (Tax, 1992). These were some of the debates that simultaneously nurtured the decolonial turn.

Scholars associated with the New Social Sciences also argued that science is usually seen as a non-human entity, objective and without interest (Tax, 1992). For them, science is intrinsically human, and therefore it responds to society's interests and structures. This critique goes particularly to the process of knowledge production, that generates an "elitization" of knowledge production. In response, they proposed a "democratization" of scientific work. This translated into including the communication of research as part of the research design itself (Rahman & Fals Borda, 1992). By doing this, their goal was to change the nature of social research by creating Participatory Action Research (PAR). PAR is proposed as a method for not only working with the communities, but also having the communities designing and actively taking part in the research process.

In a reflective text on the evolution of PAR in Latin America, the sociologist Zamosc says that the objectives of researchers using PAR are empathetic and synergic. They seek to "obtain valid knowledge that corresponds to the interest of the researched groups and joins active and direct efforts to achieve these groups goals" (Zamosc, 1992, p. 98). Overall, the purpose and reach of the movements in Latin America using these approaches to the field was to create a social science for liberation, where the role of the researcher was to "help exploited groups to achieve their revolutionary historic mission" (Zamosc, 1992, p. 98). The main method of this militant research, as they called it, is participative observation. The largest of these initial exercises was carried out by the sociologist Orlando Fals Borda in the region of Córdoba (Colombia) in the 1970s. The main accomplishment of the project was to recover historic information on class struggles in the region through interviews and exploration of personal archives. This was known as a "critical recovery of history". The result was a compilation of the history of the class struggles of the peasants in the Córdoba region and the revitalization of those fights.

It is important to mention here that for Fals Borda, a true science of the people also includes the environment, as it contains "empiric or practical knowledges, from common sense, possessed by ancestral tradition in working classes. This knowledge allows them to create, work and understand with, mainly, the resources coming from nature" (Rahman & Fals Borda, 1992, p. 213). Fals Borda subscribed to the view that a science of the people would, by itself, strengthen the claims from the working classes. A pivotal point for him was the socialization of the research findings, to give working classes the opportunity to achieve new levels of political consciousness. In short, for Fals Borda, research can only be designed with people from communities and their leaders, and with the goal of social transformation.

In the past three decades, the decolonial movement has elaborated on these and other postulates to strengthen a theoretic frame, however, that frame "does not reflect a problematization of the methodologies used in field" (Puentes, 2015, p. 2). It has been mainly the work of feminist scholars and activists to challenge this aspect and write literature on the topic. María Lugones proposes a theory of intersectionality to decolonize gender, based on the history of the groups that are being oppressed (Lugones, 2018). Ochy Curiel reflects in depth on how it is central that research on communities is led by people from those communities, by legitimizing their knowledge and problematizing the conditions of production of knowledge (Curiel, 2014). Julia Suárez-Krabbe reflects from an anthropological perspective about the necessity of using engaged research, as postulated by Fals Borda, to actively transform injustices (Suárez-Krabbe, 2011).

In specific relation to historical research, arguably the most important contribution is the work of the Bolivian sociologist Silvia Rivera Cusicanqui (De Souza Veras, 2012). In 1983 she co-created the Workshop for Andean Oral History (THOA, for its initials in Spanish), which still continues today in La Paz. Rivera is critical towards PAR as she finds it does not transform the instrumentalization of communities within the production of scientific knowledge. She also critiques the strong Marxist influences in PAR, since she considers it is a theory that does not translate to the Latin context, and it does not deal with the background differences between researcher and researched. As an alternative, through oral history in the Aymara language, the THOA aims to collect the "existence of diverse historic rationalities, with legitimizing functions" (De Souza Veras, 2012, p. 6) that went against traditional Bolivian historiography written in Spanish. With this methodology, she aims to shift the starting point of historic reflections from academia to the communities themselves within their own cosmologies and realities. THOA has been considered the main tool from which "to defend a history written from the bases with ethnic revindication where the protagonist of the history is the one to reconstruct its past" (Apaza, 2019, p. 6). THOA has also been recognized as a source of political-ideological resistance within the colonial context of Latin America (De Souza Veras, 2012).

Thus far, it has been explained how some researchers were keen on using participative methods for researching the past. However, not all historians agreed with these approaches. For example, as a response to Orlando Fals Borda's most important work, the "Historia Doble de la Costa" (Double history of the Colombian Caribbean Coast), the Latin-Americanist historian Charles Bergquist wrote a paper entitled "En Defensa de la Historia" (For the defence of history, Bergquist, 1989). In this paper, Bergquist criticized the use of PAR for historic research in depth, claiming inaccuracy, poor source criticism and neglection of the existing historiography on the topic of study. He claimed Fals Borda "constantly attempts to subvert the cause of science in the name of political engagement" (Bergquist, 1989, p. 226). Participative methods in general received further critiques, such as idealization of communities, as they are not equipped with tools to recognize the inner inequalities of gender or race within them (Salazar, 1992). While most of these critiques raised points for improvement in the use of participative methods for historic research, we subscribe to the argument that it is indeed necessary to subvert the cause of science. By subverting, we particularly refer to being critical of established methods, notions and theories that claim universal objectivity and accuracy. This applies especially when these approaches do not question the power mechanisms that are creating and reproducing social and environmental inequality.

In this and the previous sections, we have discussed how historians were looking for ways to be inclusive, and in parallel, how other social scientists in Latin America were questioning scientific methodology and epistemology as a whole in the quest for social equality. The discussions that gave room to the decolonial turn were going hand-in-hand with the development of a particular type of action research, different from its contemporary in Anglophone literature. The scholars at the centre of this debate were questioning not only the separation between researchers and communities, but also the nature of scientific knowledge itself. Action Research and Participative Action Research became useful tools in the quest to transform realities. During the last three decades these debates have changed form, questioning the matter of how the inequalities experienced nowadays are linked to the project of modernity and coloniality that is expressed in capitalism. Decolonial theory places emphasis on finding out the particular ways in which this system is oppressive; it questions relationships of power between centre and peripheries, the nature-social dichotomy, and the homogenization of people. In this process, research about the past has taken an important role, as it is regarded as an empowering tool for communities (Zamosc, 1992).

Environmental History, Decoloniality and Participative Methods

As with many other social disciplines and humanities, in the second half of the twentieth century, historians started asking how to expand historic narratives to include the non-human. The epistemological and practical consequences of the environmental turn in history have been widely researched by other authors (Hughes, 2016). Nonetheless, it is important to mention that environmental historians built a theoretical body to challenge the exclusion of the non-human in history, and a methodological praxis that reached into other sources, interpretations and disciplines to change our interpretation of the past. Some scholars have approached the task of including non-humans by focusing on their agency in history. Others have challenged established narratives on weather and culture, by reconstructing the past based on archival and biophysical information. In this way, many different disciplinary and methodological elements have been used to build the narrative of environmental history. A summary of how environmental historians have thought about methods can be found in Donald Hughes' book *What is Environmental History?*. He compiles the views of several English-speaking authors on the matter and highlights how environmental historians can collect data on societies, the biophysical world and ideas, establishing a dialogue with other disciplines and using the lens of the historic method. He concludes by saying "environmental history refuses to cut culture from nature. Equally it must not cut history from geography" (Hughes, 2016, p. 126).

In terms of how environmental history from Latin America has been influenced by the decolonial and participative approach, there is some discussion; it has been acknowledged that in Latin America environmental history has its own characteristics and there is already a strong body of production that has particularities when compared with environmental histories from other latitudes (Sánchez-Calderón & Blanc, 2019). Mark Carey coined the term "chronic deficiency" for the way Latin American history is typically told: that is, as a story of deterioration. According to Carey, this focus does not allow the seeing of other narratives or processes in the region that were part of the complex process of socio-environmental change. On the other hand, economic historian Patricia Clare of Costa Rica wrote in 2009 that there is an ideological difference between the Latin American environmental historiography and that from other places, making Latin American histories more politically engaged, especially in working towards the end of the deep inequalities of the region (Clare, 2009).

In 2019, the environmental historians Vladimir Sánchez and Jacob Blanc highlighted the close relationship between the environmental deterioration denounced in declencionist narratives, with the social inequalities in Latin America (Sánchez-Calderón & Blanc, 2019). They point out that because of this close relationship, the environmental justice movement and environmental history in the region have grown hand-in-hand; however, they also mention other types of Latin environmental historiography that have been written since 2010, with fewer declensionist narratives. Stefania Gallini wrote two pieces analysing this same topic, one in 2009 and one in 2020 (Gallini, 2009, 2020). She reflects on how the particular context of inequalities and neoextractivist economies in Latin America has made environmental historians more interested in explaining the source of these realities than in other topics. She argues that environmental history in Latin America has focused more on dialoguing with environmental and development studies than with the discipline of history. She also points out that, unlike in other places, environmental history in Latin America is written with the goal of having an impact on policy and social transformation.

In line with these arguments, we suggest that environmental history from Latin America is influenced by the development of the decolonial literature and can contribute to create narratives from the Global South. This is because it aims to support socio-environmental justice claims, challenges the preference of written sources over others, and has especial interest in the history of extractivist economies. Accordingly, within the claims of the environmental justice movement, the work of environmental historians should be able to recognize dynamics of exclusion, which are particularly acute in the Latin American case. This, in turn, is why inclusive methods for environmental history have been more popular in this part of the world. In short, this influence is visible starting with the selection of research topics that have a denunciation tone towards the quest for participative methodologies. It extends to theoretical approaches, use of concepts and the proximity it has with other disciplines like political ecology or development studies.

One great example of these developments is the publication, in 2015, of a manual on how to practice environmental history in Colombia by Gallini et al. (Gallini et al., 2015). The manual was part of a collection of volumes, published with the support of the Colombian government, to produce technical data and recommendations to guide the conservation of strategic ecosystems. The government wanted to create conservation areas, taking into account ecological data and considering the information from the communities that had inhabited the space. As pointed out by the authors of the manual, to be part of this process opened the way for environmental history towards activism and forced it out of the "dangerous ivory tower of academia" (Gallini et al., 2015, p. 7). The idea of the authors was to illustrate "the imperious need of believing in the transformative capacity of a historic research" (Gallini et al., 2015, p. 64), made ideally through co-creation with communities and being appropriated by locals and decision-making stakeholders. The manual warns about the importance of history, since not being aware of it diminishes the ability of adaptation for communities and governments. Another point in the manual refers to the power environmental history can play in granting access to natural resources to different stakeholders, emphasizing that researchers have the responsibility of offering the opportunity of knowledge to build a more equal society. The methodological proposal contained within the manual on how to make environmental history is openly participative, asking researchers to use in-depth interviews and collect archives from local inhabitants. Here, it is important to highlight that this publication came not only from one of the front figures in environmental history in Latin America, namely Stefania Gallini, but also that it was published by a governmental institution that had in mind to affect policy. The origin of the manual, speaks of how these are tools for transformation, not only for academia.

Another critical reflection on the work of including oral testimonies in doing environmental history in Latin America was written by Emilio Vargas Mena (Vargas Mena, 2014). His travels through the territory object of study, had the goal of "trying to read the printed prints that human experience left behind" (Vargas Mena, 2014, p. 230). Vargas Mena problematizes methodological issues around oral sources, including for example, on how to do sampling, how to be prepared for an in-depth interview, and how to build an archive with these sources. Vargas Mena concludes emphasizing the transdisciplinary nature of environmental history. While it needs to take into account non-human dynamics (from plants, animals and others), for environmental history it is still "fundamental to approach civil society…in order to contribute to the urgent tasks of environmental protection and the social and political transformations that make history possible" (Vargas Mena, 2014, p. 257). Another author that reflects on the contributions of Latin American environmental history is Katherine Mora. She calls for going beyond the "declensionist" and tragic stories of the human/non-human relationship, while actively using historical approaches (especially from environmental history) in the construction of adaptative strategies for the present and the future (Mora Pacheco, 2018).

Memory

Memory has been the main concept that history has drawn upon to use participative methods. Memory is a complex concept and has many interpretations. For example, it is often asked where historiography ends and where memory starts (Traverso, 2007). As the historian and expert in cultural studies Mario Rufer conceptualizes it, memory is composed of public references to the past under a non-official or scholar context (Rufer, 2009). Another academic who has explored the concept of memory in depth is the historian Enzo Traverso. For him, memory can only be understood from the present, as it is composed of representations of the past at the individual or collective level, with historic continuity and filled with meaning and direction (Traverso, 2007). Historians do not have control over this type of knowledge, as explained by Gonzalo Pasamar, since they "share the public space with lots of other actors" (Pasamar Alzuria, 2003, p. 240). For that reason, memory (hand-in-hand with oral sources) was long deemed not valuable as an historic source. However, as explained earlier, by the turn of the century it was already widely accepted among historians of all latitudes that oral history had "freed us from the cul-de-sac question of veracity-falsehood" (Gallini, 2004) of testimonies. By then it was clear that oral sources responded to different questions from written ones. In addition, the interpretation of these sources also changed, because conventional history asks for what happened, whereas oral history asks for meaning (Portelli, 2009).

We argue that memory should be considered a necessary step of societal transformation processes within environmental activism, particularly constructed from a participative and decolonial approach. Engaging in the use of the concept of memory and decolonial participative approaches can also help in uncovering "the existence of diverse historical rationalities that fulfil legitimizing functions within conflicts" (Cusicanqui, 2008, p. 59). As Rivera Cusicanqui argues, in situations of inequality, the discipline of history can focus on unveiling the different interpretations of reality, not on determining the objective truth about what occurred in the past (Cusicanqui, 2008). In other words, to understand the meaning attached to the past "in terms of justice within a cause… making historic research a collective exercise of misalignment" (Cusicanqui, 2008) for the researcher and the communities that are involved in the research. Within the struggles faced by human and environmental rights defenders in the twenty-first century, unveiling the multiple, co-existent and sometimes contradicting memories among stakeholders has an impact in determining the route towards environmental justice.

There can be multiple uses of the concept of memory for a decolonial and participative approach to environmental history. If the goal of the researcher is to commit to an engaged exercise of transformation, memory is a fundamental part of change. In a volume on historical memory in Africa, the anthropologist and historian of Africa oral history, Mamadou Diawara asserts that memory can be a "a rich and powerful tool for orientation in the present and for opening future perspectives on human action. Even on the most abstract level, the way in which we remember has consequences" (Diawara et al., 2010, p. 2). From a pedagogic studies perspective, Gabriela Ortiz says that memory:

> can move in the quotidian life and be a powerful tool for fighting when brought to the present. Through the lived experience... memory is brought up to life by constituting transmitted experience that is commemorated in the collective, forming identities and the ways of being in life. (Ortiz Zambrano, 2019, p. 29)

Contrastingly, just as it is important to analyse what is remembered, one of the most important components of memory is *forgetting*, since "forgetting is the norm and remembering is the exception, even though constructed forgetfulness may exist" (Gallini et al., 2015, p. 54).

An illustration of the co-existence of multiple narratives of the past that determine identities is the 1927 massacre of banana workers by the United Fruit Company in Northern Colombia. Official reports limit the number killed to a maximum of 1,000 people, according to the then US ambassador (Bucheli, 2002). In popular memory, that number has been highly influenced by the publication of the Nobel laureate novel *Hundred years of solitude* (where it was suggested there were some 3,000 people killed) by Gabriel García Márquez (García Márquez, 1967). For current processes of labour and environmental inequalities in the area of the massacre, where bananas are still being planted, the popular memory of peasants, conservation workers and social movement gives higher priority to the collective memory than the archival sources (De la Rosa Solano, 2018). The *remembering* of this event has, for decades, fuelled the union movements in Colombia in the fight for proper work conditions.

Another example we can offer here is drawn from fieldwork in the city of Coventry (UK) (undertaken by the first author, De la Rosa Solano), while researching the history of urban waterways. During the post-war period, the main river in the city was diverted underground, disconnecting it from the historic centre. While doing a social cartography exercise on a local festival, different inhabitants of the city informed De la Rosa Solano that they did not know about the existence of the river, a part of which still runs above surface not more than 200 m away from the festival. When asking environmental and local authorities about the problems around the river, the lack of citizen engagement in conservation and care was one of the main issues cited. In her research, De la Rosa Solano finds a connection between the lack of citizen engagement with the literal lack of visibility the river has in the city. In this case, one of the processes in the history of the river has been a constructed forgetting and the subsequent processes of remembering within artistic and environmental activities.[7]

Accessing individual and collective memory as a researcher can be done in multiple ways. Interviews, life stories, focal groups, social cartography, photo elicitation and other exercises are valuable in the quest for meaning about the past. Particularly, methods that push for interaction with the lived environment and exploration of the researched areas are fundamental in including the non-human in environmental history.

Finally, we would like to end this section by highlighting some reflections on the overall role of the researcher when using the concept of memory to engage in participative and decolonial methods. One useful piece of advice is given by Portelli, when he says that asking the interviewees for their life story, even if some of it is not directly related to the field of study, is of utmost importance. For him, oral history cannot be done, "unless your interest is focused on the person with whom you are talking" (Portelli, 2009, p. 29). The informant experiences are constituent of the ways remembering (and forgetting) is constructed. Silvia Rivera Cusicanqui makes two other notes. The first is to understand that there is cultural and linguistic *untranslatability* that is a natural part of different individuals with different backgrounds, like scholar

[7] For more information on this research, see www.recoms.eu.

researchers and communities. These need to be accepted and recognized. They can even bring opportunities for communities to build their narratives in their own terms and for their own purposes. Second, Rivera also reflects on the "negotiation" between the researcher and researched; this should be given in terms of "two subjects that reflect together around their experiences and about the vision each has of the other" (Cusicanqui, 2008, p. 60). These reflections are pointed towards the de-colonizing and construction of truly participative narratives of the past. There will still be limitations though, as previously discussed, depending mainly on each context where a participative and decolonial methodology is used.

Conclusions

This chapter has explored how participatory and decolonial methods of research can contribute to the practise of an inclusive and communicative environmental history that contributes to the cause of environmental justice. Discussed first, was how historical research has looked at inclusion, and then in turn, at participative approaches. The review of Latin American literature offered an introduction to an extensive body of scholarship that explores these topics, providing references for the one interested in taking this exploration further. From undertaking this review we argue that decoloniality and participation can be powerful allies of environmental history research. Namely, the decolonial approach helps reading the past through a critical lens that connects specific cases with larger phenomena, such as imperialism and capitalism, highlighting the spaces for change within them. Similarly, participation challenges historical research to go beyond inclusion and to place people's knowledge in the centre of scientific production. Environmental history brings together these elements to ask questions of the social sciences and humanities about non-human actors and to contextualize the discipline of history within the socio-environmental present day challenges.

The contributions of this chapter are three-fold. First we have studied the integration of social sciences methods for historical data collection in fieldwork; second, we have pointed out the relevance of people's knowledge for environmental history; and thirdly, we have highlighted the links between research practice of environmental history and environmental justice. This contribution was accompanied by a reflection on the fundamental task of including the non-human as part of that collective and individual narrative that forms the collective and individual identity. By way of illustration, the use of memory as a conceptual tool in applying this approach has also been explored. Likewise, by understanding the limits of participative and decolonial methods, the discussions in this chapter are intended to act as a foundation for further explorations on this way of doing environmental history. In this way, we can find spaces of convergence with social and environmental movements, similar to what is done in political ecology, but here, from an historical perspective.

This English language exploration of the evolution of participation in environmental history research in Latin America hopes to open the dialogue for exchanges between different traditions on methods in this field. More importantly, the particularities of the environmental history in Latin America, as they have been highlighted throughout this text, hope to inform research in other regions. Environmental history has already influenced policies in conservation in Latin America, and this is a clear route for more politically committed academic exercises in other latitudes. This type of research could be used for understanding the long-term temporality of social and ecological transformation and inform civil society of how inequities are shaped through time. Awareness of these issues can be created, and actively used, giving room to a politically committed academic practice from history. As many of the environmental historians from Latin America we have quoted, we strongly believe our field has a role to play in facing challenges of the twenty-first century. Among others, environmental history can be used to create awareness, push for inclusion, creating strategies for facing socio-environmental injustices, and have an impact on public policies.

Acknowledgements This research received funding from the European Union's Horizon 2020 research and innovation programme under the Marie Sklowdoska-Curie Grant Agreement No. 765389.

References

Álvarez, L., & Coolsaet, B. (2020). 'Decolonizing Environmental Justice Studies: A Latin American Perspective. *Capitalism, Nature Socialism, 31*(2), 50–69. https://doi.org/10.1080/10455752.2018.1558272

Apaza, I. R. L. (2019). Y qué con la historia oral? Un balance aproximado, perspectivas y cuestiones pendientes. *Voces Educativas. Reflexiones desde la experiencia pedagógica.*

Archila Neira, M. (2005). Voces subalternas e historia oral. *Anuario Colombiano de Historia Social y de la Cultura, 32*, 293–308. https://revistas.unal.edu.co/index.php/achsc/article/view/8196. Accessed 15 December 2020.

Bergquist, C. (1989). En nombre de la historia: una crítica disciplinaria de la Historia doble de la costa de Orlando Fals Borda. *Anuario Colombiano de Historia Social y de la Cultura, 0*(16–17), 205–229.

Bucheli, M. (2002). *The united fruit company in Colombia: Labor, local elite and multinational enterprise 1900–1970*. Standford University.

Burke, P. (1984). Historia popular o historia total. In R. Samuel (Ed.), *Historia popular y teoría socialista*. Crítica.

Carey, M. (2009). Latin american environmental history: Current trends, interdisciplinary insights, and future directions. *Environmental History, 14*(2), 221–252. https://doi.org/10.1093/envhis/14.2.221

Clare, P. (2009). Un balance de la historia ambiental latinoamericana. *Revista Historia*. https://www.revistas.una.ac.cr/index.php/historia/article/view/3474/3331. Accessed 12 November 2020.

Curiel, O. (2014). Construyendo metodologías feministas desde el feminismo decolonial. In M. Franco (Ed.), *Joirnadas de metodología e investigación feminista y aplicación en el ámbito de los derechos humanos* (pp. 32–51). San Sebastián Donostia.

Cusicanqui, S. R. (2008). El potencial epistemológico y teórico de la historia oral: de la lógica instrumental a la descolonización de la historia. *Teoria crítica dos Direitos Humanos no século XXI. Porto Alegre: EDIPUCRS*, 154–175.

De la Rosa Solano, S. (2018). *La maldita circuntancia del agua por todas partes. The evolution of the waterscape of the Ciénaga Grande de Santa Marta between 1950–2010*. Universiteit van Amsterdam. https://scripties.uba.uva.nl/search?id=658039. Accessed 16 December 2020.

De Souza Veras, J. J. (2012). A história oral e sua dimensão epistemológica. In *XI Encontro Nacional de História Oral*. Rio de Janeiro.

Diawara, M., Lategan, B., & Rüsen, J. (Eds.). (2010). *Historical memory in Africa: Dealing with the past, reaching for the future in an intercultural context (Vol. 12)*. Berghahn.

Fernandez, R., Lauxmann, C. T., & Facundo Trevignani, M. (2014). Emergencia del Sur Global. Perspectivas para el desarrollo de la periferia latinoamericana. *Economia e Sociedade, 23*(3), 611–643.

Gallini, S. (2004). Problemas de métodos en la historia ambiental de América Latina. *Anuario AHES, 19*, 147–171.

Gallini, S. (2009). Historia, ambiente, política: El camino de la historia ambiental en América Latina. *Nómadas, 30*, 92–102.

Gallini, S. (2020). Qué hay de histórico en la Historiografía ambiental en América Latina? *Historia y Memoria*, Especial, 179–233.

Gallini, S., De la Rosa Solano, S., & Abello, R. (2015). Historia ambiental. In P. Ungar (Ed.), *Hojas de ruta. Guías para el estudio socioecológico de la alta montaña en Colombia*. Instituto de Investigación de Recursos Biológicos Alexander von Humboldt.

García Márquez, G. (1967). *Cien Años de Soledad*. Sudamericana.

Guha, R. (1982). *Las voces de la historia y otros estudios subalternos*. Crítica.

Hobsbawm, E. (2011). *On history*. Hachette.

Hughes, J. D. (2016). *What is environmental history?* Polity Press.

Kemmis, S., McTaggart, R., & Nixon, R. (Eds.). (2014). *The action research planner: Doing critical participatory action research*. Springer.

Kindon, S., Pain, R., & Kesby, M. (2007). *Participatory action research approaches and methods: Connecting people*. Routledge.

Kraikovski, A., & Lajus, J. (2017). Living on the River over the Year: the significance of the Neva to Imperial Saint Petersbury. In M. Knoll, U. Lübken & D. Scholl (Eds.), *Rivers lost, rivers regained: Rethinking city-river relations* (pp. 235–254). University of Pittsburgh Press. https://doi.org/10.2307/j.ctt1qnw8gv.16

Lander, E. (2000). *La colonialidad del saber: Eurocentrismo y ciencias sociales. Perspectivas latinoamericanas*. CLACSO.

Lewin, K. (1946). Action research and minority problems. *Journal of Social Issues, 2*(4), 34–46.

Lugones, M. (2018). Hacia metodologías de la decolonialidad. In A. Escobar (Ed.), *Prácticas otras de conocimiento(s)*. CLACSO.

McNeill, J. R. (2003). Observations on the nature and culture of environmental history. *History and Theory, 42*(4), 5–43. https://doi.org/10.1046/j.1468-2303.2003.00255.x

Mora Pacheco, K. (2018). Pensar el pasado para adaptarse al cambio climático. El aporte necesario de la historia ambiental latinoamericana. *Letras Verdes, Revista Latinoamericana de Estudios Socioambientales, 24*, 8–26.
Ortiz Zambrano, G. B. (2019). Cómo nos construimos desde la memoria? *Revista Andina de Educación. Universidad Andina Simon Bolivar, 2*(2), 28–33. https://doi.org/10.32719/26312816.2019.2.2.4
Pasamar, G. (2003). Los historiadores y el "uso público" de la historia: Viejo problema y desafío reciente. *Ayer, 49*, 221–248.
Portelli, A. (2009). What makes oral history different. In L. Del Giudice (Ed.), *Oral history, oral culture, and Italian Americans* (pp. 21–30). Palgrave Macmillan. https://doi.org/10.1057/9780230101395_2
Puentes, J. P. (2015). Descolonización metodológica e interculturalidad. Reflexiones desde la investigación etnográfica. *Revista Latinoamericana De Metodología De Las Ciencias Sociales RELMECS, 5*(2), 6.
Rahman, A., & Fals Borda, O. (1992). La situación actual y las perspectivas de la IAP en el mundo. In M. C. Salazar (Ed.), *La investigación-acción participativa: inicios y desarrollos* (pp. 205–233). Popular.
Rufer, M. (2009). *La nación en escenas: memoria pública y usos del pasado en contextos poscoloniales*. El Colegio de México.
Salazar, M. C. (1992). *La investigación-acción participativa: inicios y desarrollos.* Popular.
Sánchez-Calderón, V. (2017). A "slum river": The unequal urganization of Bogota (Colombia) and the transformation of the Tunjuelo River in the twentieth century. In M. Knoll, U. Lübken, & D. Schott (Eds.), *Rivers Lost* (pp. 123–139). University of Pittsburgh Press.
Sánchez-Calderón, V., & Blanc, J. (2019). La historia ambiental latinoamericana: Cambios y permanencias de un campo en crecimiento [Latin American environmental history: Change and continuity in a growing field]. *Historia Crítica, 74*, 3–18. https://doi.org/10.7440/histcrit74.2019.01
Suárez-Krabbe, J. (2011). En la realidad. Hacia metodologías de investigación descoloniales. *Tabula Rasa, 14*.
Tax, S. (1992). Antropología-acción. In M. C. Salazar (Ed.), *La investigación-acción participativa : inicios y desarrollos* (pp. 27–36). Popular.
Thompson, E. P. (1963). *The making of the English working class.* Penguin.
Traverso, E. (2007). Historia y memoria. Notas sobre un debate. In M. Franco & F. Levín (Eds.), *Historia Reciente. Perspectivas y desafíos oara un campo en construcción* (pp. 67–96). Paidós.
Vargas Mena, E. (2014). Problemas metodológicos de la historia ambiental. Autocrítica de una experiencia de investigación con fuentes orales en el

volcán Barva -Sacramento y Paso Llano-, Costa Rica. *Revista de Historia, 70*, 229–257.

Wallerstein, I. (1994). *Open the social sciences: Report of the gulbenkian commission on the restructuring of the social sciences*. Standford University Press. Available at: https://www.sup.org/books/title/?id=792. Accessed 16 December 2020.

Zamosc, L. (1992). Campesinos y sociólogos: Reflexiones sobre dos experiencias de investigación activa. In M. C. Salazar (Ed.), *La investigación-acción participativa: inicios y desarrollos*. Popular.

5

An Ethos and Practice of Appreciation for Transformative Research: Appreciative Inquiry, Care Ethics, and Creative Methods

Angela Moriggi

Introduction

Over the last decade, a growing debate has emerged in the sustainability science community around the need for transformative research. The latter refers to research approaches that aim at producing impact-oriented knowledge through the co-creation of solutions with societal stakeholders, driven by researchers' commitment to partake in interventions seeking to enact and support change (Fazey et al., 2018). On the one side researchers are interested in investigating how transformational change happens, what are its main drivers and barriers, and how it can lead to a radical reshaping of human and environmental interactions in socio-ecological systems (Olsson et al., 2014). On the other, they are

A. Moriggi (✉)
Department of Land, Environment, Agriculture and Forestry, University of Padova, Padua, Italy
e-mail: angela.moriggi@unipd.it

A. Franklin (ed.), *Co-Creativity and Engaged Scholarship*,
https://doi.org/10.1007/978-3-030-84248-2_5

increasingly experimenting with action-oriented modes of knowledge co-production. In doing so, they contribute to revolutionizing the scientific paradigm towards transdisciplinary and participatory approaches that embrace uncertainty and exploration when dealing with the complex, multi-dimensional nature of socio-ecological problems (Abson et al., 2017; Fazey et al., 2020).

The content expounded in this chapter stems from a Ph.D. project (2016–2021) aimed at understanding transformational change, as well as contributing to transformative research. The study focused on Green Care practices in Finland, nature-based activities with a social innovation purpose, and their significance for pathways of place-based sustainability transformations. The data collection process was carried out over the span of three years, engaging three communities of Green Care practitioners by means of a participatory action research (PAR) approach. The conceptual building blocks of the research drew extensively from care-inspired understandings of sustainability (Pulcini, 2009; Tronto, 2013) and place-based and resourceful approaches to participatory co-production of knowledge (Gibson-Graham, 2008; Horlings, 2016).

In line with Fazey et al., (2018, p. 56) the study followed four preconditions believed to be crucial to practice transformative research: (1) it took into account the real world of politics, values, and ethics in societal change; (2) it included both practical and academic forms of knowledge; (3) it embraced creativity, innovation, and imagination as forms of knowledge production; (4) it was explicit about my position towards society and what kind of impact I expected my research to have.

The conceptual and practical understanding of change that underlined the transformative engagement in the study was inspired by the tenets of Appreciative Inquiry (AI). The latter refers to a form of action research long used in the field of organizational change and management (Busche, 2013). AI is commonly known as a strengths-based and positive approach to change. It provides a framework for anticipatory learning that supports collective processes of envisioning the future in a company, organization, or community. It can go hand-in-hand with a resourceful-oriented approach to participatory engagement (Franklin, 2018). At the same time, AI's philosophical groundings are in tune with a relational view of human agency and a celebration of life in all its forms, which is in line with a care-based understanding of sustainability. According to Zandee and Cooperrider (2008, p. 196), AI is grounded in an 'ethos of appreciation'. Up to now, however, the latter has been seldom discussed in relevant literature, and little account exists that explains how these philosophical tenets play out in practice.

The aim of this chapter is twofold. First, I wish to bring attention to the 'ethos of appreciation' underlying AI, and highlight its promising contribution for reinforcing a care-based approach to transformative research and a resourceful approach to participatory practice. In doing so, I draw particularly from the contribution of Zandee and Cooperrider (2008) mentioned above. The second aim of this chapter is to showcase how an 'ethos of appreciation' can be embodied and applied in practice, detailing five kinds of creative and arts-based methods used in the Ph.D. study. For each of the five methods, I explain the context of use, the purpose, the design, the modes of implementation, and the outcomes achieved. I also link each method to a specific dimension of AI's 'ethos of appreciation', to give a tangible account of how I interpreted it in practice.

In the discussion part of this chapter, I draw some reflections about the methodological potentials and limitations of using creative methods in this study, and the challenges and outcomes they yield when doing transformative research that aims to enable care-*full* and resourceful engagement processes. The chapter ends with concluding remarks about possible avenues for future research.

A Caring and Resourceful Approach to Transformative Research: Insights from the Literature

A growing number of sustainability researchers are looking at transformative change from a relational perspective, one that moves away from focusing on interactions between entities, and rather emphasizes continually unfolding processes and relationships (West et al., 2020). A major source of inspiration in relational thinking is characterized by scholarship on care ethics. In a recently published joint work, we have explored the potential of care ethics as a relational ontology to contribute to sustainability transformation theory and practice. As a result, three caring dimensions emerged as particularly relevant: *ethically-informed practices, relational response-ability,* and *emotional awareness* (Moriggi et al., 2020). Many researchers committed to explore and support resourcefulness in sustainability-oriented community pathways, embrace these three dimensions in their work, in more or less conscious ways. We see their goal as twofold. On the one hand they investigate how bottom-up local initiatives may contribute to "multi-fold social, cultural, environmental and economic value-creation at a community scale" (Franklin, 2018, p. 271). Examples include research on community food initiatives, on alternative forms of health and social care provisioning, on sustainable natural resource management, etc. (Franklin, 2018). At the same time, they also engage in collaborative processes that can nurture the inherent (and more or less latent) potential of the community to sustain and enhance its own resourcefulness and resilience (Franklin, 2018). This is often done by resorting to PAR and co-creative approaches of knowledge co-production, similar to what has been done in the Ph.D. study presented in this chapter. The remainder of this section will elaborate on the meaning of each of the three caring dimensions mentioned above, while giving examples of its application in participatory and resourceful research practice.

As far as the first dimension is concerned, seeing research—and participatory engagement in particular—as an *ethically-informed practice* inspired by caring principles, implies three main conditions: (a) attentive engagement to context and its interdependencies; (b) willingness to experiment; (c) attention towards empowerment (Moriggi et al., 2020). From a care perspective, context matters greatly. Issues cannot be understood only through a universal, standardized lens, nor can they be judged through abstract moral norms (Held, 2006). Embeddedness plays an important role in caring. Similarly, we contend that engaging in care-*full* research practices implies fostering deep relationships with specific contexts and realities, understanding and learning from them (Warren, 2000) as opposed to imposing sterile top-down knowledge or extracting useful data only for the sake of it. Much participatory and place-based researchers embrace embeddedness, relationality, and attention to context in their work (Brown et al., 2017; Giambartolomei et al., 2021; Horlings et al., 2020; see also Franklin, this book). Relating to context and its peculiarities also means becoming aware of its complexity and of the multifold relationships that constitute its socio-ecological system. This ideally implies the recognition of human-nature interdependence, and an openness to appreciate many forms of life, with an eco-centric rather than an anthropo-centric approach (Kimmerer, 2014). Arts-based research and transformative learning approaches offer meaningful examples of this kind of inquiry in practice (Harmin et al., 2017; Pearson et al., 2018) Notably, Harmin et al. (2017) resort to 'epistemological stretching' during a graduate level seminar course on environmental decision-making. They describe it as "a pedagogical orientation which focuses on expanding the ways of knowing that someone respects, understands, and/or engages with" (Harmin et al., 2017, p. 1). During the course, students were asked to combine course readings and lectures with personal experiences in nature, recorded through painting, sketching, prose poems, and photographs.

A second condition of research when seen as a caring and ethically-informed practice has to do with willingness to experiment. This is based on the idea that caring is an iterative practice, grounded on intensified involvement and knowledge (Noddings, 2013). For virtuous transformations to happen, things need to be done over and over again. Iteration does not merely (or necessarily) lead to betterment; however, it does create the space to adapt to the needs and capacities of those who are involved in the practice with an intentional and purpose-driven approach (Mol et al., 2010; Valencia-Sandoval et al., 2010). It also requires experimentation, tinkering, trial-and-error, and eventually, failure. Experimentation and iterative learning are considered essential factors in transformative research (Fazey et al., 2018; Giambartolomei et al., 2021). To carry out a Participatory Learning and Action Research (PLAR) project in Uganda, Sanginga et al. (2010) went through four iterative and complementary stages, including bottom-up experimentation and learning, sharing between communities, involvement of policy-makers and local administrations, and of district policy stakeholders.

Another example is provided by Foster (2016, p. 112) who narrates of the successes and failure of carrying out experimental collaborative arts-based research in order to promote social justice.

The third condition we identified in caring practices is tension towards empowerment. From a care perspective, empowerment goes hand-in-hand with recognizing the agency of both sides of the caring spectrum. Both sides must be given a voice, by re-framing relationships of power and by focusing on what people *can* do throughout the research process (Barnes, 2008). This resonates deeply with the call for co-creation and co-production of knowledge animating the transformative research debate in sustainability science today (Norström et al., 2020). Likewise, in PAR the desire to empower participants has motivated decades of attempts of inclusive and generative forms of engagement of communities and individuals (Evans et al., 2010; Reason & Bradbury, 2008). Notably, Masterson et al. (2018) describe practical and ethical advantages and challenges of using Photo-voice to engage local communities in Kenya and South Africa and to foster deep learning about human well-being in relation to socio-ecological systems dynamics (Masterson et al., 2018).

A care-based approach to empowerment is valuable as it also prompts us to see non-humans on the other side of the caring spectrum, recognizing their agency and dignity (Puig de la Bellacasa, 2011; Spretnak, 1997). The starting point to refuse objectification and domination is to explore with curiosity the rhythm and needs of non-human beings, recognizing them as sentient and communicative (Harmin et al., 2017; Kimmerer, 2014). There are examples of this practice in place-based experiential learning, where storytelling is used to support students of field philosophy to engage with nature affectively, embracing feelings such as wonder and mystery (Goralnik & Nelson, 2017).

Moving on to the second dimension of a care-based approach to transformative research, we can talk of *relational response-ability*, understood as the ability to respond to the needs we see around us (Haraway, 2016). Earlier I highlighted the importance of relationality and embeddedness to context. Close interactions and embodied experiences create bonds, connections, and responsibilities. Most importantly, they enable the possibility to notice and understand the needs of others (Tronto,

2013). This is typical of PAR work, as researchers refuse detachment and neutrality, and take a pro-active committed stance in relation to the community involved. Heras and Tàbara (2014) review around 20 examples of community-based research that used performative methods as an integrative research approach drawing on elements from the performing arts to support individual, community, and institutional reflexivity and transformation. Many researchers also feel the need to train their capacity for attentiveness (and consequently, response-ability) by learning to be "present – in the moment – and also open to what is not yet known" (Foster, 2016, p. 129). This can be done through mindfulness (Wamsler et al., 2018), spirituality (Kaufman, 2017), reflexivity (Robertson, 2000), and a general willingness to "dig in, to develop meaning, make connections, be honest and vulnerable, and seek growth" (Goralnik & Nelson, 2017, p. 15).

Finally, the third component of a care-based approach to transformative change is *emotional awareness.* For a long time, emotions have been fenced from the research arena. Recently, the humanities and social (sustainability) sciences have started to appreciate the centrality of emotions—both 'negative' and 'positive'—for change agency (Leys, 2011). Feelings such as anger, joy, fear, and hope can orient one's self towards the future and guide transformative actions (Pearson, 2021). Emotions are also deeply connected to our value systems, and greatly influence our moral compass, and the decisions we choose to take (Held, 2006). As a result, transformative researchers are slowly experimenting with the practice of bridging emotional and rational dimensions in processes of collective co-creation (Galafassi, 2018; Pearson et al., 2018). The dramatic urgency of socio-environmental issues, exemplified by the climate crisis, cannot be purely discussed through the medium of sterile modelling forecasts. By engaging with emotions, people can foster imagination, creativity, and intuition, and project themselves into the future in hopeful and liberating ways (Pearson, 2021). To this aim, novel ways of generating knowledge are being pursued, including visioning techniques that help people to embrace uncertainty and vulnerability (Evans et al., 2010; Tschakert et al., 2014), and that tap into existing positive and inspirational initiatives to explore alternative pathways to the future (Pereira et al., 2019).

Appreciative Inquiry and an Ethos of Appreciation

Since its introduction in the 1980s, AI has been used extensively around the world to promote transformative change in organizations and groups (Ludema & Fry, 2008). The idea underlying AI is not to implement change towards a goal. Rather, it is "about chang*ing* … convening, conversing and relating with each another in order to tap into the natural capacity for cooperation and change that is in every system" (Ludema & Fry, 2008, p. 281). AI is based on the assumption that each organization or group has a positive core that provides a source of 'renewable' energy for both personal and organizational transformation. Often, this positive source of energy remains untapped due to a long-standing reliance on a problem-solving approach. However, accounts from research and practice have demonstrated how focusing on problems to search for solutions often leads to ineffective and disappointing results (Hung et al., 2018). In contrast, AI builds on the idea of 'generativity'. Rather than being stuck in conversations about gaps and challenges, or getting trapped in reductionist thinking about one solution versus another, AI leverages the capacity for generative dialogue between individuals (Busche, 2011). It empowers people to build new knowledge, spur inventiveness, create energy, and enhance co-operative capacity, through curiosity, wonder, and surprise (Ludema & Fry, 2008; Zandee & Cooperrider, 2008). This triggers a virtuous circle, where inspiration, joy, and strength feed into each other towards "ascending spirals of co-operative action" (Ludema & Fry, 2008, p. 282).

The tendency to focus on the positive characterizing AI has also been viewed with criticism. Practitioners who favour exclusively positive narratives at the expense of negative experiences and feelings, may reduce AI to a simplistic, mechanical, and even manipulative form of engagement (Zandee & Cooperrider, 2008). Moreover, many using AI mostly rely on its 4-D cycle of inquiry (*discovery—dream—design—destiny*), without truly understanding the origins of the practice and the philosophical principles inspiring it (Ludema & Fry, 2008). To countervail risks of trivialization of AI, Zandee and Cooperrider (2008) elaborate on five dimensions that lie at the heart of its 'ethos of appreciation'. These dimensions are not only meaningful as they support AI practitioners in substantiating and elevating the generative capacity of their work. They are also valuable lenses that illuminate over the similarities and complementarity between AI, a care-based approach to transformative change, and participatory and resourceful approaches to engagement. As such, they were particularly valuable in the Ph.D. study object of this chapter. I will now briefly explore each of the five dimensions (also shown in Fig. 5.1).

Illuminating the miracle of life is the first dimension proposed by Zandee and Cooperrider. It is based on the assumption that life is mysterious, and as such must be appreciated with wonder and "childlike openness in inquiry" (Zandee & Cooperrider, 2008, p. 193). Accepting mystery also implies embracing uncertainty—for a long time banished from the scientific realm, and now increasingly called for by transformative research proponents (Keeler et al., 2017; West et al., 2020). As far as

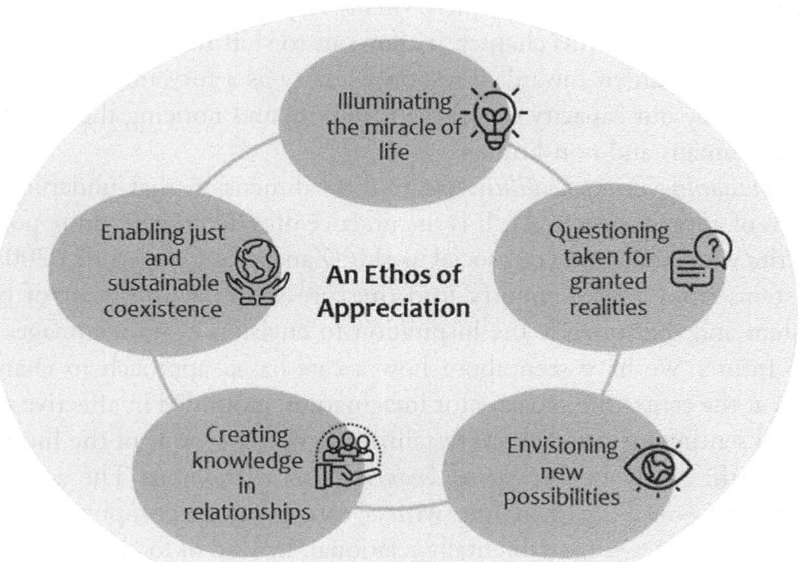

Fig. 5.1 Five dimensions of an ethos of appreciation in AI (*Source* Developed following Zandee & Cooperrider, 2008)

wonder is concerned, from a care perspective, it is absolutely crucial. It is a way to appreciate the Earth's beauty, as well as its suffering; to see, feel, and sense empathetically *with* it (Kimmerer, 2014). Nurturing a sense of wonder is not merely an aesthetic exercise, but also a moral virtue, as Kathleen Dean Moore beautifully explains, drawing from Rachel Carson and her masterpiece *The Edge of the Sea* (2007). As Carson wrote: "I believe that the more clearly we can focus our attention on the wonders and realities of the universe about us, the less taste we will have for destruction" (Moore, 2005, p. 28).

The second dimension characterizing an ethos of appreciation is *questioning taken for granted realities.* AI invites us to rethink the questions we ask, and reframe the topics of inquiry. The goal is to break free of habituated patterns of thinking and acting, and unleash curiosity, imagination, and fresh thinking and deliberation (Zandee & Cooperrider, 2008). Care ethics scholars have long questioned taken for granted ontologies, engrained in Western philosophical thinking. The idea of relational response-ability is one of many examples. As explained in the previous section of this chapter, it allows us to shift the focus on responsibility as a burden towards a response-ability as a forward-looking act, triggered by our capacity of being in relation, and noticing the needs of other humans and non-humans.

Envisioning new possibilities is the third dimension that underlies an ethos of appreciation in AI. It is the practice of welcoming infinite possibilities when imagining our social worlds (Zandee & Cooperrider, 2008). It stems from a deep inquiry into the core values at the heart of our system and that provide the inspiration to envision evocative images of the future. We have seen above how a care-based approach to change puts at the centre the capacity for imagination, grounded in affective and moral sentiments, needed to crystallize alternative visions of the future.

Fourth, AI is about *creating knowledge in relationship.* The assumption here—once again in line with a care ethics philosophy—is that human existence is fundamentally relational. Instead of focusing on individualistic accounts of human agency, we should focus on relationships and "see others as vital co-creators of our mind, our self, and our society" (Sampson in Zandee & Cooperrider, 2008, p. 195). The process of inquiry is supposed to nurture this relational knowledge by creating the conditions for interpersonal connection and sharing.

Finally, the fifth dimension is about *enabling just and sustainable coexistence*. Relationships are not to be nurtured solely with other human beings, but should embrace other species as well. Engaging in AI processes should therefore also remind us of "our own embodied participation in a spirited, biological realm" and "appreciate our sensuous participation in a more-than-human world" (Abram, 1996 in Zandee & Cooperrider, 2008, p. 195). This eco-centric approach resonates well with awareness of interdependence animating a care-based approach to transformative change.

In the remainder of this chapter, my goal is to showcase how an ethos of appreciation can be put into practice, by detailing five kinds of creative methods I have employed during the collaborative engagement part of my Ph.D. project. For each method, I will explain the purpose, the sources of inspiration, the way it was implemented, and the outcomes it produced. To each method is associated one of the five dimensions of an ethos of appreciation presented above. By elaborating on the practical applicability of each dimension in detail, my goal is to complement and enrich the conceptual assumptions sketched by Zandee and Cooperrider (2008), offering additional interpretations of *how* an ethos of appreciation can be understood and operationalized.

An Ethos of Appreciation in Practice: An Account from the Field Using Creative Methods

This section provides a methodological and empirical account of the application of selected methods during the course of my Ph.D. study. Before introducing each method in detail, I provide some background information about the research and a brief overview of the various stages of participatory engagement.

Empirical Study: A Three-Year Collaboration with Green Care Practitioners in Finland

The overall aim of my Ph.D. project was to analyze and appreciate place-based practices of Green Care in Finland and their possible significance and contribution to processes of transformative change (Moriggi, 2021). Green Care is an umbrella term used to describe a wide range of activities in nature aimed at health and social care, social inclusion, pedagogy, and recreation (Sempik et al., 2010). In this study I explored three diverse examples of Green Care practices. The first case, a care farm, involves a group of mentally disabled people in sheep husbandry and farming activities for therapeutic purposes. The second case, a biodynamic farm, engages different target groups (e.g., long-term unemployed, children with special needs) in farming practices for social inclusion and pedagogy. The third case, a nature-tourism company, offers outdoor sports, wellbeing, educational, and recreational activities to a variety of users, including company employees, people with disabilities, and the elderly.

The main practitioners of the two farms and of the company (seven people) were engaged over the span of three years on a continuous basis (2016–2019). Most stages of fieldwork also involved other stakeholders, such as the staff of the three enterprises, their clients, the external networks of collaborators (e.g., civil servants, business partners, buyers, etc.), as well as experts in the field of Green Care. Around 75 people were involved in total. The collaboration aimed at not only gathering relevant data, but also at fostering a process of critical reflection and capacity-building for the three communities of practitioners, appreciating their assets and capacities, in line with a resourceful approach to participatory practice. The study focused on people's values and motivations to initiate Green Care activities, on the caring relationships enacted through the practices (see Moriggi et al., 2020), and on the role of place-based resources in sustaining the process of change agency at both individual and collective levels (Moriggi, 2019).

The empirical work relied on an in-depth qualitative research informed by place-based, transdisciplinary sustainability science, enriched by the principles and techniques of Participatory Action Research (PAR). In line with these traditions, all methods were designed and implemented following principles of inclusiveness, transparency, reflexivity, and empathy. The methods of data collection in the study stem from an 'eclectic pluralism' of approaches and techniques, borrowing from both academic and non-academic fields (Chambers, 2008, p. 311). As often happens in action research, my methodological approach was that of a 'bricoleur', as I integrated and made sense of various perspectives with the evolving of the research process, and of my understanding of the issues under study (Wicks et al., 2008, p. 26). As a result, more conventional data collection activities, such as semi-structured interviews and participant observation, were coupled with visual and creative ones, such as Photo-voice and arts-based methods.

Figure 5.2 provides an overview of the various stages of fieldwork and the methods used.

The methods detailed in the following section are creative techniques that were specifically designed and used during the co-creation workshops. The first workshop, called 'Sharing and Reflecting', was carried out in August 2018, and brought together the practitioners of the three cases (nine people)—including the main entrepreneurs and some of their staff. The objectives were twofold: (a) to present and discuss preliminary results and the conceptual framework of the research work; (b) to provide an opportunity for sharing and reflection, highlighting both commonalities and differences of the various approaches to Green Care

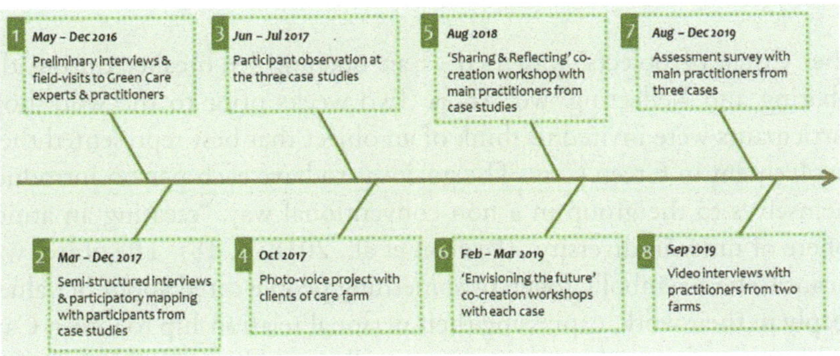

Fig. 5.2 Stages of fieldwork and related methods of data collection (*Source* The author)

across the three cases. The workshop lasted seven hours and was structured loosely following the tenets of Theory U, a facilitation framework particularly used in organizational management and change (Scharmer, 2007). Different techniques were used, some borrowed from organizational management, others inspired by system thinking, others adapted from arts-based research and experiential learning.

The second series of workshops, called 'Envisioning the future', involved each of the three cases separately, to support practitioners in crystallizing future visions of their Green Care practices, and the wider development of their community and place. In total, sixteen people participated in the three workshops. Also in this case, the workshop lasted seven hours, and was designed combining Theory U with the 4-D model of Appreciative Inquiry (*discovery—dream—design—destiny*). The methods used borrowed from system thinking, design thinking, and arts-based research.

I acted as a facilitator in both workshops and had the support of a Masters' student (Finnish mother-tongue), who provided logistical help and interpretation assistance when needed.

Below I introduce five methods used during the two workshops, following the chronological order with which they were implemented.

Method No. 1—'Circle of Objects': Creating Knowledge in Relationships

The 'Circle of objects' was used as an opening ice-breaker during the 'Sharing and Reflecting' workshop. Two weeks prior to the workshop, participants were invited to think of an object that best represented their involvement in Green Care. The goal was to have each person introduce themselves to the group in a non-conventional way, "creating an atmosphere of unity in diversity" (Pearson et al., 2018, p. 18). The object was meant to be a symbolic token of something people cared about or valued deeply in their work, expressing their personal relationship to Green Care and learning from others in an emotionally-sensed way. "Objects have a great evocative and aesthetic power" and "enable people to communicate tactically and metaphorically" (Pearson et al., 2018, p. 18). The method

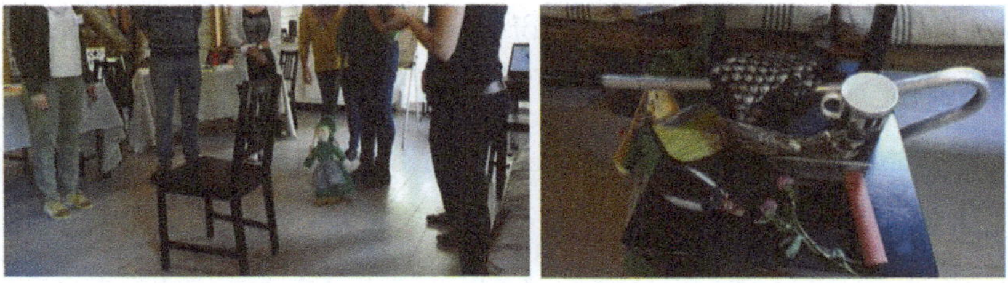

Fig. 5.3 'Circle of objects': people sharing in circle; all the objects collected on a chair (*Source* The author)

was adapted from a version previously designed by the author of this chapter jointly with other colleagues (from the collective Re.imaginary[1]) for a workshop not related to this Ph.D. study (see Pearson et al., 2018).

After a few welcoming words at the start of the workshop, participants were invited to join in a circle, holding the objects in their hands. One after the other, people introduced themselves, briefly narrating the story behind their object, and placed it at the centre of the circle, as a symbolic gesture signalling their belonging to the community of people participating to the workshop. As shown in Fig. 5.3, some of the objects were pieces of equipment used daily by people in their work, such as a shovel—reminding them of the importance of caring for the soil, crucial source of life for all practices happening on the farm. Others brought a nail clipper for rabbits and a sheep cane, telling of daily practices of caring for animals, needed 'partners' in the rehabilitation and social inclusion activities. Other objects related to the different aspects or roles taken in Green Care practices. Notably, one participant brought an enamel cup with an image of the Moomins, fantastic characters designed by Tove Jansson (Finnish writer of children's literature), widely popular and appreciated across the country. The object symbolized her role as a storyteller when working as a nature guide, taking groups into the forest and narrating of traditional livelihoods and human-nature relationships. Another person brought a broom with a puppet of a Nature Witch, to introduce her 'alias' during Green Care activities. She expressed her wish to convey wonder and magic to people, via experiential learning in

[1] See: https://www.reimaginary.com/.

nature. A woman brought a pair of hand mittens, to signify the combination of softness and strength required of a professional Green Care practitioner. Cloves flowers reminded a person of the regenerating power of forest walks, that she herself experienced living on the biodynamic farm.

The objects were rich in meaning and acted as a medium to people's personal stories and experiences. In line with AI's principle of *creating knowledge in relationships*, they allowed participants to connect as a group and share their sources of inspiration and accomplishment (Zandee & Cooperrider, 2008). These 'narrative rich' short introductions brought smiles to people's faces, and created a sense of kindness and mutual empathy that set the tone for the remainder of the workshop.

Method No. 2—'Creating with the Soil': Enabling Just and Sustainable Coexistence

The second method, called 'Creating with the soil', was used about two-thirds of the way through the 'Sharing & Reflecting' workshop. The core part of the workshop combined my presentations of the main findings of the Ph.D. study (up to that point in time), with discussions in pairs, and sharing in plenary. It then involved an individual mapping exercise focused on Green Care practitioners' resourcefulness, also followed by collective sharing. All these activities required substantial intellectual effort from participants, with a great deal of information being conveyed and thoughts expressed. As such, I deemed it necessary to include a somatic break in the workshop, to allow people to reconnect with their bodies, and rest their brains for a while (Evans et al., 2009).

Participants were therefore invited to 'create with the soil'. A week before the workshop, I asked the practitioners from the biodynamic farm (located only 5 km away from the workshop's venue) to collect a bucket of fresh soil from their farm, and bring it along on the workshop day, together with a thick wood branch. The 'Creating with the soil' method is about working with clayish material in a freestyle, letting shapes emerge and crystallize without previous planning. The goal is to enable a playful and relaxed atmosphere, away from complex and articulated discussions, engaging with a simple, almost child-like, artful expression. The method also aims at facilitating generativity and intuition, by allowing people to 'think with their hands'. As such, it disrupts normal patterns of thought, and can lead to unexpected wisdom and insights during the process of creation. The method is also an inclusive one, as it does not require the communication of articulated thoughts verbally, nor to be 'proficient' in any creative forms of expression, as people can make shapes freely and at their own pace, without any expectations of the outcome.

'Creating with the soil' was inspired by artist Lotte Kravitz, who facilitated a similar exercise during the international conference 'Transformations to Sustainability', held at the University of Dundee in September 2017. When designing the co-creation workshop for my research, I found this method particularly fit, knowing the participants involved, and their natural attitude to 'get their hands dirty'. Moreover, the act of shaping the soil has strong symbolic connotation. The soil of the biodynamic farm is far from being mere dirt; rather, it is composed of lively organic matter, rich in dead and living organisms. For the practitioners there, it is a crucial source of life on the farm, something to attentively care for, and the subject of various sacred rituals. Although the other participants did not follow anthroposophical principles, all expressed great respect for the soil and the natural elements they engage with daily. As such, I felt that 'creating with the soil' could well express AI's proposition to be reminded of "our own embodied participation in a spirited, biological realm & appreciate our sensuous participation in a more-than-human world" (Abram in Zandee & Cooperrider, 2008, p. 195). This is linked to the fifth dimension of an ethos of appreciation expounded earlier, namely *enabling just and sustainable coexistence*. The idea underlying it is to nurture relationships with both humans and non-humans, avoiding an anthropocentric approach, and bringing to the room and the co-creation process other living elements. According to Zandee and Cooperrider (2008), one good way to do so is by engaging in bodily exercises, sensing and feeling one's own rootedness in the larger ecology. This may also enable reflections and insights that allow people to reconnect to the values and sources of motivation that inspire their everyday undertakings.

During the 'Sharing and Reflecting' workshop, I placed the wood branch on a table in the middle of the room, together with the bucket of soil and a bowl of water. As shown in Fig. 5.4, people were invited to join on a voluntary basis, and start working with their hands. When and if a shape formed, they could place it on the wood branch for display. Participants responded enthusiastically, and a few of them immediately joined the table. As they started to play with the soil, they engaged in relaxed conversations, laughing and smiling. However, it was not easy to get everyone to join at the same time; this made the process not as fluid and lively as I had experienced it as a participant in Dundee. Finnish people are very respectful of other people's space, and therefore some of the participants preferred to 'take turns' rather than mingle with the whole group. In the end, everyone ended up with a little creation, and placed it on the wood branch.

The shapes were mostly related to beings from the natural world, including a mushroom, a rabbit, a pig, a bug, and a horseshoe. Other shapes included some alien forms, and a small human with open arms. The meaning of the creations was not discussed, but simply shared in plenary to foster a feeling of collectiveness and unity in diversity, and anchor key impressions from the exercise.

Reflecting back on the effectiveness of the method, I can say that it worked as a relaxing and aesthetically pleasant break. However, it required more time than planned (including the time needed for people to wash their hands and reconvene in the room), and I therefore had to

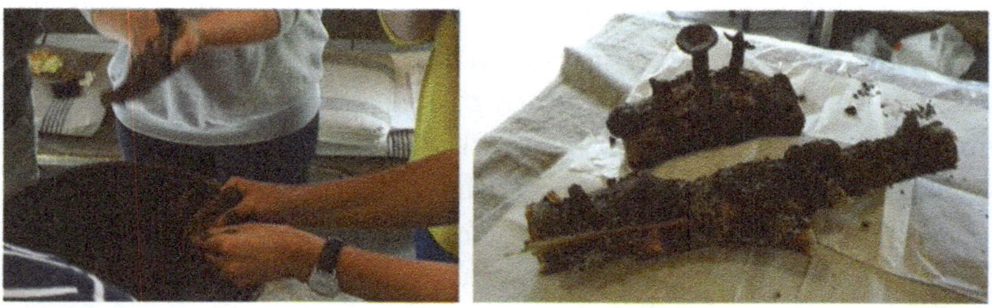

Fig. 5.4 'Creating with the soil': people shaping the clay; the forms created displayed on the wood branch (*Source* The author)

rush it towards the end. Moreover, the exercise that was scheduled afterwards did not build on the feeling of more-than-human co-existence, nor did it engage further with the somatic intelligence of the participants. As such, the potential to fully put into practice an ethos of appreciation was partially lost on this occasion.

Method No.3—'Council of Beings': Questioning Taken for Granted Realities

The third method, called the 'Council of beings', was used during the 'Envisioning the future' workshops, in close combination with method No. 4—'Letters from the future'—which I will present next. Three separate 'Envisioning the Future' workshops were held for this study, and I will therefore reflect on Methods Nos. 3 and 4 thinking of all the three events.

The 'Council of beings' is a combined adaptation of two techniques (respectively called 'Inviting non-human stakeholders' and 'Expanding time') previously designed by a group of colleagues and myself for a workshop not related to this Ph.D. project (see Pearson et al., 2018). It is directly inspired from Joanna Macy's 'Council of all beings' (Macy & Brown, 1998). The latter is a communal ritual in which participants are asked to step aside from their human identity and speak on behalf of another life-form, in order to gain stronger awareness of our interdependence with other living beings, and trigger emotions of care and wonder for them. In this sense, the exercise lends itself to *questioning taken for granted realities*, one of the five dimensions of AI's ethos of appreciation. Instead of approaching an issue exclusively from an anthropocentric perspective, as it is normally done, this method forces people to reframe the questions to be asked, and break free of habituated ways of thinking, to take on non-human perspectives.

The 'Council of beings' (together with 'Letters from the future') formed the 'Dream' part of my workshops, which were structured along AI's 4-D cycle of inquiry. During that phase, participants were asked to give voice to desires and wishes for the future, when thinking of the development of their place and practices. As a first step, I invited them to embrace a different perception of time by showing them a timeline, portraying the time span of different human and non-human beings. The beings on each timeline were closely related to each case. In the one shown in Fig. 5.5—designed for the care farm—I included the picture of a building, namely the guided-living unit where the mentally disabled clients live, imagining it would be there for about 200 years. Human beings included a child, a disabled person, and an elderly lady (living on average 85 years)—representing the current and possible future dwellers of the farm (all the pictures were taken from the internet and did not represent any real person living on the farm, for confidentiality reasons). The remaining pictures portrayed the miniature pigs living on the farm (living 12 years on average), one of the many sheep they raise (living 10 years on average), and a butterfly (living 1 month on average). By showing the different time span of the various beings, the aim was to make people reflect about our usual perception of time, often mostly focused on our short-term needs as humans, and highlight other time perceptions as well. When thinking about the future development of a place, it is important to become aware of the needs of different beings, including elements of the socio-ecological system (Pearson et al., 2018).

As a second step, I laid out seven cards on a table (one more than the number of participants present), each representing one of the beings on the timeline. I asked people to observe the cards, and pick one they would want to give voice to, in this exercise and in the following one

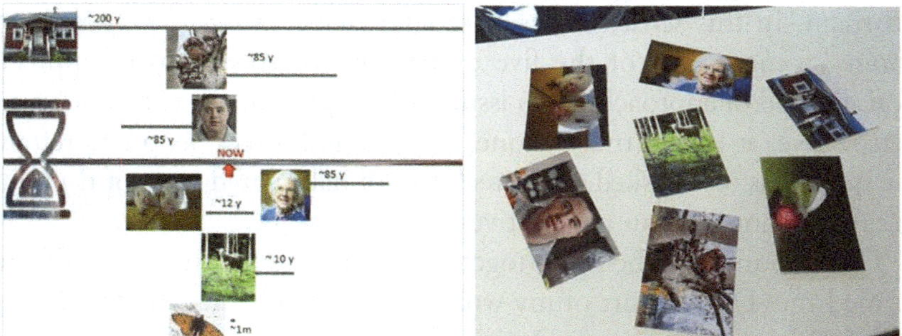

Fig. 5.5 'Council of beings': timeline of change; cards with different beings (*Source* The author)

of imagining the future. Once everyone had a card, people introduced their 'new self', and named a characteristic or something they loved about their character. This moment of sharing was meant to symbolically invite the new stakeholders' identities to the discussion, and trigger people's imagination and capacity to step out of their comfort zone. Both steps worked well in all the workshops I held. People could immediately relate to the beings introduced, as they were part of their everyday life and work. They also quickly 'embraced' their new self, and got into the playful atmosphere of pretending to be a different being. Knowing the participants in advance certainly helped me to select the most appropriate choice of human and non-human beings, in a way that would speak to each participant's experiences and aspirations.

Method No. 4—'Letters from the Future': Illuminating the Miracle of Life

The method 'Letters from the future' followed straight after the 'Council of beings'. Once people went back to their seats, I distributed a nice piece of paper, resembling the texture of a letter, and a pen. People were then invited to write a short letter to themselves, thinking from the perspective of their 'new' being. This person or animal or thing would speak to them from the future, in the year 2039, 20 years from the date of the workshop. Their future selves had the capacity to see their place—the care farm, the biodynamic farm, or the location of the nature-tourism company—in its future and most ideal development state. The following guiding questions were given as prompts: *(1) What do you see in the place? What does it look like? (2) What activities are happening? Who is there? (3) How do you feel? What sparks your joy?* People were then given ca. 20 minutes to write their letters, choosing either English or their mother tongue (Finnish or Swedish), as the preferred language of writing (see Fig. 5.6). Once everyone had finished their letters, participants were asked to read them out loud, sharing their visions with the group. Language interpretation was provided by my assistant, in case the letter was not in English, to allow me to understand the content fully.

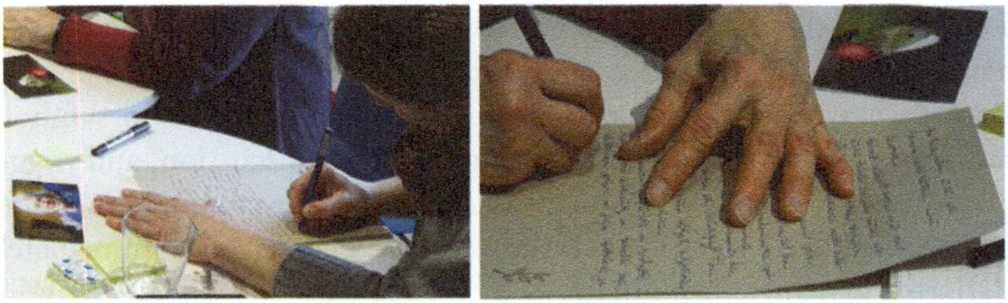

Fig. 5.6 'Letters from the future': participants writing letters from the perspective of other beings (*Source* The author)

The main goal of this method was to elicit a dream-like situation in which people could picture the best possible scenario for their place. In AI, it is important to go through this stage, rather than moving straight from the present situation to a future one. The *Dream* phase allows people to connect to their deepest sources of motivation and to give voice to their wishes, without being held back by cynicism or caution. It answers the question *What could be*? and prompts people to envision multiple possibilities. It is followed by the *Design* phase, when space is given to building the necessary steps to realize the ideal vision, answering the question *What should be?* (Busche, 2011). Zandee and Cooperrider suggest using artful creations, such as drawing, poems, and songs, in the *Dream* phase, to "express latent images of ideal futures" and discover and communicate shared meaning (2008, p. 194). According to them, this is conducive to *illuminate the miracle of life*. Artful creations introduce a sense of wonder and childlike inquiry to the discussion, allowing people to access a more 'intuitive' and 'sensuous' understanding of organizational life (Zandee & Cooperrider, 2008).

In the case of the workshops I held, my hope was to heighten the sense of wonder, by taking the perspective of a different being. In that way, participants were asked to disrupt their normal patterns of thinking, and crystallize thoughts previously untapped. For instance, when thinking as a butterfly, participants imagined flying over their place, and got a landscape view of what was happening and how, noticing smells or seeing things that usually went unnoticed. Additionally, people were asked to picture an ideal and positive future in the mind of an animal or a plant. As such, more chances were created to portray regenerative possibilities—conducive for both human's and non-human's flourishing and well-being—and therefore not only illuminate, but also celebrate and nurture the miracle of life.

It is important to note that not all workshops led to the same outcomes. In one of them, I encountered resistance by one of the participants—who held an important role in the organization—who refused to embrace a positive perspective on the future. Rather, they chose to depict a dystopian future, portraying ecological destruction and loss of the human-nature connection. This partially jeopardized the process, as it created a sense of awkwardness and mismatch with the visions presented by the other participants. It also toned down the collective energy in the group, and somehow trivialized the imaginative and dream-like efforts in the other letters. In response to this, I slightly adapted the remaining part of the workshops, trying also to give voice to risks and challenges when building realistic steps towards the future vision. However, looking back at the workshop now, I can say that a more flexible structure and stronger experience as a facilitator from my side could have helped to welcome the resistance in a more fruitful way, and re-shape the workshop most appropriately.

As far as the other workshops were concerned, the 'Letters from the future' method yielded very positive results. There was a collective sense of empathy and heartfelt connection while people read their letters. People later said that hearing others depict positive visions of their places, made them feel hopeful and energized to further pursue their plans and wishes. This exercise also gave them confidence for the next part of the workshop, in which concrete activities and needed resources had to be envisioned.

Method No. 5—'Vision Tree': Envisioning New Possibilities

The last method I would like to introduce is called 'Vision tree' and takes direct inspiration from a manual that tells of experiences of using AI with rural Indian communities (Ashford & Patkar, 2001). The tree is a visual metaphor that helps to brainstorm, crystallize, and prioritize thoughts in the group. In the case of my workshops, I used it in conjunction with 'Letters from the future', explained in the previous section. While participants were reading their letters out loud, I recorded as many keywords as possible on post-its, responding to the guiding questions given earlier: (1) *What do you see in the place? What does it look like?* (2) *What activities are happening? Who is there?* (3) *How do you feel? What sparks your joy?* I then clustered the keywords into three main areas: *core elements*, representing the roots of the tree; *main activities*, to be placed on the trunk of the tree; and *values & emotions*, manifested on the fruit or branches of the tree.

Once all participants had finished their letters, as can be seen in Fig. 5.7, I placed the various post-its on the Vision tree—painted by myself prior to the workshop on thick paper, hanging on one of the walls of the workshop's venue. In the case of the care farm, words like horses, visitors, butterflies, wool, water streams, new buildings, etc., appeared at the roots of the tree. On the trunk, there were yoga courses, musical gigs, farmers' markets, horse care, etc. The branches were populated by feelings of joy, beauty, love, community, trust, etc.

My main goal with this adaptation of the exercise was to acknowledge participants' visions in a way that would be immediately visible to

Fig. 5.7 'Vision Tree': the tree with and without post-its (*Source* The author)

everyone in the group. By seeing all the keywords in one image, people could easily spot similarities or resonances, but also differences. Moreover, having keywords clustered on the different parts of the tree, allowed for analytical clarity, leading naturally to the next stage in the workshop, in which concrete plans and steps had to be taken into account—focusing on future practices, ways of working, and needed resources. The goal was also in line with AI's ethos of appreciation, and in particular with the idea of *envisioning new possibilities*. This process had already started with the 'Letters from the future', and was now further consolidated as people could actually *see* black on white the most important ingredients that made new possibilities alive. As Zandee and Cooperrider put it, 'words create worlds' (2008, p. 194). It is important to co-create the positive imagery collectively and to highlight the connections between ingredients and inspiration, so that the image of the future feels like a shared one.

Discussion

In this section I draw some methodological reflections, focusing on design and execution of the techniques detailed above, and reflecting on their added value for transformative research aimed at enabling care-*full* and resourceful processes of engagement. I also briefly elaborate on the kind of research data and more general outcomes the methods yielded, highlighting challenges and limitations. These reflections draw from my own observations and the notes in my research diary, as well as the feedback given by the participants about the methods—right after the workshop, as well as via a questionnaire survey administered to the main practitioners (seven people) at the end of the research project (Moriggi, 2021).

As far as design is concerned, a lot of preparation and thorough planning was dedicated to the methods. I followed an informed rationale, and aimed at achieving specific objectives. Only at a later stage, once the methods had been tested multiple times, did I gain stronger awareness of its strengths and challenges. Notably, only when trying out the methods with different audiences did I realize how they could allow me to put specific AI principles into practice and facilitate transformative processes.

In terms of execution, I believe creative techniques substantially enriched the processes of co-creation during the workshops. They also enhanced the care-*full* and resourceful approach I was trying to embody. Notably, they facilitated greater empathy and connection within the group (and with non-human beings), spurred people's imagination and out-of-the-box thinking, they helped participants to access their inner wisdom and emotions, they disrupted habituated patterns of thought and action, and allowed for experimentation and tinkering. The visual artefacts acted as useful 'boundary objects'—prompts that facilitated communication and understanding around a certain issue (Home & Rump, 2015). These outcomes are in line with what is expected of action-oriented and transdisciplinary forms of research. A crucial pre-condition to their effectiveness was that most participants could *trust the process*, accepting to play and participate, without knowing the outcome. On the other hand, even when some of the techniques were not used to their full potential, or when I encountered the resistance of some of the participants, I could see the long-term benefits of using these methods for the engagement process in its entirety. A certain degree of flexibility was important to adapt the methods to different circumstances. On the other hand, the structure of the workshops was rather tight, and extra space could have been made for improvisation and serendipity. Flexibility and adaptation are crucial for care-*full* and resourceful research: they allow the facilitator/researcher to tap into the full (and perhaps unexplored) potential of the method, while helping the group to feel empowered and thrive along the process.

For the purpose of my Ph.D. study, during the workshops I used creative methods in combination with more 'analytical' ones (e.g., SWOT analysis, system mapping). These methods were particularly appropriate to gather text-rich information, and allowed participants to rely on more conventional and familiar forms of learning and collaborating. By combining different techniques, I was able to elicit different modalities of knowledge generation—engaging brains, hearts, and hands—and facilitate both individual and collective learning.

In terms of outcomes, looking back at the whole research process, I did not merely rely on co-creative workshops and creative methods, but also on more conventional forms of data collection, such as semi-structured interviews and participant observation. These provided in-depth information that cannot always be accessed through group events lasting only a few hours. Interviews transcripts have been a crucial in-depth source of empirical findings for my Ph.D. papers and thesis. I also analyzed some of the data obtained through creative methods. To this aim, it was very important to document the process during the workshops, by taking pictures, recording people's observations and thoughts (either with the help of a note-keeper, or with a tape recorder), and transcribing the information written on post-its, maps, letters, etc. The data obtained were mostly used for triangulation purposes, namely to achieve greater rigour when interpreting different datasets and enhancing the validity of the formulated findings. To some extent, the process of triangulation mitigated risks of 'deference' and 'social desirability' effects, namely when participants tell the researcher what they want to hear, or what makes them look good in front of the group (Galafassi, 2018). Moreover, presenting preliminary findings and conceptualizations to participants was extremely valuable to validate their accuracy and relevance. Documenting the process during the workshops was also valuable as it provided so-called 'presentational knowledge', useful for communication purposes beyond fieldwork (Gearty et al., 2015, p. 61). In the weeks following each workshop, I compiled a 'Learning Portfolio', a short document where I collected pictures, slides, and thoughts discussed during the workshop. The Portfolios were then sent in both hardcopy and electronic format to the participants, as a record of what had happened during the workshops, and as a resource they could tap into for their future development steps. Moreover, I also used the pictures in presentations and events to communicate findings in a more effective and memorable way.

Conclusions

In this chapter, I have discussed the need to foster transformative research, and presented Appreciative Inquiry (AI) as a framework for organizational change and management. AI provides both conceptual and practical tools that can enrich care-*full* and resourceful transformative research practice. In particular, I presented five dimensions of AI's 'ethos of appreciation', laying out their philosophical meaning, as well as their practical application, by giving a detailed account of five creative methods I employed during my Ph.D. study.

In conclusion, it can be said that the methods proved very valuable to facilitate care-*full* and resourceful processes of co-creation. They also revealed, especially over time and with multiple applications, how an 'ethos of appreciation' can be put into practice. However, challenges and limitations were also present. Additional empirical testing of these methods is needed, to explore their possible application and potential in various contexts of action, and in combination with different techniques. Moreover, the five dimensions of an ethos of appreciation deserve further attention and elaboration, at both conceptual and methodological levels. AI offers meaningful perspectives that have so far been only partially taken on by proponents of transformative sustainability research. The hope is that this chapter can provide inspiration for other researchers and practitioners to embrace an 'ethos of appreciation', and to foster care-*full* and resourceful engagement processes for transformative change.

Acknowledgements The Ph.D. project object of this chapter has received funding from the European Union's Horizon 2020 research and innovation programme under the Marie Sklowdoska-Curie Grant Agreement No. 674962. Additional support has been provided via a personal working grant funded by Kone Foundation, project identification no. 201801752.

References

Abram, D. (1996). *The spell of the sensuous: Perception and language in a more-than-human world*. Random House.

Abson, D. J., Fischer, J., Leventon, J., Newig, J., Schomerus, T., Vilsmaier, U., von Wehrden, H., Abernethy, P., Ives, D. C., Jager, N., & Lang, D. J. (2017). Leverage points for sustainability transformation. *Ambio, 46*, 30–39. https://doi.org/10.1007/s13280-016-0800-y

Ashford, G., & Patkar, S. (2001). *The positive path: Using appreciative inquiry in rural Indian communities*.

Barnes, M. (2008). Care, deliberation and social justice. In J. Dessein (Ed.), *Community of practice Farming for Health, 6–9 November 2007, Ghent, Belgium* (pp. 27–37). ILVO.

Brown, K., Eernstman, N., Huke, A. R., & Reding, N. (2017). The drama of resilience: Learning, doing, and sharing for sustainability. *Ecology and Society, 22*(2). https://doi.org/10.5751/ES-09145-220208.

Busche, G. R. (2011). Appreciative inquiry: Theory and critique. In D. Boje, B. Burnes, & J. Hassard (Eds.), *The Routledge companion to organizational change* (pp. 87–103). Routledge.

Busche, G. R. (2013). Generative process, generative outcome: The transformational potential of Appreciative Inquiry. *Advances in Appreciative Inquiry, 4*, 89–113. https://doi.org/10.1108/S1475-9152(2013)0000004003

Chambers, R. (2008). PRA, PLA and pluralism: Reflections and theory. In P. Reason & H. Bradbury (Eds.), *The Sage handbook of action research: Participative inquiry and practice* (pp. 297–319). Sage.

Evans, J., Davies, B., & Rich, E. (2009). The body made flesh: Embodied learning and the corporeal device. *British Journal of Sociology of Education, 30*(4), 391–406. https://doi.org/10.1080/01425690902954588.

Evans, K., de Jong, W., Cronkleton, P., & Nghi, T. H. (2010). Participatory methods for planning the future in forest communities. *Society & Natural Resources, 23*(7), 604–619. https://doi.org/10.1080/08941920802713572.

Fazey, I., Schäpke, N., Caniglia, G., Hodgson, A., Kendrick, I., Lyon, C., Page, G., Patterson, J., Riedy, C., Strasser, T., Verveen, S., Adams, D., Goldstein, B., Klaes, M., Leicester, G., Linyard, A., McCurdy, A., Ryan, P., Sharpe, B., …Young, H. R. (2020). Transforming knowledge systems for life on Earth: Visions of future systems and how to get there. *Energy Research & Social Science, 70*(101724). https://doi.org/10.1016/j.erss.2020.101724.

Fazey, I., Schäpke, N., Caniglia, G., Patterson, J., Hultman, J., van Mierlo, B., Säwe, F., Wiek, A., Wittmayer, J., Aldunce, P., Waer, A. H., Battacharya, N., Bradbury, H., Carmen, E., Colvin, J., Cvitanovic, C., D'Souza, M., Gopel, M., Goldstein, B., ...Wyborn, C. (2018). Ten essentials for action-oriented and second order energy transitions, transformations and climate change research. *Energy Research & Social Science, 40*, 54–70. https://doi.org/10.1016/j.erss.2017.11.026.

Foster, V. (2016). *Collaborative Arts-based research for social justice*. Routledge.

Franklin, A. (2018). Spacing natures: Resourceful and resilient community environmental practice. In T. Marsden (Ed.), *The Sage handbook of nature* (pp. 267–285). Sage.

Galafassi, D. (2018). *The transformative imagination: Re-imagining the world towards sustainability*. Ph.D. thesis. Stockholm University.

Gearty, M. R., Bradbury-Huang, H., & Reason, P. (2015). Learning history in an open system: Creating histories for sustainable futures. *Management Learning, 46*(1).

Giambartolomei, G., Franklin, A., & Fried, J. (2021). Supporting institutional transformations: experimenting with reflexive and embodied cross-boundary research. In A. Franklin (Ed.), *Co-creativity and engaged scholarship. Transformative methods in social sustainability research*. Palgrave-Macmillan.

Gibson-Graham, J. K. (2008). Diverse economies: Performative practices for "other worlds." *Progress in Human Geography, 32*(5), 613–632. https://doi.org/10.1177/0309132508090821.

Goralnik, L., & Nelson, M. P. (2017). Field philosophy: Environmental learning and moral development in Isle Royale National Park. *Environmental Education Research, 23*(5), 687–707. https://doi.org/10.1080/13504622.2015.1074661.

Haraway, D. (2016). *Staying with the trouble: Making Kin in the Chthulucene*. Duke University Press.

Harmin, M., Barrett, M. J., & Hoessler, C. (2017). Stretching the boundaries of transformative sustainability learning: On the importance of decolonizing ways of knowing and relations with the more-than-human. *Environmental Education Research, 23*(10), 1489–1500. https://doi.org/10.1080/13504622.2016.1263279.

Held, V. (2006). *The ethics of care: Personal, political, and global*. Oxford University Press.

Heras, M., & Tàbara, J. D. (2014). Let's play transformations! Performative Methods for Sustainability. *Sustainability Science, 9*(3), 379–398. https://doi.org/10.1007/s11625-014-0245-9.

Home, R., & Rump, N. (2015). Evaluation of a multi-case participatory action research project: The case of SOLINSA. *Journal of Agricultural Education and Extension, 21*(1), 73–89. https://doi.org/10.1080/1389224X.2014.991112.

Horlings, L. G. (2016). Connecting people to place: Sustainable place-shaping practices as transformative power. *Current Opinion in Environmental Sustainability, 20*, 32–40. https://doi.org/10.1016/j.cosust.2016.05.003.

Horlings, L. G., Nieto-Romero, M., Pisters, S., & Soini, K. (2020). Operationalising transformative sustainability science through place-based research: The role of researchers. *Sustainability Science, 15*, 467–484. https://doi.org/10.1007/s11625-019-00757-x.

Hung, L., Phinney, A., Chaudhury, H., Rodney, P., Tabamo, J., & Bohl, D. (2018). Appreciative inquiry: Bridging research and practice in a hospital setting. *International Journal of Qualitative Methods, 17*, 1–10. https://doi.org/10.1177/1609406918769444.

Kaufman, P. (2017). Critical contemplative pedagogy. *Radical. Pedagogy, 14*(1), 1–20.

Keeler, B. L., Chaplin-Kramer, R., Guerry, A. D., Addison, P. F. E., Bettigole, C., Burke, I. C., Gentry, B., Chambliss, L., Young, C., Travis, A. J., Darimont, C. T., Gordon, D. R., Hellmann, J., Kareiva, P., Monfort, S., Olander, L., Profeta, T., Possingham, H. P., Slotterback, C., & Vira, B. (2017). Society is ready for a new kind of scienceis academia? *BioScience, 67*(7), 591–592. https://doi.org/10.1093/biosci/bix051.

Kimmerer, R. (2014). Returning the gift. *Minding Nature, 7*(2), 18–24. https://www.humansandnature.org/filebin/pdf/minding_nature/May2014_Returning_the_Gift.pdf.

Leys, R. (2011). The turn to affect: A critique. *Critical Inquiry, 37*, 434–472. https://doi.org/10.1086/659353.

Ludema, J. D., & Fry, R. E. (2008). The practice of appreciative inquiry. In P. Reason & H. Bradbury (Eds.), *The Sage handbook of action research: Participative inquiry and practice* (pp. 280–297). Sage.

Macy, J. R., & Brown, M. Y. (1998). *Coming back to life: Practices to reconnect our lives, our world*. New Society Publishers.

Masterson, V. A., Mahajan, S. L., & Tengö, M. (2018). Photovoice for mobilizing insights on human well-being in complex social-ecological systems:

Case studies from Kenya and South Africa. *Ecology and Society, 23*(3), 13. https://doi.org/10.5751/ES-10259-230313.

Mol, A., Moser, I., Piras, E. M., Turrini, M., Pols, J., & Zanutto, A. (2010). Care in practice. On normativity, concepts, and boundaries. *Tecnoscienza. Italian Journal of Science & Techonology Studies, 2*(1), 73–86.

Moore, K. D. (2005). The truth of the barnacles: Rachel Carson and the moral significance of wonder. *Environmental Ethics, 27*(3), 265–277. https://doi.org/10.5840/enviroethics200527316

Moriggi, A. (2019). Exploring enabling resources for place-based social entrepreneurship: A participatory study of green care practices in Finland. *Sustainability Science, 15*(2), 437–453. https://doi.org/10.1007/s11625-019-00738-0.

Moriggi, A., Soini, K., Franklin, A., & Roep, D. (2020). A care-based approach to transformative change: Ethically-informed practices, relational response-ability, & emotional awareness. *Ethics, Policy & Environment, 23*(3), 281–298. https://doi.org/10.1080/21550085.2020.1848186.

Moriggi, A. (2021). *Green care practices and place-based sustainability transformations: A participatory action-oriented study in Finland.* Ph.D. thesis. Wageningen, NE: Wageningen University. https://doi.org/10.18174-544553.

Noddings, N. (2013). Why care about caring? In *Caring: A feminine approach to ethics and moral education* (pp. 7–29). University of California Press.

Norström, A. V., Cvitanovic, C., Löf, M. F., West, S., Wyborn, C., Balvanera, P., Bednarek, A. T., Bennett, E. M., de Bremond, A., Campbell, B. M., Canadell, J. G., Carpenter, S., & R., Folke, C., Fulton, E. A., Gaffney, O., Gelcich, S., Jouffary, J.-B., Tissier, M. L., Leach, M., … Österblom, H. . (2020). Principles for knowledge co-production in sustainability research. *Nature Sustainability, 9*. https://doi.org/10.1038/s41893-019-0448-2.

Olsson, P., Galaz, V., & Boonstra, W. J. (2014). Sustainability transformations: A resilience perspective. *Ecology and Society, 19*(4), 1. https://doi.org/10.5751/ES-06799-190401.

Pearson, K. R., Backman, M., Grenni, S., Moriggi, A., Pisters, S., & Vrieze de, A. (2018). *Arts-based methods for transformative engagement. A toolkit.* SUSPLACE.

Pearson, K. R. (2021). Imaginative leadership: A conceptual frame for the design and facilitation of creative methods and generative engagement. In A. Franklin (Ed.), *Co-creativity and engaged scholarship: Transformative methods in social sustainability research.* Palgrave-Macmillan.

Pereira, L. M., Bennett, E., Biggs, R. (Oonsie), Mangnus, A., Norström, A. V., Peterson, G., Raudsepp-Hearne, C., Sellberg, M., Vervoort, J. (2019). Seeding change by visioning good Anthropocenes. *Solutions, 10*(3).

Puig de la Bellacasa, M. (2011). Matters of care in technoscience: Assembling neglected things. *Social Studies of Science, 41*(1), 85–106.

Pulcini, E. (2009). *Care of the world: Fear, responsibility and justice in the global age*. Springer.

Reason, P., & Bradbury, H. (Eds.). (2008). *The Sage handbook of action research: Participative inquiry and practice* (2nd Ed.). Sage. https://doi.org/10.1177/14767503030012006.

Robertson, J. (2000). The three Rs of action research methodology: Reciprocity, reflexivity and reflection-on-reality. *Educational Action Research, 8*(2), 307–326. https://doi.org/10.1080/09650790000200124.

Sanginga, P. C., Kamugisha, R. N., & Martin, A. M. (2010). Strengthening social capital for adaptive governance of natural resources: A participatory learning and action research for bylaws reforms in Uganda. *Society & Natural Resources, 23*(8), 695–710. https://doi.org/10.1080/08941920802653513.

Scharmer, O. (2007). *Theory U: Leading from the future as it emerges*. Society for Organizational Learning.

Sempik, J., Hine, R., & Wilcox, D. (2010). *Green care: A conceptual framework. A report of the working group on the health benefits of Green Care, COST Action 866, Green Care in agriculture. Loughborough University*. Centre for Child and Family Research, Loughborough University.

Spretnak, C. (1997). Radical nonduality in ecofeminist philosophy. In K. Warren (Ed.), *Ecofeminism: Women, culture, nature* (pp. 425–436). Indiana University Press.

Tronto, J. C. (2013). *Caring democracy: Markets, equality, and justice*. New York University Press.

Tschakert, P., Dietrich, K., Tamminga, K., Prins, E., Shaffer, J., Liwenga, E., & Asiedu, A. (2014). Learning and envisioning under climatic uncertainty: An African experience. *Environment and Planning a: Economy and Society, 46*(5), 1049–1068. https://doi.org/10.1068/a46257.

Valencia-Sandoval, C., Flanders, D. N., & Kozak, R. A. (2010). Participatory landscape planning and sustainable community development: Methodological observations from a case study in rural Mexico. *Landscape and Urban Planning, 94*(1), 63–70. https://doi.org/10.1016/j.landurbplan.2009.07.018.

Wamsler, C., Brossmann, J., Hendersson, H., Kristjansdottir, R., McDonald, C., & Scarampi, P. (2018). Mindfulness in sustainability science, practice, and teaching. *Sustainability Science, 13*, 143–162. https://doi.org/10.1007/s11625-017-0428-2.

Warren, K. (2000). *Ecofeminist philosophy: A Western perspective on what it is and why it matters*. Rowman & Littefield.

West, S., Haider, L. J., Stålhammar, S., & Woroniecki, S. (2020). A relational turn for sustainability science? Relational thinking, leverage points and transformations. *Ecosystems and People, 16*(1), 304–325. https://doi.org/10.1080/26395916.2020.1814417.

Wicks, P. G., Reason, P., & Bradbury, H. (2008). Living inquiry: Personal, political and philosophical groundings for action research practice. In R. Peter & H. Bradbury (Eds.), *The Sage handbook of action research: Participative inquiry and practice* (pp. 15–31). Sage.

Zandee, D. P., & Cooperrider, D. L. (2008). Appreciable worlds, inspired inquiry. In P. Reason & H. Bradbury (Eds.), *The Sage handbook of action research: Participative inquiry and practice* (pp. 190–199).

6

Imaginative Leadership: A Conceptual Frame for the Design and Facilitation of Creative Methods and Generative Engagement

Kelli R. Pearson

Introduction

> At the very least, participatory involvement with the many forms of art can enable us to see more in our experience, to hear more on normally unheard frequencies, to become conscious of what daily routines have obscured, what habit and convention have suppressed—Maxine Greene (1995: 123)

Complex sustainability challenges can only be understood and addressed via ambiguous subjective judgements, which are shaped by the inner dimensions of individuals and groups, such as their world-views, imaginaries, interests/motivations, values, and ideologies (Rittel & Webber, 1973). Effective change processes must therefore include

K. R. Pearson (✉)
Department of Education and Learning Science, Wageningen University, Wageningen, The Netherlands

A. Franklin (ed.), *Co-Creativity and Engaged Scholarship*,
https://doi.org/10.1007/978-3-030-84248-2_6

cultural transformations and move beyond an exclusive focus on data-driven, technical, policy-oriented, and biophysical solutions (Boyden, 2001). However, even taking the inner-dimensions into account, actuating change is often constrained by the power and inertia of entrenched ways of thinking and perceiving, habituated everyday practices, and social/contextual norms and conventions (Ajzen, 1991; Dewey, 1922; Greene, 1995; Kagan, 2011). Moreover, the human psyche is hardwired to disengage when faced with information that appears overwhelmingly difficult or disturbing and can result in apathy and eco-anxiety (Lertzman, 2015; Pihkala, 2020).

With the aspiration to support transformative change and disrupt habits of thinking and doing, many scholars argue for a 'humanistic' (Hulme, 2011) or 'artistic' (Kagan, 2017) turn in sustainability transformations.[1] A humanistic turn calls for drawing from the arts and humanities and from the fields of psychology, cognitive sciences, theology, philosophy, and cultural studies. In fact, the arts have played a vital role in social transformations throughout history (*see* Belfiore & Bennett, 2008), and many studies point to the potential role of arts and culture in supporting sustainability transformations specifically (Hawkins et al., 2015; Kagan, 2011; Kepes, 1972; Rathwell, 2016).

Research suggests that arts-based and creative practices are well-suited for engaging with the inner dimensions of sustainability (Horlings, 2017). One such approach can be termed 'generative engagements' (Eernstman et al., 2021); these include experiences or events that evoke multiple forms of intelligence (Gardner, 2011) and enable emotional, aesthetic, cognitive, somatic, and social processing (Eisner, 2002; Gardner, 2011). The process of physically creating 'practical–aesthetic' artefacts, for example, enables a process of 'thinking with our hands' (Groth, 2017; Sheridan et al., 2014) and gives us multimodal experiences that support meaning-making processes, individually and as

[1] 'Sustainability transformations' is understandably a flexible and fuzzy term as it frequently makes its way back and forth between various academic disciplines and the world of practice and policy. At its core, however, it is a way to distinguish transformative change (i.e., change that alters the fundamental properties of a system) from transitional change (processes that emphasize incremental change). For a systematic literature review of sustainability transformations, see Salomaa and Juhola (2020).

social and cultural beings (Gulliksen, 2017). Generative engagements can facilitate and trigger the exchange and co-creation of knowledge through making and sharing artefacts (Groth, 2017), by spanning and connecting knowledge systems (Rathwell et al., 2015), through embodied learning and knowing (Gulliksen, 2017), and through playful experimentation (Nørgård et al., 2017). They can also support people to reflect on their deepest values, ethics, and motivations—what they care about and why it is worth taking action (Eernstman & Wals, 2013). While addressing heavy, potentially overwhelming topics, generative engagements also include the motivating and vitalizing affective elements of pleasure (Hammond et al., 2018), humour and light-heartedness (Eernstman et al., 2021), and joy (see Morrigi, this volume).

Generative engagements focused on sustainability can take many forms including, for example, collective artist residencies (Eernstman et al., 2021), immersive/interactive art installations or performances (see Weintraub, 2012), or learning environments and workshops that make use of creative methods (Galafassi, 2018; Taylor & Ladkin, 2009). Creative practices are also widely used in research processes, particularly in participatory action research (PAR) and transdisciplinary research (TDR) (Kagan, 2011; Wang et al., 2017). Still, there is a gap in understanding how creative methods can be designed *specifically* to evoke and support mindsets that are conducive to sustainability transformations.

Therefore, in the spirit of generative, playful, and 'exuberant' experimentation (Hollings, 2004), this chapter addresses the question: How can creative methods be operationalized (via generative engagements) to support the imaginative leadership capacities of researchers and practitioners working in the arena of sustainability?[2] I use the term *imaginative leadership* (see below) to describe the ability to understand and consciously influence the symbolic/metaphorical dimensions of self and others that are linked to specific values, mindsets, worldviews.

In addressing the above question, this chapter reflects on the process of co-designing and facilitating two different workshops grounded in

[2] Although they can generally be used interchangeably, this chapter uses the more inclusive term *creative methods* instead of *arts-based methods*. Is the process of cooking together, for example, an arts-based method? That is debatable, but if used in the context of a workshop or residency, it could certainly be considered a creative method of participatory engagement.

creative practices and methods. It proposes a conceptual frame that links creative methods to specific transformative mindsets. Both workshops aimed to support the imaginative leadership of sustainability researchers and practitioners by (a) activating specific conceptual frames and processes of self-reflection with the potential to open new spaces of possibility for sensing, perceiving, feeling, and acting, and (b) inviting participants to disrupt default anthropocentric worldviews and timescales and to draw more deeply and consciously from their own values and motivations in their work as sustainability professionals or researchers.[3] This chapter focuses primarily on the process of designing the methods and workshops—the theoretical inputs and practicalities that shaped them—rather than on the methods themselves (for a detailed description of all the specific methods used during workshops, see Pearson et al., 2018 or Pearson n.d.).

First, this chapter gives an overview of the workshops and the methodology of the research process. Second, it introduces the key sensitizing concepts of *transformative imagination* and *imaginative leadership*, and third, it presents a preliminary list of *transformative mindsets* that emerged from literature, semi-structured interviews, and the co-design process. Fourth, it describes the design and implementation of the workshops, including limitations. Fifth, emerging from the co-reflection process, it proposes an updated set of transformative mindsets for use in developing a framework for imaginative leadership moving forward, and then ends with concluding thoughts.

[3] I follow the school of thought that centres the role of physical, institutional, social, and cultural structures and systems in perpetuating unsustainability, as opposed to focusing on a pro-sustainability behaviour change of individuals (e.g., reducing carbon footprint or making sustainable consumption choices).

Developing Creative Methods Workshops to Support Imaginative Leadership: An Overview of the Research Process

> The only way to approach such a period in which uncertainty is high and one cannot predict what the future holds, is not to predict, but to experiment and act inventively and exuberantly via diverse adventures in living—C. S. Holling (2004: 8)

The workshops described in this chapter enabled collaborative development and experimentation with unconventional methods for sustainability leadership within the conventional form of a workshop. The aim was to support the agency and self-efficacy of key individuals/systems entrepreneurs already working towards sustainability transformations, as a leverage point for systemic and cultural change.

First, the *Action Hub: Arts-based methods for transformative design* (referred to henceforth as 'Action Hub') was a 90-min practice session with approximately 30 participants conducted during Transformations 2017, a transdisciplinary conference that took place in Dundee, Scotland. Co-designers included a cohort of six researchers from the SUSPLACE Innovation Training Network.[4] The co-designers chose this conference as the arena for our experimentation as it is known for encouraging non-traditional conference contributions, it includes both academics and practitioners, and it is supportive of creative and experiential methods for sustainability transformations.

Second, *Imaginative Leadership: Co-producing with nature and communities* (referred to henceforth as 'Imaginative Leadership') was a full day workshop for sustainability professionals in the Welsh Government working in the area of community engagement. The concept was initiated together with a representative of the Welsh government specializing in leadership and sustainability. Additional co-designers

[4] SUSPLACE was an EU Horizon 2020 funded Marie Skłodowska-Curie Actions Innovation Training Network (2015–2019) focused on understanding 'sustainable-place shaping' from multiple perspectives.

included a professional performance artist working at the intersection of art and sustainability and transformative practices and a social entrepreneur working with Natural Resources Wales. The artist was hired as the primary facilitator of the events and the other two co-designers participated as participant-observers. The same workshop structure was repeated with two different groups of approximately 40 people each (one in northern and one in southern Wales) on two separate days.

The focus of these experiments was not to track the impact of specific methods, but to use the design and implementation process as an arena for reflection, for reality testing the use of creative methods in the process of developing a theoretical framework for designing and applying creative methods, and to probe promising pathways for future practice and research. The learning process can be broken into four (non-linear) phases that incorporated iterative loops throughout: (1) exploration, (2) collaborative workshop design, (3) execution, (4) reflection.

Phase 1: Exploration

The exploration phase combined semi-structured expert interviews with a wide, cross-disciplinary sampling of literature related to the inner dimensions of sustainability transformations. In total, I conducted 14 semi-structured interviews in the Netherlands and the UK with people who work at the intersection of arts-based or creative practices in facilitation, community engagement, and sustainability (identified via snowballing). They were intended to give insight into how and why professional practitioners use creative methods, as well as what makes them successful and/or challenging (in their perspective). Literature guided the direction of interview questions, and the interviews, in turn, pointed to additional arenas of relevant academic research and theory.

Early influences that shaped my conceptualization of creative practices for sustainability transformations included academic literature in the arenas of art and aesthetic experience (e.g., Dewey, 1934), art-based environmental education (Mantere, 1998; van Boeckel, 2013), and multiple intelligences theory (Gardner, 2011). Together these strains of literature emphasize the role of art for sense-making, engaging diverse styles of learning and knowing, processing information through multiple senses and somatic-cognitive processes, re-sensitizing ourselves to the environment (and specific places), releasing conditioned perceptions, and engaging with sustainability issues (and each other) based on depth of emotional experience. From a practice perspective, I was influenced by my experience with Joanna Macy's *Work the Reconnects* (Macy & Brown, 2014) and the social and earth-based practices found in permaculture (Macnamara, 2012). Both experiment with new, transformative ways of relating to the natural world, and both incorporate creative and pragmatic practices that highlight attentiveness to emotions, to interdependencies, to inter-relationships between people, and to the details, rhythms, and cycles of natural systems. Finally, the methodological frameworks of Appreciative Inquiry (AI) (Cooperrider & Whitney, 2001) and Participatory Action Research (PAR) influenced my overall approach. AI shifts attention from 'solving' problems, to strengthening what is already working, including re-appreciating more intangible place-based resources (i.e., Horlings et al., 2020) and can be linked to designing creative methods for sustainability transformations (covered more thoroughly in Moriggi, this book). PAR acknowledges and highlights the dual role of the researcher as a scientist and social change agent, particularly in light of the need for urgent sustainability transformations (Wittmayer & Schäpke, 2014; see also De La Rosa, this volume).

Phase 2: Collaborative Workshop Design

The concept of the workshop format was first described by Osborn in *Applied Imagination* (1953) in which he outlined methods for creative group problem-solving. From their inception, workshops were intended to spark imagination and collective creativity (Isaksen et al., 2010). Workshops were chosen as the arena for experimentation in part because the format is highly accessible to a range of participants, it requires a relatively low-time commitment on the part of participants, and it is a familiar, 'safe' structure, which is important when people are working outside of their comfort zones (Sol et al., 2013).

Each workshop was co-designed with a different constellation of collaborators (co-researchers, stakeholders, and practitioners). Co-design is understood here as a joint team effort to initiate, develop and implement a participatory process. Although the workshops were different in terms of collaborators, size, participant profiles, content, and length, there were elements common to both. First, they were targeted towards people who already work in sustainability-related arenas and therefore the intent was *not* to change participants' minds or even to persuade or influence them, but instead help them access mindsets that they already value. Second, in order to set the stage for productive collaboration, each experiment began with a series of discussions around workshop goals and parameters, including personal goals, research goals, motivational goals (e.g., planetary health or 'islands of sanity'[5]) and participant-centred goals (i.e., what would be most useful and generative from the perspective of targeted participants?). In each case, the final step was to design the overall workshop concept and the specific methods. Data collected from this phase consisted of meeting notes, workshop design drafts (that included goals for each activity), and detailed final agendas, together with guidelines for spoken scripts, room set-up, and materials. Still, the nature of collaboration is often ad hoc and messy, and the chaotic demands of practice often subsume tidy categories and intentionality of theory.

Phase 3: Execution

The execution phase included the actual set-up, production, and facilitation of each event. Data collected included various documents, and artefacts were produced including presentations and notes, written instructions for participants, photographs, short video clips in some cases, and creative outputs/artefacts resulting from specific methods (collages, poems, etc.).

[5] Margaret Wheatley (2017) proposes that whether or not humans can stem the tide towards unsustainability, we have the possibility to contribute to 'islands of sanity' that evoke the "conditions for our basic human qualities of generosity, contribution, community and love" (p. 8).

Phase 4: Reflection

To reflect on the design process and resulting workshops, I drew from both practice-led research and art-based and qualitative methods. Practice-led research is widely used in the context of creative arts and performance studies. It employs iterative cycles of doing/creating and reflection (Candy, 2006) that contribute either to a body of theory or to a more pragmatic concept of social usefulness (Smith, 2009) or new knowledge gained (Mäkelä, 2007). The practitioner-researcher assesses the value and potential of a practical engagement in the world (i.e., the making of an object or a creative process) through reflection and evaluation. In this case, the 'practice' consisted of the creative development and implementation of the workshops. Reflection consisted of discussion sessions with co-collaborators and participant-observers, and was also informed by end-of-session evaluations and follow-up questionnaires. It also involved extended periods of interaction with co-designers engaged in what Clifford Geertz (1998) terms 'deep hanging out'—spending formal and informal time together reviewing and revisiting insights and learnings again and again.

AI also influenced the reflection process by focusing attention on what worked and what contributed to the successes and areas of vitality in the process of designing and executing the workshops (not, of course, to the exclusion of critical discernment—see Morrigi, this book). Finally, reflection was supported by the process of synthesizing findings from both workshops into a toolkit and open-source database (reimginary.com) for researchers and practitioners, which describes each specific method and our overall approach in great detail (see Pearson et al., 2018; Pearson, n.d).

Sensitizing Concepts: Imaginative Leadership and Transformative Mindsets

> Ideas are as important as facts and nowhere is it evident that they are inducable from them. We need imagination not rules; intuition not technique; warm ideas not cold facts; inventive people not conformists, fertile thinking not rigid rules to follow—Arthur P. Bochner (2009: 363)

In 1954, Herbert Blumer argued for the value of 'sensitizing concepts' in social sciences research. In contrast to 'definitive concepts', sensitizing concepts "merely suggest directions along which to look…they rest on a general sense of what is relevant" (p. 7). In accordance with Blumer, this research was guided by the concept of the *transformative imagination* (Galafassi, 2018); this assumes that imagination, and therefore the arts, have an important role to play in sparking and strengthening people's individual and collective capacity to create fundamentally new social-ecological systems.

Both 'imagination' and 'imaginaries' shape our sense of reality and possibility as we encounter the world. Imagination can be understood as a social and individual cognitive process by which we are able to conceptualize something beyond that which is immediately in front of us. It is a capacity that enables us to envision fantastical scenarios, but also more pragmatic possibilities for both what *could* happen and what *should* happen in reality (Bøttcher, 2020; Vadeboncoeur & Vellos, 2016). Imagination is central to human agency because it orients people to future possibilities that require actions in the present (Appadurai, 1996; Zittoun & Cerchia, 2013). At the same time, when people feel that a present situation is urgently untenable, it can stimulate a leap in their ability to imagine new scenarios, which then results in novel behaviours (Sannino, 2015).

Imaginaries, on the other hand, are less process and more structured, existing as deep, often unconscious, symbolic matrices that filter and mediate our experience of the world. As Kagan (2017) describes it, "the imaginary is like a cognitive and cultural hummus from which more articulate cultural constructs such as visions, narratives, discourses and utopias can grow and where they can take root" (p. 161). He points out that humans do not simply create and impose imaginaries on reality, but rather that they result from imaginative relational encounters between humans and the rest of the 'more-than-human' world.[6] The term 'social imaginary' is therefore used to describe how groups of people collectively imagine and shape the parameters of society—in terms of aspirations and priorities and in terms of institutional and social structures (Taylor, 2004). Within the social imaginary, a plurality of paradoxical and conflicting interpretations or landscapes exists, but the overall sense of the possible is bounded by the scope of the imaginary.

Grounded in the above, the concept of the *transformative imagination* (as used by Galafassi, 2018) is a way of describing how individuals and groups can alter the social imaginary (or evoke different dimensions of it) by activating fundamentally new ways of seeing, sensing, feeling, encountering, and envisioning the world (Galafassi, 2018). Galafassi argues that *transformative imagination* supports change agency because it alters the underlying paradigms and worldviews that create the conditions for unsustainability.

Use of the terms paradigm or worldview, however, often implies that people have one dominant and consistent perspective that they apply in all situations. On the contrary, individuals, like societies themselves, are a plurality—not so coherent or consistent. Even if we have a strong, conscious preference for a particular worldview, the majority of people have multiple, often conflicting, conceptual frameworks (linked to different worldviews) that can be activated at any given time (Lakoff & Johnson, 1980). With this understanding, the term *mindset* is used here to describe a mental model or conceptual frame/metaphor that is triggered by a specific metaphorical stimulation. A specific mindset, when triggered, defines the overall 'common sense' regarding a specific situation and therefore the scope of possibilities for decision-making and sense-making. Studies about norms (Ariely & Jones, 2008), framing (Lakoff & Johnson, 1980), and priming (Molden, 2014; Nijland, 2016), suggest that, depending on the circumstances and relevant frames or

[6] More-than-human (coined by Abram, 1996) is used to describe other biological beings (e.g., animals, plants, fungi) and non-animate natural systems or entities (e.g., rivers, mountains, ecosystems).

triggers, a different mindset, and therefore a different set of possibilities and norms, will arise, largely based on the way cognition is rooted in metaphorical thinking (Lakoff, 2010). It can be thought of as a pair of glasses that allows the wearer to see certain colours or opportunities more clearly. Therefore, *transformative mindsets* can be defined as specific cognitive lenses or frames that are helpful for orienting and motivating people specifically towards sustainability transformations (Pearson et al., 2018).

Emerging from this line of reasoning, I propose the concept of *imaginative leadership*; this can be broadly defined as the ability to influence, evoke, or shape the mental models, metaphors, and cultural narratives that people (both self and others) use to make sense of the world. My conceptualization is influenced by Bourdieu's (1991) understanding of symbolic power and Geertz's understanding of culture as a semiotic universe (1976). Lakoff (2010) makes two important points for understanding how imaginative leadership might support people's transformative capacity: (a) the inner dimensions are not static and consistent, but rather subject to ongoing fluctuation and emergent dynamics related to changing external and internal stimuli (also see Nijland, 2016); and (b) repetition of a particular metaphorical frame actually physically strengthens the synapses of specific neural circuits related to a particular ideological perspective (or mindset), which sets the parameters of possibility (in imagination and in action). Therefore, the imaginative leader develops the capacity to identify and evoke specific transformative mindsets (in both self and others) that activate conceptual frames with the potential to expand possibilities for transformative actions towards sustainability (see Pearson, 2021).

Identifying Transformative Mindsets

Creativity is an amoral capacity (Gardner, 1993; Katz, 2018): it can just as easily be used to design an astounding piece of machinery that destroys a forest as it can to spark a radical social-technical innovation that helps the forest and its inhabitants thrive. With this in mind, the methods used in the workshops were intended to evoke creativity *for* sustainability transformations by incorporating triggering metaphors and 'ideological language' (Lakoff, 2010) linked to specific mindsets. Therefore, the question became: which mindsets and which metaphors have the potential to spark mutually beneficial relationships between (and within) the human and more-than-human realms? And then: how can we intentionally evoke, anchor, and strengthen these mindsets and metaphors (in this context, through creative methods and generative engagements)?

During the exploration phase of the *Action Hub*, the co-design cohort identified a limited set of transformative mindsets (see Table 6.1) that was subsequently validated by the *Imaginative Leadership* co-designers. The list was derived via triangulation with input from literature, initial fieldwork (including expert interviews), and previous work experience related to sustainability transformations. It was not intended to be a definitive or comprehensive list of *all* transformative mindsets, but rather provide a reasonable starting point for experimentation. In the post-event reflection process, the conceptualization of these transformation mindsets was expanded and reconfigured, as presented in Section 6.4.

The concept of *regenerative sustainability* (see Table 6.1) deserves particular emphasis, because although it is designated as a mindset, the co-designers of both workshops also considered it as an overarching normative aim of sustainability transformations. In regenerative sustainability, human activities have the potential to have positive, beneficial impacts on the biosphere and all of its inhabitants, which is distinct from discourses on sustainability that primarily emphasize attempts to minimize harm (Mang & Reed, 2020; Wahl, 2016). In fact, although the term sustainability is ubiquitous in academic literature, policy, and popular culture, its usefulness in supporting the scale of social transformation required by the complexity and urgency of global challenges is contested (Wahl, 2016). Herbert Girardet of the World Futures Council, for example, argues that the word sustainability is inadequate, and that regeneration or regenerative development is both a more realistic and a more compelling paradigm (Girardet, 2014).

Table 6.1 Summary of the first iteration of 'transformative mindsets' that informed the design of the workshops and methods

Mindset	Core Concept
Regenerative Sustainability	The possibility that human activity could increase the biodiversity and health of social-ecological systems, as distinct from minimizing ecological or social harm (Mang & Reed, 2020; Wahl, 2016)
Sense of Time	The ability to consider longer perspectives (both past and future) and multiple time-scales have the potential to change the way of conceptualizing both problems and solutions (Boylston, 2019; Macy & Brown, 2014; Stewart, 2020)
More-than-Human Perspectives	De-centring anthropocentrism through imaginative consideration of 'more-than-human' (Abrams, 1996; Macy & Brown, 2014) perspectives, including biological beings (e.g., animals, plants, fungi) and non-animate natural systems or entities (e.g., rivers, mountains, ecosystems)
Care for Place	Developing a sense of willing responsibility and caring for specific places, and with that an emotional connection (Altman & Low, 1992; McEwan & Goodman, 2010)
Complexity/ Uncertainty	Sensitization to the reality of dynamic complex systems and problems requires an openness to uncertainty and a willingness to experiment (Hollings, 2004; Kagan, 2011, 2017)

Note See Fig. 2 and table 3 in section xx for revised list
Source Own conceptualization (CCBY)

Putting Theory in to Practice: Designing and Facilitating Creative Methods for Transformative Engagement

> Art is an adventure playground of the heart, where we can explore, discover, share and become who we are, in relative safety, alone and together—Francois Matarasso (2019: 43)

To give context to the reflections and insights that follow, here I return to the process of designing and executing the two workshops. Some decisions made in the design and execution of the *Action Hub* carried over to the *Imaginative Leadership,* so are covered in more detail in the description of the former.

Action Hub: Arts-based Methods for Transformative Design

The cohort of co-designers for the *Action Hub* originally came together around a shared academic interest in theory and methodologies related to creative methods, but we also shared a more personal interest in using methods that make us (and our research participants) feel 'energized' and 'inspired'. Our collective objective was to put theory into practice and experiment 'exuberantly'. We were also motivated to share practical applications of our research that change-makers, action-researchers, and local leaders could use in their work.

As a first step for organizing our design, we collectively chose the change management framework of Theory U (see Scharmer, 2009) to structure the workshop (see Fig. 6.1 below). We selected Theory U for many reasons—expedient, intuitive, and logical. Expediently, it was already familiar to the co-design cohort; intuitively, it is easy to understand, communicate, and use even in its simplest form (as described here); and logically, it is backed by academic and philosophical rigour and the layers and nuances of the theory resonated with our overall AI and PAR approach. Moreover, Theory U balances a clear linear structure with space for iterative looping, for spontaneity, and for indeterminacy. In several of the expert interviews, practitioners emphasized that establishing a stable, predictable framework for facilitation processes can help participants to leave their comfort zones and engage with unorthodox practices. It can also help consolidate outcomes and transitions into action. Leaving space for indeterminacy, on the other hand, is vital for cultivating serendipity, intuition, and lateral thinking, and therefore for sparking creativity, 'generative engagement', and new ways of perceiving. Theory U also highlights a balance between interpersonal processes of

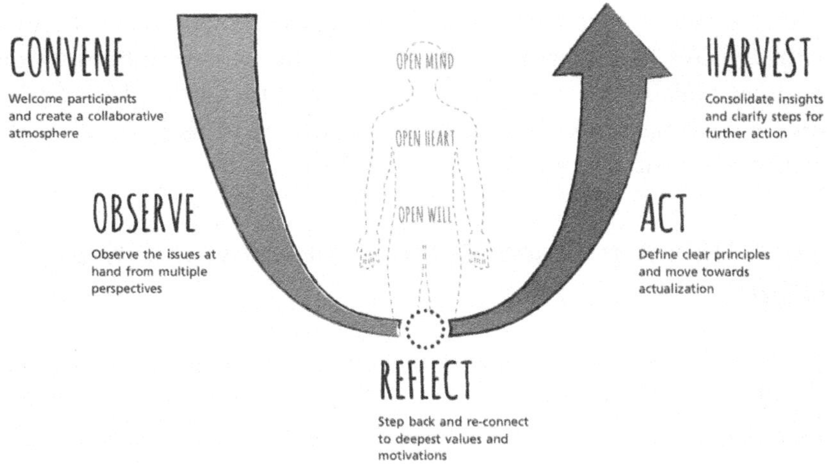

Fig. 6.1 Theory U process of observing, reflecting, acting, harvesting (*Source* Pearson et al. [2018] as adapted from Scharmer [2009])

collaboration and individual or introspective processes of reflection. It acknowledges the importance and role of emotional intelligence and values, and is explicitly intended to open spaces of possibility, or in Scharmer's terms 'seeing with fresh eyes' and 'sensing the field'. Theory U also includes a phase for reflecting,[7] or 'presencing'. This creates time for participants to intuitively connect with their deepest values and motivations; this is often missing from academic, community, governance, and corporate work on sustainability issues. Scharmer (2009) refers to this as 'the blind spot of leadership'.

In parallel with anchoring our design process in a clear structure, we identified key transformative mindsets (Table 6.1) that would be woven into our methods design and overall approach. To demonstrate the practical application of the methods, we decided to focus on three different specific design challenges (Table 6.2) based on real cases that were familiar to the co-designers. Next, before choosing, adapting, and

[7] Note: As a strategic decision for communicating clearly and accessibly to our target audiences, we chose to change the term 'presencing' used by Scharmer (2009) and the Theory U practice community) to 'reflecting' to describe the bottom of the U.

Table 6.2 Design challenges and stimulating questions: Action Hub

Design Challenge	Description	Guiding Question
(1) Dismissed Military Area in Italy	Bottom-up cultural, economic, and ecological regeneration of a dismissed military area in Northern Italy	(Group A) How can we imagine the distant future?
		(Group B) How can the dismissed military area include more-than-human perspectives?
(2) Abandoned Farmstead in the Netherlands	Re-imagining the potential uses of an abandoned farm in Overijssel, The Netherlands	(Group A) How can the farm regeneration project include more-than-human perspectives?
		(Group B) How is a farm like a church?
(3) Moving the city centre in Kiruna Sweden	Moving and re-designing a new city centre in Kiruna Sweden due to the expansion of mining operations	How can the new town square incorporate more-than-human perspectives?

Note Challenges 1 and 2 were split into two groups with different guiding questions. Challenge 3 was addressed by only one group
Source Own conceptualization (CCBY)

designing specific methods, we identified the overall goals and stimulating questions related to each design challenge, and each phase of the Theory U. Eventually, we settled on a design that enabled 5 small groups of 4–6 people to follow a set structure in terms of timing, but within the context of different pre-prepared design challenges, different stimulating questions, employing different creative methods, and emphasizing different transformative mindsets within each group.

The creative methods were then designed with the intention to root and anchor transformative mindsets via sticky metaphors and multisensory experiential learning, making them more auto-accessible and increasing participants' self-efficacy. At the same time the methods were intended for uptake by the participants—to support them in using creative methods (based in transformative mindsets) in their own research and work.

To illustrate the workshop in more concrete terms, Table 6.3 outlines the Dismissed Military Area challenge, including the methods used and the related transformative mindsets. Figure 6.2 shows some of the methods in action.

In the execution phase, we made sure that materials were well-organized for smooth transitions between activities, that instructions were available verbally and in writing, and that the room was aesthetically pleasing and had a welcoming atmosphere. We used nature-based images and objects (e.g., flowers, pinecones, rocks) to stimulate a sense of biophilia.

In order to get feedback on the workshop, three academic colleagues acted as participant-observers in the smaller groups and reported back their observations. In addition, each of the table facilitators reported on their experience and the 'harvest' with their respective participant groups, and we sent out a follow-up survey. Overall, the feedback was overwhelmingly enthusiastic, with some small technical suggestions (more time being the primary request) and ideas for further experimentation, such as putting more emphasis on establishing trusting group dynamics.

Imaginative Leadership: Co-producing with Nature and Communities (for Frontline Staff in Welsh Government)

Wales has been a global leader in creating leading-edge policy agendas to support sustainability transformations (Jones et al., 2020), and many people are now working to figure out how to accelerate implementation in various arenas (see Giambartolomei et al., this book). The co-designers of the *Imaginative Leadership* workshop were all interested in supporting the Welsh agenda, specifically with leadership development in the Welsh Government.

As with the development of the *Action Hub*, before designing specific methods, we started with the overall objectives of the workshop from the perspective of the participants and different participating stakeholders. We aimed to (a) introduce the concept of creative methods and transformative mindsets, (b) demonstrate the use (and usefulness) of specific

Table 6.3 Action Hub workshop: Sample itinerary—Dismissed military area

Time	Theory U Phase	Methods	Description	Transformative Mindsets
10 min	Convene (in plenary)	Embodiment: Regenerative Paradigm	Participants were asked to physically demonstrate the feeling of reducing their ecological footprint, then asked to demonstrate the feeling of increasing their positive (regenerative) impacts and comment on the difference	Regenerative Sustainability
5 min	Convene (in small groups)	Circle of Objects	Participants picked one of a group of presented objects (natural and human-made), explained why they picked it, and then strung it on a common thread	n/a
15 min	Observe	Storytelling, Evoking the Senses & Silent Conversation	The host read a pre-written story of the dismissed military area that included geological time, more-than-human perspectives, and layers of cultural, ecological, and *biophysical meaning*. Participants were invited to listen to the story using all of their senses and to write down what they imagined experiencing (smell, look, feel, etc.) onto sticky notes. Silently, as a group, participants then organized notes into affinity groups	Care for Place, Sense of Time, More-than-human Perspectives, Complexity/ Uncertainty

(continued)

Table 6.3 (continued)

Time	Theory U Phase	Methods	Description	Transformative Mindsets
20 min	Reflect	Expanding Time & Inviting Non-human Stakeholders	The host shared a visual representation of different time scales depicted in the story of the case. Participants then chose a card representing a more-than-human stakeholder (bear, mountain, river, etc.) and were invited to quietly reflect on the story from that perspective. Next, participants repeated the Circle of Objects exercise above, but from the perspective of their character	Sense of Time, More-than-human Perspectives
30 min	Act	Collage & Predicting Future Headlines	The host invited participants to imagine the place 100 years in the future and asked several guiding questions. Participants then created a collage envisioning the future from the perspective of their more-than-human character. Finally, in pairs, and then in the group, they extrapolated the most evocative qualities and messages of each collage and combined them into a 'future headline' (or headlines) for an imagined newspaper in the future	Sense of Time, More-than-human Perspectives,, Regenerative Sustainability

Time	Theory U Phase	Methods	Description	Transformative Mindsets
15 min	Harvest	Learning & Commitment	Individually, then in pairs, and with the group, participants reflected on what they learned, what they could apply to their own lives and in their work, and what they felt grateful for	n/a

Note Each of these methods can be found in Pearson et al. (2018) and the associated open source database reimgainary.com (Pearson, n.d.)
Source Own conceptualization (CCBY)

Fig. 6.2 Photographs from *Action Hub* (*Source* Photographs taken by Action Hub co-designers with permission)

creative methods for uptake by participants to employ in their own projects, and (c) provide the opportunity for participants to work on actual challenges from their work through the lens of specific transformative mindsets. With consensus from the *Imaginative Leadership* co-design group, the structure of Theory U was carried over from the *Action Hub.*

We used a hypothetical design challenge based on the town of Treherbert in Wales, which the local co-designers identified as emblematic of communities whose economic livelihood used to depend on the now-defunct mining sector. In the post-mining era, many towns and villages have struggled to re-invent themselves and re-define economic (and ecological) well-being for themselves. For the first half of the day, the workshop design focused on re-framing possible futures for Treherbert, evoking an expanded sense of time and more-than-human perspectives, using methods such as the Timeline of Transformation, Storytelling, and Inviting More-Than-human Stakeholders (Pearson et al., 2018). For the second half of the day, building on these new perspectives, we structured a form of peer-to-peer mentoring that looked at specific challenges faced by participants, while still including more-than-human stakeholders.

The *Action Hub* event venue was predetermined, but for *Imaginative Leadership* we were able to choose the locations. Based on her experience in place-responsive performative arts and sustainability, the artist/facilitator emphasized the importance of establishing relationality between the physical space of the workshop (including its history and its symbolic/cultural dimensions) and the design and methodology of the workshop. We looked for spaces that had access to nature, that aligned with our sustainability values (i.e., minimal disposable plastics, availability of sustainability produced food), and that had some cultural/symbolic significance. Once again, we put attention on creating a warm, welcoming ambiance in setting up the room. We also provided a participant workbook that included instructions for each method, key references, and space to take notes.

Feedback from participants was gathered during the harvesting phase of the event and was, again, overwhelmingly enthusiastic. Co-designers and two colleagues acted as participant-observers, and I subsequently conducted follow-up interviews (together with many informal conversations) with co-designers in the months following the workshop.

Acknowledging Limits

Here I highlight three decisions that limited the scope of these experiments in substantive ways. First, the duration of the workshops was limited to 90 minutes and a full day respectively, with no follow-up or ongoing engagement. This was due to the constraints of the context in which the research took place, and was not an intentional part of the design. Sustained generative engagements, such as extended ongoing training in which people meet regularly over a longer period of time or multi-day intensive 'collective artist residencies' (Eernstman et al., 2021), could involve more iterative processes and yield rich data and more detailed insights about the potential of creative methods to support transformative mindsets.

Second, this research was deliberately focused on increasing the self-efficacy and leadership capacity of people already engaged with sustainability. This decision stemmed from my research parameters, but also from ethical and practical considerations. Ethically, there is a fine line between persuasion and manipulation (Noggle, 2020), which I preferred not to approach, and, practically, within the short time scope of the engagements, starting from a place of common understanding and shared values saved time and effort in terms of setting the groundwork for willing and enthusiastic participation. In the future, it would be interesting to invite people with less familiarity or commitment to sustainability issues to experiment with some of these practices.

Third was the decision to avoid controversial topics or areas of conflict and avoid processes of decision-making. This was intentional; it allowed for a relatively simplistic approach to designing and facilitating the workshops and enabled us to focus on developing our concept of using creative methods to support/spark transformative mindsets. Moreover, it was not realistic or appropriate to surface deeper, potentially traumatic issues given the time constraints. Within different parameters, however, creative methods that are rooted in the deeper common values of participants have the potential to engage generativity with the reality of conflict, power-dynamics, eco-anxiety, and other hidden dimensions such as conflicting goals, values, and agendas (e.g., the value of surfacing conflict in social learning for sustainability, Wals & Heymann, 2004).

Putting Practice into Theory: Another Look at Transformative Mindsets for Imaginative Leadership

The initial list of transformative mindsets, while incomplete, provided a jumping off point for experimentation. Upon revisiting them during the post-workshop reflection process, both co-design cohorts agreed that they were indeed useful and valuable and that they stretched our own creativity and transformative imagination. They supported our novel approach to facilitating generative engagements rooted in creativity *for* transformations towards regenerative sustainability. At the same time, we identified areas for fine-tuning and some gaps. The revised list (see Fig. 6.3) proposes a new starting point for further experimentation and the expansion of a framework for supporting imaginative leadership through generative engagements and creative methods. Notably, the list, as it is presented here, is meant for uptake in the field, and is therefore framed for simplicity and clarity with the lay reader in mind. Moreover, it is with humility that I emphasize that each of these mindsets has been studied extensively across disciplines and each represents a vast arena of interconnected literature; they have been framed in many different ways in literature and in practice. In accordance with the parameters of this chapter, the following discussion represents only a brief and limited

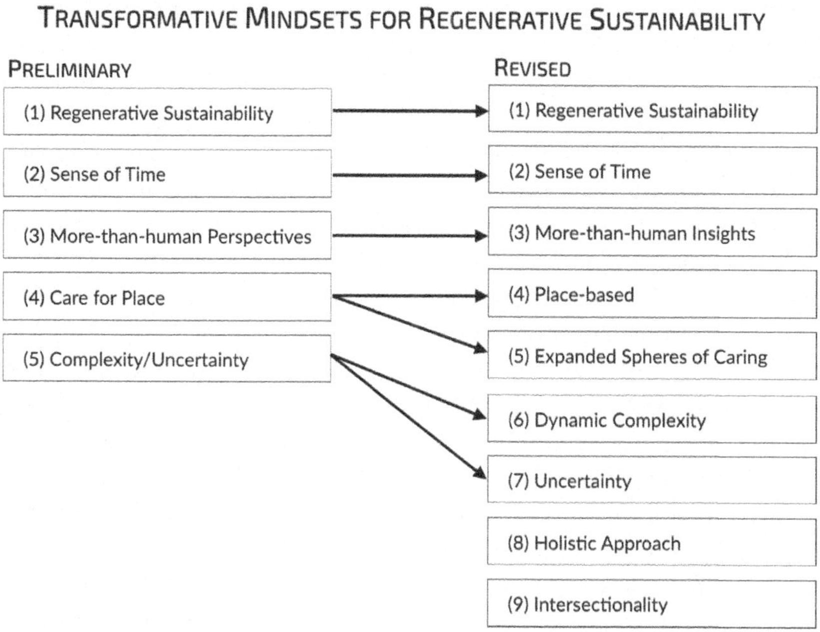

Fig. 6.3 Revised list of transformative mindsets (*Source* Own conceptualization [CCBY])

summary of each (for a more detailed account of each, see Pearson, 2021).

First, *regenerative sustainability* (1) was validated by both co-designers and participants as a foundational concept for imaginative leadership. As a normative aim and as a transformative mindset, it represents a generative evolution in the concept and application of sustainability (Mang & Reed, 2020; Wahl, 2016). The importance of *sense of time* and consideration of *more-than-human perspectives* were also confirmed. Although participants worked in the field of sustainability, in general they found it challenging to imagine 100 years or even 20 years into the future and they appreciated the chance to reflect through the lens of multiple time-scales. Likewise, people valued the opportunity to engage with the design challenges and their own projects through an imaginatively more-than-human lens.

The more-than-human can be considered empathically and ethically (Abrams, 1996; de La Bellacasa, 2017), in planning and decision-making (Macy & Brown, 2014) and from a legal rights-of-nature perspective (Boyd, 2017), but also more instrumentally as inspiration for innovation. The practice of biomimicry (Benyus, 1997), for example, aims to learn from and appreciate the design intelligence ("3.8 billion years of research and development") inherent in natural systems as input for innovations, not only for technology and infrastructure, but also for social and economic innovations (i.e., what could an economic system learn from a forest?). The term *more-than-human 'perspectives'* was therefore modified to the more expansive term *more-than-human 'insights'* (3).

The mindset of *caring for place* in the initial list was indeed a useful lens for designing methods that evoke an emotional, sensory connection to specific places. Upon reflection, however, *caring* as a stand-alone concept was woven into so many dimensions of the design process that it emerged as foundational to our approach on multiple levels. Therefore, we split this mindset into its two components: *place-based* (4) and *expanded spheres of care* (5). *A place-based* lens (see Massey, 2015) emphasizes an attentiveness to the specificity, assemblage of relationships, and the 'situatedness' of what makes a place *a place* (biophysical, symbolic, cultural, relational, etc.); places are 'where things happen' in terms of sustainability transformations (Horlings et al., 2020). It also implies a felt mutuality or attachment; this can be both affective/emotional (Altman & Low, 1992; McEwan & Goodman, 2010) and pragmatic, appreciating our (inter) dependence with tangible and intangible place-based resources, for example (Horlings et al., 2020).

Expanded spheres of care (5) highlights both an expanding circle (Singer, 1981/2011) of ethical concern (*who* is being cared for) and the attitudes and practices for expressing care (i.e., *how* to care). The expanded sphere moves beyond self and immediate kin to include humans 'others', the more-than-human, and even future (and past) generations. A broad scope of caring is seen as an essential component of leadership for regenerative sustainability (see Schein, 2017 for an overview of the caring/ecological worldview).[8] In terms of attitudes and practices, during the workshops we aimed to be attentive to and inclusive of diverse (and overlooked) voices and perspectives and to respectfully support the physical and mental well-being of participants (and co-designers). Notably, we observed the value of a caring intentionality in designing the workshop 'container', i.e., the physical place (from acoustics to aesthetics to temperature and light), the relationality among the participants and facilitators, and other, often 'invisible' supportive elements such as the food or even the organization of materials.

In our initial list, *complexity* and *uncertainty* were considered as one mindset. Both were present and played important roles in shaping our approach, but in practice they were quite distinct. Much has been written about how the ability to respond to *dynamic complexity*[9] (5) is an underdeveloped capacity (Kagan, 2011; Schein, 2017). Complex adaptive living systems (a watershed for example) are often not predictable or rationally knowable in terms of observable relationality between cause and effect as they are in 'complicated' mechanistic systems (Burns et al., 2015; Holling, 2004); they therefore require a probing and experimental approach to problem solving. In conceptualizing complex living systems,

[8] Moriggi et al. (2020) propose an in-depth framework of caring in relation to sustainability transformations that includes ethically informed practices, emotional awareness, and relational response-ability (Haraway, 2016) i.e., the ability to responsibility respond to the context at hand.

[9] Burns et al. (2015), for example, identify complex living systems as an overarching paradigm in sustainability leadership (in opposition to the Newtonian mechanistic worldview).

queer ecology adds another dimension, in which diversity is appreciated, and essentializing or reductionist categories placed on self and others are problematized ('freaked out'), and instead considered more fluidly (Kagan, 2011, 2017).

Uncertainty (6), on the other hand, can be thought of as an essential attitude in the face of complexity. The capacity to be open to 'not knowing' emerged as a golden thread frequently emphasized by practitioners, artists, participants, and the co-designers in both projects and in literature (see Kagan, 2017). It can be linked to the ability to look at problems through new imaginative perspectives (e.g., more-than-human), to weakening the static hierarchy of the expert/audience duality, to opening the scope of possibilities for action, to communicating in new ways, and to re-defining constellations of collaboration (Arora, 2019; Clampitt et al., 2001; Kagan, 2017). Uncertainty can also be characterized as 'beginner's mind'; this has been central to many mindfulness traditions and, in modern applications, has been applied widely, for example in diagnosis and care in medical practices (Epstein, 2003) and in pedagogy (Kochhar-Lindgren, 2001). In contrast to a static destination, Kagan (2011) frames sustainability as a dynamic 'search process', emphasizing that people do not fully understand complex living systems, or even what a regenerative or sustainable society should or could look like in the future.

In addition to revising the original list of transformative mindsets, during the design, execution, and reflection processes two key gaps became evident. First, was the importance of a *holistic approach* (7) to knowledge, places, and people. A holistic approach takes into account context and relationality, including historical, biophysical, cultural, social, psychological, and symbolic dimensions; it acknowledges both the embeddedness and embodiedness of both social imaginaries and physical realities (Haraway, 2016).[10] Through this lens, knowledge must be grounded in context and specific places (Horlings et al., 2020). Importantly, all participants (in the broadest sense possible) were considered with a 'whole-person approach' that considered their well-being,

[10] See Warm Data Lab (n.d) for a promising approach to addressing the deep relationality and complexity inherent in social science research.

thoughts, emotions, motivations, perceptions of place, and constellation of relationships through time.[11] From a holistic perspective, the methods themselves were embedded in the context of the process (or the 'container'). A holistic approach can be woven into the fabric of an event, as demonstrated in the process of incorporating a relational response to our event location in *Imaginative Leadership*.[12] In addition, the twin concepts of mutuality and interdependence are vital to a holistic approach and they were emphasized repeatedly in our design process from a philosophical perspective. Notably, we did not link the concept of interdependence to specific methods—perhaps because it was not a part of the initial list. The concept of interdependence has long roots in indigenous and non-occidental philosophies, knowledge, and worldviews (Avalos Cisneros, 2015), but has only more recently been mainstreamed in western positivist sciences such as ecology (Callenbach, 2008).

The second gap that we identified was a mindset of *intersectionality* (8); this is not only foundational for supporting transformational change, but must also be explicitly highlighted. It is crucial to strengthen our collective and individual conceptual frames that connect social issues, such as racism, gender issues, wealth inequality, colonialism, or oppressive violence and dominance-based power dynamics with issues of ecological destruction and degradation.[13] Moving away from an anthropocentric perspective can help to disrupt default assumptions about humans' right to dominate other species (as in the workshops described in this chapter), but within the scope of our workshops and methods, we did not address the topic directly. Indeed, there is potential for exploring and surfacing these connections with a guided application of creative methods within the conceptual framework of imaginative leadership. For a critical literature review on intersectionality and sustainability education see Maina-Okori et al. (2018), an intersectional perspective on

[11] 'Whole person approach' has been applied in many contexts, such as medical care (Thomas et al., 2018) and pedagogy (Fadeeva et al., 2010).

[12] It also points to research about the way metaphors can be embodied, or grounded in physical environments.

[13] Environmental racism and the genocide of indigenous people, for example, cannot, in reality, be separated from the so-called 'ecological dimensions' of unsustainability, such as biodiversity loss and pollution/degradation of natural environments.

climate change see Kaijser and Kronsell (2014), and reflection on intersectionality in light of the life and murder of Berta Cáceres see Méndez (2018).

The revised list of mindsets is summarized in Table 6.4, together with a short statement of key transformative aspects and suggestions for further reading.

Table 6.4 Revised transformative mindsets

Mindset	Transformative Aspect
(1) Regenerative Sustainability	From minimizing harm to generating resilience and vitality for the biosphere and its inhabitants (Mang & Reed, 2020; Wahl, 2016)
(2) Sense of Time	From chronic short-termism, to an expanded ability to think in multiple time-scales, especially incorporating long-term perspectives (Macy & Brown, 2014; Boylston, 2019; Steward, 2020)
(3) More-than-human Insights	From anthropocentrism to attentively, imaginatively, and ethically including more-than-human perspectives in processes of knowledge co-creation (Abrams, 1996; Benyus, 1997; Boyd, 2017; de La Bellacasa, 2017)
(4) Place-based	From universalist approaches to 'emplacement'—grounded and contextualized and emerging from a relational approach to place-specificity (Massey, 2005; Macnamara, 2012; Horlings et al., 2020)
(5) Expanded Spheres of Care	Expanded spheres of ethical concern for humans, places, and our ecological selves (de La Bellacasa, 2017; Moriggi et al., 2020; Schein, 2017; Singer, 1981/2011; Haraway, 2016)
(6) Dynamic Complexity	Limitations of mechanistic mindset for problem solving and knowledge creation; De-essentializing living systems, diversity and queer conviviality (Boylston, 2019; Burns et al., 2015; Holling, 2004; Kagan, 2011, 2017)

(continued)

Table 6.4 (continued)

Mindset	Transformative Aspect
(7) Uncertainty	From a 'need-to-know' model of expertise to comfortability with not knowing; framing sustainability as 'a search process' instead of a destination (Arora, 2019; Kagan, 2017; Clampitt et al., 2001; Epstein, 2003; Kochhar-Lindgren, 2001)
(8) Holistic Approach	From abstracted, to embedded (physically, relationally, and semiotically), situated and contextual (often place-based), and interdependent (from compartmentalization to mutuality). Includes a 'whole-person' approach to design and facilitation (Avalos Cisneros, 2015; Callenbach, 2008; Fadeeva et al., 2010; Haraway, 2016; Thomas et al., 2018)
(9) Intersectionality	The way humans interact with other species and the biosphere with violence and extractive motivations is intertwined with dysfunctions in intra-human dynamics (Kaijser & Kronsell, 2014; Maina-Okori et al., 2018; Méndez, 2018)

Source Own conceptualization (CCBY)

Conclusion

> The survival of civilization and the well-being of humankind in the future will require a dramatic shift in the dominant cultures of global society—a veritable cultural renaissance—Boyden (2001: 112)

The poet, philosopher, artist, and storyteller in each of us shape our sense of what is important, worthwhile, and possible. When we are touched and moved by the emotional resonance or compelling aesthetic of an artistic endeavour, new pathways emerge in the landscapes of our imagination, which counters the stifling, fatalistic perception that 'there is no alternative'. Within the broad landscape of our collective social imaginary, specific worldviews, metaphors, and mental models invisibly "channel attention, filter information, categorize experience, anchor interpretation, orient learning, establish moods, secrete norms, and legitimates narratives, ideologies, and power structures" (Gladwin et al., 1997, p. 245). In fact, the structures, institutions, and technologies created and deployed by a society reflect its culture, its worldview, and how it understands and engages with natural and more-than-human systems (Mang & Reed, 2020). Imaginative leadership through the arts can nourish a cultural renaissance towards regenerative sustainability by sparking new stories, metaphors, and practices that support transformative mindsets and open new spaces of possibility.

The design of generative engagements that employ creative methods, which are consciously and explicitly linked to transformative mindsets, is one arena among many for playful experimentation. It is worth re-emphasizing that because mindsets are not constant, the point of this experimentation is to practice *deliberately* evoking *specific* mindsets, based on people's own values.

The loose experimental nature of the learning process recounted in this chapter leaves significant room for future exploration and discovery. The revised list of *transformative mindsets* reflects the direct experience of the co-designers and is intended to act as a starting point for the next iteration of exploration and experimentation with creative methods, the transformative imagination, and the development of imaginative leadership. Future research could more fully consider:

- the quality and typologies of participation during the design process and during the event;
- the role of the 'container' and how it connects to a *holistic approach* and a deep commitment to caring as practice;
- the validity, interpretation, and range of *transformative mindsets* could be co-explored and contextualized with participants or compared with other aligned frameworks.

On one hand, it would be interesting to design a research experiment that looks at the influence of specific mindsets on tangible design outcomes in processes of planning or the design of specific initiatives. On the other, it could be fruitful to problematize and explore the instrumentalism of creative approaches and the focus on solution-oriented strategic development, in contrast to more open-ended and ontological explorations. Creative methods are certainly not a panacea and they can be applied more or less skilfully, and used more or less appropriately in different contexts, for different aims; it would certainly be illuminating to look in more depth at how and when creative methods fail or even backfire and increase resistance and conflict *(see* van der Vaart et al., 2019).

Although sustainability is an ongoing 'search process' (Kagan, 2011) rooted in productive uncertainty, transformations towards just and ecologically healthy societies will always involve a reflection on what we value, accept, reject, love, care for, are passionate about, what we find just, fair, and sensible. They are also shaped and constrained by path-dependent contexts and systemic structures and accepted norms. As Herbert Marcuse has said: "The truth of art lies in its power to break the monopoly of established reality to define what is real…Art cannot change the world, but it can contribute to changing the consciousness and drives of the men and women who could change the world" (Marcuse, 1978: 9/33).

Acknowledgements This research was supported by SUSPLACE, a Marie Skłodowska-Curie Actions Innovative Training Network funded by the European Union's Horizon 2020 research and innovation programme under the Marie Skłodowska-Curie grant agreement No. 674962.

References

Abram, D. (1996). *The spell of the sensuous: Perception and language in a more-than-human world*. Vintage.

Ajzen, I. (1991). The theory of planned behavior. *Organizational Behavior and Human Decision Processes, 50*(2), 179–211.

Altman I., & Low S. M. (Eds) (1992). *Place attachment.* Plenum Press.

Appadurai, A. (1996). Disjuncture and difference in the global cultural economy. In *Modernity at large: Cultural dimensions of globalization* (pp. 27–47). University of Minnesota.

Ariely, D., & Jones, S. (2008). *Predictably irrational*. New York, NY: Harper Audio.
Arora, S. (2019). Admitting uncertainty, transforming engagement: Towards caring practices for sustainability beyond climate change. *Regional Environmental Change, 19*(6), 1571–1584.
Avalos Cisneros, N. (2015). *Interdependence as a lifeway: Decolonization and resistance in transnational native American and Tibetan communities* (Doctoral dissertation, UC Santa Barbara).
Belfiore, E., & Bennett, O. (2008). *The social impact of the arts—An intellectual history*. Palgrave MacMillan.
Benyus, J. M. (1997). *Biomimicry: Innovation inspired by nature*. Morrow.
Blumer, H. (1954). What is wrong with social theory? *American Sociological Review, 19*(1), 3–10.
Bøttcher, L. (2020). Supporting unusual development through moral imagination. *Learning, Culture and Social Interaction,* 100384.
Bochner, A. P. (2009). Warm ideas and chilling consequences. *International Review of Qualitative Research, 2*(3), 357–370.
Boyd, D. R. (2017). *The rights of nature: A legal revolution that could save the world*. ECW Press.
Boyden, S. (2001). Nature, society, history and social change. *Innovation: The European Journal of Social Science Research, 14*(2), 103–116.
Boylston, S. (2019). *Designing with society: A capabilities approach to design, systems thinking and social innovation*. Routledge.
Bourdieu, P. (1991). *Language and symbolic power*. Harvard University Press.
Burns, H., Diamond-Vaught, H., & Bauman, C. (2015). Leadership for sustainability: Theoretical foundations and pedagogical practices that Foster change. *International Journal of Leadership Studies, 9*(1).
Candy, L. (2006). Practice-based research: A guide. *CCS report, 1*(2).
Callenbach, E. (2008). Interdependence. In *Ecology, revised and expanded* (pp. 83–85). University of California Press.
Clampitt, P., Williams, M. L., & DeKoch, R. (2001). Embracing uncertainty: The executive's challenge. *Journal of Change Management, 2*(3), 212–228.
Cohen, I. R., & Harel, D. (2007). Explaining a complex living system: Dynamics, multi-scaling and emergence. *Journal of the Royal Society Interface, 4*(13), 175–182.
Cooperrider, D. L., & Whitney, D. (2001). A positive revolution in change: Appreciative inquiry. *Public Administration and Public Policy, 87*, 611–630.
Dewey, J. (1922). *Human nature and conduct: An introduction to social psychology*. Henry Holt

Dewey J. (1934/2005). *Art as experience*. Penguin.

Eernstman, N., van Boeckel, J., Sacks, S., & Myers, M. (2012). *Inviting the unforeseen: A dialogue about art, learning and sustainability* (pp. 201–212). Wageningen Publishers.

Eernstman, N., & Wals, A. E. (2013). Locative meaning-making: An arts-based approach to learning for sustainable development. *Sustainability, 5*(4), 1645–1660.

Eernstman, N., Pearson, K. R., de Vrieze, A., Wals, A., & Bjurström, A. E. (2021). Designing collective artist residencies: Cultivating imaginative disruptions and lightheartedness in times of gravity. *Arts and Interdisciplinary Research*, (3), 17–34.

Eisner, E. W. (2002). *The arts and the creation of mind*. Yale University Press.

Epstein, R. M. (2003). Mindful practice in action (I): Technical competence, evidence-based medicine, and relationship-centered care. *Families, Systems, & Health, 21*(1), 1.

Fadeeva, Z., Mochizuki, Y., Podger, D. M., Mustakova-Possardt, E., & Reid, A. (2010). A whole-person approach to educating for sustainability. *International Journal of Sustainability in Higher Education.*

Galafassi, D. (2018). *The transformative imagination: Re-imagining the world towards sustainability* (Doctoral dissertation, Stockholm Resilience Centre, Stockholm University).

Gardner, H. (1993). *Creating minds*. Basic Books.

Gardner, H. (2011). *Frames of mind: The theory of multiple intelligences.* Hachette.

Geertz, C. (1976). Art as a cultural system. *Modern Language Notes, 91*(6), 1473–1499.

Geertz, C. (1998). Deep hanging out. *The New York Review of Books, 45*(16), 69–72.

Gergen, K. (2009). *Relational being: Beyond self and community.* Oxford University Press.

Girardet, H. (2014). *Creating regenerative cities*. Routledge.

Gladwin, T. N., Newberry, W. E., & Reiskin, E. D. (1997). Why is the northern elite mind biased against community, the environment, and a sustainable future? *Environment, Ethics, and Behaviour,* 234–247.

Groth, C. (2017). *Making sense through hands: Design and craft practice analysed as embodied cognition.* Aalto University, Helsinki.

Greene, M. (1995). *Releasing the imagination: Essays on education, the arts, and social change.* Jossey-Bass.

Gulliksen, M. S. (2017). Making matters? Unpacking the role of practical aesthetic making activities in the general education through the theoretical lens of embodied learning. *Cogent Education, 4*(1), 1415108.

Hammond, C., Gifford, W., Thomas, R., Rabaa, S., Thomas, O., & Domecq, M. C. (2018). Arts-based research methods with indigenous peoples: An international scoping review. *AlterNative: An International Journal of Indigenous Peoples, 14*(3), 260–276.

Haraway, D. J. (2016). *Staying with the trouble: Making kin in the Chthulucene.* Duke University Press.

Hawkins, H., Marston, S. A., Ingram, M., & Straughan, E. (2015). The art of socioecological transformation. *Annals of the Association of American Geographers, 105*(2), 331–341.

Hedlund-de Witt, A. (2013). Worldviews and their significance for the global sustainable development debate. *Environmental Ethics, 35*, 133–162.

Holling, C. S. (2004). From complex regions to complex worlds. *Ecology and Society, 9*(1).

Horlings, L. G. (2017). The role of artists and researchers in sustainable place-shaping. *Culture in Sustainability*, 131.

Horlings, L. G., Roep, D., Mathijs, E., & Marsden, T. (2020). Exploring the transformative capacity of place-shaping practices. *Sustainability Science, 15*(2), 353–362.

Hulme, M. (2011). Meet the humanities. *Nature Climate Change, 1*(4), 177–179.

Isaksen, S. G., Dorval, K. B., & Treffinger, D. J. (2010). *Creative approaches to problem solving: A framework for innovation and change*. Sage.

Jones, N. A., Ross, H., Lynam, T., Perez, P., & Leitch, A. (2011). Mental models: An interdisciplinary synthesis of theory and methods. *Ecology and Society, 16*(1), 46–46.

Jones, R., Goodwin-Hawkins, B., & Woods, M. (2020). From territorial cohesion to regional spatial justice: The well-being of future generations act in Wales. *International Journal of Urban and Regional Research, 44*(5), 894–912.

Kagan, S. (2011). *Art and sustainability: Connecting patterns for a culture of complexity*, Transcript Verlag.

Kagan, S. (2017). Artful sustainability: Queer-convivialist life-art and the artistic turn in sustainability research. *Transdisciplinary Journal of Engineering & Science,* 8.

Kaijser, A., & Kronsell, A. (2014). Climate change through the lens of intersectionality. *Environmental Politics, 23*(3), 417–433.

Katz, L. (2018). Dark dreams and malign creativity. *Knowledge Cultures, 6*(02), 64–75.

Kearns, L. L. (2015). Subjects of wonder: Toward an aesthetics, ethics, and pedagogy of wonder. *Journal of Aesthetic Education, 49*(1), 98–119.

Kepes, G. (1972). *Arts of the environment*. Aidan Ellis.

Kochhar-Lindgren, G. (2001). Beginner's mind: Opening the open in the classroom. *Pedagogy, 1*(2), 410–415.

de La Bellacasa, M. P. (2017). *Matters of care: Speculative ethics in more than human worlds* (Vol. 41). University of Minnesota Press.

Lakoff, G. (2010). Why it matters how we frame the environment. *Environmental Communication, 4*(1), 70–81.

Lakoff, G., & Johnson, M. (1980). *Metaphors we live by*. University of Chicago press.

Lertzman, R. (2015). *Environmental melancholia: Psychoanalytic dimensions of engagement*. Routledge.

Macnamara, L. (2012), *People & permaculture: Caring & designing for ourselves, each other & the planet* (1st ed.). Permanent Publications.

Macy, J., & Brown, M. Y. (2014). *Coming back to life: The guide to the work that reconnects*. New Society Publishers.

Maina-Okori, N. M., Koushik, J. R., & Wilson, A. (2018). Reimagining intersectionality in environmental and sustainability education: A critical literature review. *The Journal of Environmental Education, 49*(4), 286–296.

Mäkelä, M. (2007). Knowing through making: The role of the artefact in practice-led research. *Knowledge, Technology & Policy, 20*(3), 157–163.

Mang, P., & Reed, B. (2020). Regenerative development and design. *Sustainable Built Environments*, 115–141.

Mantere, M. H. (1998). Art and the environment: An art-based approach to environmental education. In L. Rubinstein Reich (Ed.), *Rapporter om utbildning* (Vol. 3, pp. 30–35). Lärarhögskolan.

Marcuse, H. (1978). *The aesthetic dimension*. Beacon Press.

Massey, D. (2005). *For space*. Sage.

Matarasso, F. (2019). *A restless art. How participation won, and why it matters*. Digital edition. Calouste Gulbenkian Foundation. Viitattu, 15, 2020.

McEwan, C., & Goodman, M. K. (2010). Place geography and the ethics of care: Introductory remarks on the geographies of ethics, responsibility and care. *Ethics, Place and Environment, 13*(2), 103–112.

Méndez, M. J. (2018). "The river told me": Rethinking intersectionality from the world of Berta Cáceres. *Capitalism Nature Socialism, 29*(1), 7–24.

Molden, D. C. (2014). Understanding priming effects in social psychology: An overview and integration. *Social Cognition, 32*(Supplement), 243–249.

Moriggi, A., Soini, K., Bock, B. B., & Roep, D. (2020). Caring in, for, and with nature: An integrative framework to understand green care practices. *Sustainability, 12*(8), 3361.

Nijland, H. J. (2016). *Disentangling the domestic contract: understanding the everyday-life construction of acceptability-or non-acceptability-of keeping and killing animals for food* (Doctoral dissertation, Wageningen University).

Noggle, R. (2020). The ethics of manipulation in *The Stanford Encyclopedia of Philosophy* (Summer 2020 Edition).

Nørgård, R. T., Toft-Nielsen, C., & Whitton, N. (2017). Playful learning in higher education: Developing a signature pedagogy. *International Journal of Play, 6*(3), 272–282.

Osborn, A. F. (1953). *Applied imagination: Principles and procedures of creative thinking*. Charles Scribner's Sons.

Pearson, K. R. (n.d.). *Reimaginary*. Retrieved April 01, 2021, from https://www.reimaginary.com/.

Pearson, K. R., Bäckman, M., Grenni, S., Moriggi, A., Pisters, S., & de Vrieze, A. (2018). *Arts-based methods for transformative engagement: A toolkit.* SUSPLACE.

Pearson, K. R. (2021, Forthcoming). *Imaginative leadership: A conceptual framework for operationalizing creative practices in support of regenerative sustainability.* Unpublished Thesis. Wageningen University, The Netherlands.

Pihkala, P. (2020). Eco-anxiety and environmental education. *Sustainability, 12*(23), 10149.

Rathwell, K., Armitage, D., & Berkes, F. (2015). Bridging knowledge systems to enhance governance of environmental commons: A typology of settings. *International Journal of the Commons, 9*(2).

Rathwell, K. J., & Armitage, D. (2016). Art and artistic processes bridge knowledge systems about social-ecological change: An empirical examination with Inuit artists from Nunavut, Canada. *Ecology and Society, 21*(2).

Rittel, H. W., & Webber, M. M. (1973). Dilemmas in a general theory of planning. *Policy Sciences, 4*(2), 155–169.

Salomaa, A., & Juhola, S. (2020) How to assess sustainability transformations: A review Global. *Sustainability, 3.*

Sannino, A. (2015). The emergence of transformative agency and double stimulation: Activity-based studies in the Vygotskian tradition. *Learning, Culture and Social Interaction,* (4), 1–3.

Scharmer, C. O. (2009). *Theory U: Learning from the future as it emerges.* Berrett-Koehler Publishers.

Schein, S. (2017). *A new psychology for sustainability leadership: The hidden power of ecological worldviews.* Routledge.

Sheridan, K., Halverson, E. R., Litts, B., Brahms, L., Jacobs-Priebe, L., & Owens, T. (2014). Learning in the making: A comparative case study of three makerspaces. *Harvard Educational Review, 84*(4), 505–531.

Singer, P. (1981/2011). *The expanding circle: Ethics, evolution, and moral progress.* Princeton University Press.

Sol, J., Beers, P. J., & Wals, A. E. J. (2013). Social learning in regional innovation networks: Trust, commitment and reframing as emergent properties of interaction. *Journal of Cleaner Production, 49*(8), 35–43.

Smith, H. (Ed.) (2009). *Practice-led research, research-led practice in the creative arts.* Edinburgh University Press.

Stewart, F. (2020). Overcoming short-termism: Incorporating future generations into current decision-making. *Irish Studies in International Affairs,* 1–17.

Taylor, C. (2004). *Modern social imaginaries.* Duke University Press.

Taylor, S. S., & Ladkin, D. (2009). Understanding arts-based methods in managerial development. *Academy of Management Learning & Education, 8*(1), 55–69.

Thomas, H., Mitchell, G., Rich, J., & Best, M. (2018). Definition of whole person care in general practice in the English language literature: A systematic review. *BMJ Open, 8*(12), e023758.

Treffinger, D. J., Isaksen, S. G., & Dorval, K. B. (1994). Creative problem solving: An overview. *Problem Finding, Problem Solving, and Creativity,* 223–236.

Vadeboncoeur, J. A., & Vellos, R. E. (2016). Re-creating social futures: The role of the moral imagination in student–teacher relationships in alternative education. *International Journal of Child, Youth and Family Studies, 7*(2), 307–323.

van Boeckel, J. (2013). *At the heart of art and earth: An exploration of practices in arts-based environmental education.* Aalto Publications.

van der Vaart, G., van Hoven, B., & Huigen, P. P. (2019). 'It is not only an artist village, it is much more than that'. The binding and dividing effects of the arts on a community. *Community Development Journal, 54*(3), 446–462.

Wahl, D. (2016). *Designing regenerative cultures*. Triarchy Press.
Wals, A. E. J., & Heymann, F. V. (2004). Learning on the edge: exploring the change potential of conflict in social learning for sustainable living. In A. Wenden (Ed.), *Educating for a culture of social and ecological peace* (pp. 123–145). SUNY Press. 0–7914–6174–2.
Wang, Q., Coemans, S., Siegesmund, R., & Hannes, K. (2017). Arts-based methods in socially engaged research practice: A classification framework. *Art/research International, 2*(2), 5–39.
Warm data lab. (n.d.). Retrieved April 01, 2021, from https://warmdatalab.net/.
Wheatley, M. J. (2017). *Who do we choose to be?: Facing reality, claiming leadership, restoring sanity*. Berrett-Koehler Publishers.
Wittmayer, J. M., & Schäpke, N. (2014). Action, research and participation: Roles of researchers in sustainability transitions. *Sustainability Science, 9*(4), 483–496.
Weintraub, L. (2012). *To life!: Eco art in pursuit of a sustainable planet*. University of California Press.
Zittoun, T., & Cerchia, F. (2013). Imagination as expansion of experience. *Integrative Psychological and Behavioral Science, 47*(3), 305–324.

7

Insights and Inspiration from Explorative Research into the Impacts of a Community Arts Project

Gwenda Van der Vaart

Introduction

For researchers focusing on the question of how communities can be strengthened for a more sustainable future, working at the crossroads of science, arts and society can provide interesting opportunities. Not surprisingly, an increase of academic work related to this intersection can be witnessed (e.g., Brice & Fernández Arconada, 2018; Coemans & Hannes, 2017; Hawkins, 2011). There are several types of relationships between the fields of arts and science. Wang et al., (2017, p. 6), for instance, distinguish between "research about art, art as research, and art in research". As this book demonstrates, researchers interested in resourceful and resilient community practices use a variety of creative and/or arts-based methods in their work. In this context though, looking

G. Van der Vaart (✉)
Department of Spatial Planning and Environment, University of Groningen, Groningen, The Netherlands
e-mail: g.van.der.vaart@rug.nl

A. Franklin (ed.), *Co-Creativity and Engaged Scholarship*,
https://doi.org/10.1007/978-3-030-84248-2_7

at artistic practices themselves can also provide valuable insights for such researchers in order to learn more about ways in which communities can be strengthened.

Over the past decades, an expansion from the arts beyond galleries into society took place, with artists actively engaging with particular social contexts (Brice & Fernández Arconada, 2018). Various terms are adopted in this context, such as community-based art, socially engaged art, site-specific art, social practice, dialogic art, interventionist art, contextual art, and collaborative art (Bishop, 2012; Simoniti, 2018). The objectives and output of artists involved in this kind of art vary enormously, but, as Bishop (2006, p. 179) notes, they all share "a belief in the empowering creativity of collective action and shared ideas". In line with this, there are many examples of artists who actively work with communities and aim, through their work, to contribute to that community (Guetzkow, 2002; Matarasso, 2007).

Recently, there has also been increasing attention for and recognition of the role artists can play in contributing to a community's resilience specifically. Neal (2015), for instance, argues that artists play a crucial role in rethinking the future and can help to reinvent and reimagine our world. She regards artists as 'agents of change', whom:

> can be circuit breakers of tragedy, surprising people with alternative ways of seeing, jolting them awake from denial and speeding up a public process of seeing and feeling the "truth" of climate change [...] opening possibilities for change and renewal. (p. 7)

For communities that want to develop themselves towards a more sustainable future, especially community arts projects (in which an artist actively works together with the community members) appear to hold much potential (see Anwar McHenry, 2011; Burnell, 2012; Derrett, 2008; Mulligan et al., 2006; Stocker & Kennedy, 2011; Van der Vaart, 2018). Ferreira and Duxbury (2017, p. 46) note that participation in the arts "can be a powerful driver for individual and collective capacity to (individually and collectively) rethink values, norms, and behavioural conducts". Horlings (2017, p. 137) explains that these relate to the "'inner' dimension of sustainability" that, next to practical behaviour and politics, accelerates change towards sustainability.

In addition, community arts projects can contribute to a community's resilience because of their ability to enhance links between community members, their community and the wider surroundings. Subsequently, this can stimulate people's willingness as well as ability to work together for a common good, and empower them to engage in other types of civic activities to protect and pursue their collective interests (Anwar McHenry, 2011; Derrett, 2008; Larsen et al., 2004). In this context, McCarthy et al. (2004) speak of the role the arts can play in building a community's 'organizational capacity'. According to their literature review on the benefits of the arts at the community level, this is stimulated in three ways, by: developing local arts groups and leaders, promoting cooperation among arts and non-arts groups, and "the more general process of people organizing and getting involved in civic institutions and volunteer associations" (p. 14).

This chapter takes as its empirical focus one such community arts project: the theatre-trilogy *Grutte Pier*. This project took place in the Frisian village Kimswerd, the Netherlands, between 2014 and 2018. The chapter reflects on an explorative research project into the impact of this community arts project on the village of Kimswerd. By discussing what researching such artistic practices can bring, the chapter draws wider lessons for researchers interested in community engagement, place-based action, and community resilience.

The chapter is structured as follows: first, more background on the relationship between the arts and community resilience is provided and concerns related to evaluating the impacts of community arts projects are discussed. Then the Grutte Pier community arts project and the research project are further introduced. The next sections first provide a reflection on researching such an artistic practice, and then summarize and reflect on the reported impacts of the Grutte Pier project on the community of Kimswerd. The chapter finishes with a discussion and conclusions on what researching community arts projects can bring to researchers interested in achieving meaningful change in communities in order to prepare them for a more sustainable future.

Community Resilience and the Arts

When you are interested in the sustainable development of communities, it is nearly impossible to avoid encountering the concept of resilience. Over the past decades, despite attracting sustained critique by some (see, e.g., Kaika, 2017; Porter & Davoudi, 2012), this term has become widely used in both academia and practice in order to discuss how communities can cope with the changes and uncertainties they face (Pendall et al., 2010; Revell & Dinnie, 2020; White & O'Hare, 2014). The concept of resilience is put in the spotlight even more because of the increasingly louder calls for transformative, sustainable change in light of current pressing global issues such as climate change, but also the COVID-19 pandemic, that underscores the current unsustainable system.

There is a wide variety of interpretations of resilience to be found in the literature (see, e.g., Davoudi, 2012; Hutter & Kuhlicke, 2013; Pendall et al., 2010). At its base though, when linked to communities, an evolutionary understanding of resilience revolves around the question of how communities can shape and respond to the challenges they face in order to achieve a better future (see Davoudi, 2012). Resilient communities are considered as being able to utilize and develop their resources to respond and adapt to challenges as well as opportunities that are brought about by changes (Revell & Dinnie, 2020).

Several scholars stress that the everyday lifeworld and local knowledge of communities should be incorporated when planning for community resilience (see, e.g., Steiner & Markantoni, 2013). Traditionally, however, "the majority of work in the burgeoning field of resilience [...] has not been grounded within the everyday practices of communities of policy and practice" (Coaffee, 2013, p. 327). Top-down, managerialist approaches to resilience raise certain questions, as Brice and Fernández Arconada (2018, p. 225) rightfully point out: how are resilience objectives identified and achieved? And "by what methods [can] these initiatives [...] be kept relevant to specific places and specific communities"?

In light of the above, turning to community arts holds much potential. In using the term community arts, it is referred to as "a collective method of art-making, engaging professional artists and self-defined communities through collaborative artistic expression" (Ontario Arts Council, 1998, p. 7). This art form fundamentally depends on actively engaging people in the creative process and, therefore, unlocks their everyday, lived experiences. This aligns with resilience policies that are directed towards smaller spatial scales and everyday activities (Coaffee, 2013), and can help to take the specific socio-spatial context of a community into account (see also Christopherson et al., 2010; Hutter & Kuhlicke, 2013; O'Hare & White, 2013).

Evaluating Impact

As noted, community arts projects can provide valuable insights and inspiration to researchers interested in community engagement, place-based action, and community resilience. However, when researching such artistic practices to assess their impact, certain issues are at stake.

First, when looking at the value of such projects for community resilience, it is important to realize that they are not a panacea. Positive outcomes of community arts projects cannot be taken for granted and projects can also have negative outcomes. As Matarasso (1997, p. 75) already noted in his classic work *Use or Ornament?*: "the arts are not fast-food, predictable in content in every place and on every occasion". Therefore, a nuanced perspective on the effects of the arts is required (see also McCarthy & Jinnett, 2001; Mulligan et al., 2006). Moreover, in the literature, potential barriers to engaging in the arts are noted and, related to this, concerns around the extent to which a community is involved are expressed (see, e.g., Anwar McHenry, 2011; Balfour et al., 2018; Mulligan et al., 2006; McCarthy & Jinnett, 2001). A key lesson here is that context matters. The influence of the arts is context-dependent, with the arts, for instance, having both binding and dividing effects for different (groups of) community members at the same time (Van der Vaart et al., 2017). On their own, the arts cannot provide the solution to communities and therefore, should be considered as one of several potential means in community development processes (Burnell, 2012; Matarasso, 2007; Van der Vaart et al., 2017).

Another important issue at stake when discussing the value of community arts, especially in light of certain (community) development goals, is that there are concerns and tensions around evaluating the arts. Over the years, evidencing the impacts of arts projects has become more important. Belfiore and Bennett (2010) point out that there is a commitment of Western governments towards evidence-based policy-making: "to measure and assess the extent to which the subsidized arts have a socio-economic impact" (p. 122), and therefore whether they contribute to policy (or not). As a result, initiators of arts projects are increasingly pressurized to articulate the public value of their work in light of funding criteria and the need to effectively appeal to the general public and its legislative representatives (McCarthy et al., 2004). Belfiore and Bennet (2010) observe that this encouraged a blooming of impact studies and the development of a "toolkit mentality" (p. 122), with a search for a straightforward method of impact evaluation that can be applied in different contexts.

Many art advocates, however, have a resistance to, or negative perceptions of, evaluation (Reeves, 2002). According to Jermyn (2001), this resistance can be composed of several elements including lack of time, resources or skills, but also lack of motivation, inclination or understanding about the value of evaluation (outside the context of funding relationships). She points out that art practitioners rarely regard evaluation and monitoring as central or integral to their work. In addition, Jermyn (2001) notes there are fears connected to the appropriateness of available evaluation methods, such as that it "will fail to reflect the spirit of the arts activity, stifle creativity or somehow reduce the arts experience" or that "the utility of the arts will be overstated at the expense of less measurable benefits" (p. 9).

This latter fear connects to concerns around the 'instrumentalization' of the arts, expressed by artists as well as scholars (see, e.g., Brice & Fernández Arconada, 2018; McCarthy et al., 2004; Mulligan et al., 2006; Khan, 2010; see also Leitheiser et al., this book). These concerns boil down to an (perceived) imbalance, whereby the instrumental benefits of the arts overshadow or suppress their intrinsic benefits. Here, arts advocates express concerns with regard to only funding the arts based on their instrumental benefits. The instrumentalization of the arts also leads to debate on how exactly to understand and judge 'quality' in the arts (see, e.g., Bishop, 2006; Simoniti, 2018). There is a tension between utility on the one hand and aesthetics on the other (Brice & Fernández Arconada, 2018).

Finally, there are scholars who take a critical stance towards arts impact research and who point to tensions between "genuine research and research for the sake of advocacy" (Belfiore, 2009, p. 353; see also Belfiore & Bennett, 2010). Belfiore (2009) notes that there is an evident temptation in this field to articulate research questions in advocacy- or policy-friendly terms, noting that as a consequence:

> research has often focused on asking *how* the presumed positive social impacts of the arts might be measured or enhanced, rather than in asking *whether* the arts have social impacts of the sort claimed for them, *if* these impacts can be expected to be positive and, more generally, *whether* it is possible to generalise people's experiences of the arts within arts forms, across art forms and across the very diverse population represented by those who engage with the arts. (p. 353, original emphasis)

Consequently, some scholars argue for a separation of arts advocacy from rigorous impact evaluation research (Belfiore & Bennett, 2010), and point to the need for carefully thought-through research questions (Belfiore, 2009), with researchers proceeding "along clear lines" while making "explicit the theories underpinning [their] research" (Merli, 2002, p. 115).

PeerGrouP, Grutte Pier and the Research Project

> A story becomes stronger when it is told. Like a tree can grow when it receives sunlight and water. You came here for a story about Grutte Pier. His story got strong roots. And a tree with strong roots blossoms every year. (translated quote from "De Bezinning", 2017–2018)

These are the opening lines of the final round of performances connected to the Grutte Pier theatre trilogy. This community arts project took place in the Dutch village Kimswerd between 2014 and 2018. The project was initiated by PeerGrouP, a site-specific theatre company in the northern Netherlands. PeerGrouP describes its work as "site-specific in a socially engaged manner" (PeerGrouP, 2020). On its website, PeerGrouP's special approach to its productions is explained; this starts with PeerGrouP's employees coming to a community as outsiders:

> Integrate and infiltrate. In this approach they [the employees] are working on a site and trying to find a (temporary) place in the community. By making use of the local artistic skills and fascinations, curiosity awakens. This is a starting point: from curiosity grows complicity and from complicity a desire to cooperate might emerge. (*Ibid.*)

Actively collaborating with local people plays an important role in PeerGrouP's productions. According to their vision, they aim to make theatre *in* and *with* a community, telling stories about the place in which they are performing. Hereby they strive to achieve a lasting effect *on* the community (PeerGrouP, 2019). As expressed in PeerGrouP's policy plan, PeerGrouP regards it as their civil mission to "challenge the community, entrepreneurs, politics and science to actively contribute to a transition to a sustainable society" (PeerGrouP, 2016, translated). In light of this, PeerGrouP works on projects in which they make connections with scientists, experts, organizations, and people from the community itself. Moreover, they strive to make such connections in such a manner that a follow-up is also possible (*Ibid.*).

The Grutte Pier trilogy in Kimswerd was one of the major productions of PeerGrouP at that time. The project revolved around the life of the village's historical figure Grutte Pier. This is a nickname of Pier Gerlofs Donia (meaning 'Big Pier') and refers to his allegedly legendary size and strength. Grutte Pier lived in Kimswerd between 1480 and 1520 and led the rebellion against the oppressors of Friesland. In keeping with the above introduction to PeerGrouP, the theatre company's approach to the project involved active collaboration with the inhabitants of Kimswerd (a village of nearly 500 inhabitants), as well as other volunteers from the wider region. During the years that the project ran in the village, PeerGrouP and the community worked together towards a trilogy around three main performances[1]:

1. *De Brân*, which involved a reconstruction of Grutte Pier's farm and a one-off event on 29 January 2015 (1,534 visitors). During the performance, the reconstructed farm was symbolically set on fire. Through this act, it was commemorated that it was 500 years ago that Grutte Pier's farm was burned down, which ignited his anger and made him start the rebellion against the occupiers of Friesland.
2. *Grutte Pier fan Kimswert*, a large open air spectacle around the life and mission of Grutte Pier; this was performed 27 times during the summer of 2016 (9,094 visitors).
3. *De Bezinning*, a more intimate performance based on the last years of Grutte Pier's life (which he spent in a monastery); this was staged in the church of Kimswerd (and two other locations outside the village) and was performed 12 times in the winter of 2017–2018 (653 visitors).

During the last part of the trilogy, I was asked by PeerGrouP to conduct a small research project into the Grutte Pier project. They were curious about the impact of their long-term involvement in Kimswerd on the village, as seen from the perspective of the inhabitants. The research project was therefore commissioned on behalf of PeerGrouP, but as a researcher I had freedom in designing and undertaking the project. The objective of the research project was to explore *whether* the Grutte Pier project had an impact on the village and *if so*, to provide insight into the nature of this impact. The project involved in-depth interviews and questionnaires and was conducted between December 2017 and February 2018.

Participants for the interviews were recruited through snowball sampling and random door-to-door recruiting. In total, 12 interviews were conducted with 13 different persons (including one married couple

[1] In addition to these main performances there were several side-events/spinoffs organized in relation to the Grutte Pier trilogy, to generate further attention for the project.

that was interviewed at the same time) comprising: (a) 3 volunteers who were part of the so-called 'village team'—a group of inhabitants actively involved in the overall organization of the Grutte Pier project; (b) 3 volunteers that participated in a (small) part the project; (c) 4 inhabitants who did not participate in the project but who visited the performances (including the married couple); and (d) 3 inhabitants who neither participated in the project nor visited the performances. The interviews focused on topics such as people's connection to Kimswerd, their involvement in the Grutte Pier project, and their opinion on, and experiences with, PeerGrouP and the project.

The questionnaires were handed out after three performances of the last part of the trilogy in Kimswerd. The questionnaire consisted of mostly brief open-ended questions and covered topics such as visitors' initial response to the performance, their reasons for visiting, their connection to Kimswerd and knowledge of Grutte Pier's story, their involvement in the project themselves, and whether they experienced any effects of the project (and which). In total, 50 questionnaires were completed by the visitors.

The interviews were recorded and transcribed, and both the interviews and questionnaires were coded inductively to analyze the various effects that emerged from the data. As the interviews were conducted in Dutch, the quotes in the following sections are all translated by the researcher (Van der Vaart). Hereby, fictional names are used.

In the following sections I will first provide reflections on researching such an artistic practice and then summarize and reflect on the effects of the Grutte Pier project that emerged from the interviews and questionnaires.

Researching a Community Arts Project

Evidencing the impact of community arts projects has become more important over the years, as noted above. However, there is no consensus on what the best evaluation methods for assessing the impact of such projects are (Belfiore, 2006; Reeves, 2002). While some scholars underline the need for more definitive quantitative evidence in order to demonstrate the impact of the arts, others argue that the development of evaluation techniques beyond the quantitative is needed (see, e.g., Belfiore, 2006; Mulligan et al., 2006). In addition, Merli (2002) points to the context-dependent nature of the impact of the arts, stating that differences are likely to exist and that there is a need to know more about this. She argues that: "without knowing what the real, specific effects of the arts are, and in which circumstances they occur [...] researchers are only going to measure what they would like to be there" (p. 115). In light of this, it is important to not only 'measure' impacts, but to 'understand' people's experiences, ideas and feelings (Merli, 2002). Such an approach aligns with the 'critical research ethos' that Belfiore (2009) advocates. She points to the need for explorative research that is:

> indifferent to the requirements of advocacy [and] aims to describe, explore and illuminate complex issues around the role and condition of culture, cultural production, consumption and administration in contemporary society. (p. 354)

Although being small-scale, it is worthwhile to briefly reflect on some choices that were made in undertaking the explorative research project in Kimswerd, in order to draw lessons for future work.

First, the research project consisted of both questionnaires and interviews. Choosing this mix of methods proved to be helpful in obtaining both a more general impression of the Great Pier project from the visitors' side and a more in-depth understanding of inhabitants' opinions and experiences with the arts project in their village. As the research project aimed to explore the arts project's effects as experienced by the inhabitants of Kimswerd, I decided to only interview inhabitants themselves and not people working at PeerGrouP. Here, I also deliberately opted for interviewing both inhabitants who were involved in the project (to various degrees) and those who were not. It was interesting to have this mix of interviewees, to explore if and how this impacted the way people perceived and experienced the project.

The research project was conducted near the end of the Grutte Pier trilogy, as it was only in the autumn of 2017 that PeerGrouP approached me to conduct an explorative research project into the effects of their arts project. This rather late timing seems to support Jermyn's (2001) earlier noted observation that art practitioners rarely regard evaluation and monitoring as being integral to their work. In hindsight, while it was still possible to collect data for a small explorative review into the effects of the arts project, a more thorough, overall evaluation of the project would have requested an earlier involvement, ideally directly from the start in 2014. From an evaluation perspective, it would have been interesting to monitor the development of the theatre trilogy and be able to investigate how the inhabitants experienced the project throughout the years. Still, the explorative research project as it was conducted presents valuable insights and inspiration for researchers interested in community engagement, place-based action and community resilience, the next section, therefore, turns to the findings of the project.

A Project as Strong as Grutte Pier Himself?

Overall, a highly positive image of the Grutte Pier trilogy arose during the research project. Many interviewees regarded the Grutte Pier project as a unique project that was very successful and argued that only *this* project could have achieved the effects that it did *to this extent*. When asked to elaborate on what they saw as specific strengths of the project, the interviewees mentioned four strengths: the size of the project; the popularity and authenticity of the story of Grutte Pier; the fact that the project connected people with one another; and that it involved the community to a great extent. These factors helped PeerGrouP to engage a large portion of the inhabitants in the project. Ann (village-team), for instance, remarked:

> If you wanted to do something, you could participate, in any way whatsoever. Whether you were sitting behind the cashier, were sewing [a costume], or controlling traffic, you are doing it all together [...] I think that's the strength of PeerGrouP [...] It is being propagated very enthusiastically, so you actually feel like participating yourself too.

In addition, the interviews also revealed that the unfamiliarity of PeerGrouP's employees—their 'other set of eyes'—seemed to be related to the project's success. Interviewees noted that this also contributed to the great enthusiasm of a large number of inhabitants to volunteer for the project. Ursula (village-team) explained this as follows:

> I think the unfamiliar faces actually made everyone feel involved. For instance, Peter [...] the chairman of Dorpsbelangen [village interest group], he is a good speaker and technically he could be able to do it in the same way, but then the people would have thought 'yeah right Peter'. But I think that would be because he is a face of the village [...] now, they were new faces in the village and that gave a very good and positive feeling.

As described above, PeerGrouP works via an 'integrate and infiltrate' approach, hoping that a desire to cooperate might emerge from people's initial curiosity. Apparently, this approach paid off enormously in Kimswerd. Following on from PeerGrouP's vision, they subsequently strived to achieve a lasting effect *on* the community *in*—and *with*—which they work.

The interviewees that participated in the Grutte Pier project themselves raised several kinds of personal effects. These are related especially to people's personal growth, social life, and feelings of pride. Some indicated that they experienced personal growth and developed certain skills due to their involvement in the project. This finding corresponds to earlier work from, among others, Matarasso (2007) and Newman et al. (2003). Becoming more assertive, developing a broader social outlook, and improving planning and communication skills are examples of the personal impacts that were mentioned. Amber (who participated in the farmers' choir[2]), for instance, opened up about the following personal change she experienced:

> A change in my being [...] I am more combative. I stand up for myself more [...] and I am more daring. I am also part of a [*different*] choir, and

[2] The famers' choir was specifically formed as part of the project and played a role in the performances.

> I just dare to be myself there. Before I never dared, I was always sitting in the back corner, everywhere. So nobody would see me. Well, that is changed.

In addition to such personal effects, the interviews and questionnaires also revealed several effects at the village level. The following quote symbolizes a broader pattern that can be observed in this regard. Alice (who did not participate or visit) answered the question whether the project resulted in certain effects on the village as follows:

> I cannot really assess that properly. In terms of sense of community you mean, right? Yeah I did not feel that, because I was not part of it [...] I believe that, if you get involved in that, let's say the 'mienskip' [*note: Frisian word for community*], that it certainly has a strengthening effect.

In line with Alice's statement, not everybody among the interviewees perceived the effects of the Grutte Pier trilogy on the community of Kimswerd to the same extent. Understandably perhaps, those who were themselves actively involved in the project noted the effects, while those who were less or not at all involved mentioned the effects less often or to a lesser extent (see also Van der Vaart et al. (2017) for a similar finding in a different context). In contrast to this finding though, in the case of the questionnaire responses, visitors who themselves were not actively involved in the project noted several positive effects of the project on Kimswerd.

The questionnaire included four open-ended questions on the effects of the Grutte Pier project. Visitors were asked if and what the project brought them personally, what they thought it brought to the village, and whether the project had any negative effects in their eyes (also specified on a personal and village level). The most often indicated village level effects were that the project put Kimswerd on the map and generated attention regarding its history. These findings were supported by the interviews. The interviews, in turn, revealed that these effects also fed into a sense of pride among the inhabitants and could result in a boost of one's identity. The interviewees expressed a certain degree of pride, both with regard to the Grutte Pier project and their village. They noted that they were proud that the project took place in their village and that the inhabitants—together—made it into such a success.

The village level effects that were most often mentioned during the interviews were that the project created and strengthened bonds between Kimswerd's inhabitants. Participants expressed that, thanks to the project, they got to know more people and/or got to know people better. Here, many interviewees also remarked that the project strengthened the general sense of community in Kimswerd. Phil (village-team), for instance, noted:

> the people who have been working with each other in the village in the past years [*during the project*], they also meet each other more often and work together more often. So it has done a lot for the social bonding in the village. Every night you had a large amount of volunteers on the move [...*this*] certainly contributed to a closer bond between a lot of people in Kimswerd.

As a subsequent result of these effects, some interviewees also saw a smoother way of collaborating as an effect of the Grutte Pier project. They noted that people could find each other quicker when they need help and had a shared 'success experience', which gives confidence. Thanks to the success of the project, people discovered how much is possible to accomplish by working together. In this sense, the project contributed to a certain awareness of, and confidence in, people's own ability. This all feeds into the 'organizational capacity' of the community that McCarthy et al. (2004) speak of, and is highly promising in light of future community development initiatives.

Hardly any negative effects of the Grutte Pier project were mentioned, either in the questionnaires or in the interviews. This absence of negative effects being mentioned corresponds to Newman et al.'s (2003) literature review on community-based arts projects, in which they found that only a few negative consequences of projects were ever mentioned. It could be that the participants might have perceived it as being inappropriate to share any negativities on the Grutte Pier project due to the overall wide support it received in the village—even though they could complete the questionnaire anonymously and the interviews were treated confidentially. There was just one interviewee who noted that he had faced some critical comments from colleagues, who were questioning the costs of the project and whether this spending was justified. Some of the interviewees, however, did raise that there might have been certain negative noises about the project, even though they did not encounter these themselves.

For the community in Kimswerd an important question is how long, and to what extent, the above-mentioned effects will continue to have an influence on their village community. At the time the interviews were conducted, the very last performances of the trilogy were taking place or had just finished. At that time, the interviewees found it hard to predict how and to what extent the project might continue to influence the community in the future. Many at least regarded the project as something that will be talked about for a long time and that will be a precious memory for many inhabitants. An entrance ticket for the first part of the trilogy, a wooden slice specially made for the project, got a place in the homes of several inhabitants and forms a tangible artefact in memory of the project.

While one interviewee expressed that she was "afraid it will slowly simmer away in the village" (Ursula—village-team), the other interviewees had higher expectations with regard to the longevity and strength of the project's effects. In addition to the experience of the project being regarded as a dear memory that would long be cherished, people mentioned three effects of the Grutte Pier project that they expected to be longer lasting. First, some people felt that the personal growth they had experienced from taking part would last for the rest of their lives. Second, interviewees expressed their expectation that the boosting of the community's 'organizational capacity' would be long lasting. In their eyes, the project contributed to a smoother way of collaborating in the village and stimulated a certain trust and belief in people's ability to accomplish things together. In light of this, people, for instance, expected that it would be easier to find volunteers for future activities, also when organized by the villagers themselves. Finally, some interviewees indicated that they expected that the Grutte Pier project would serve as an inspirational influence for future activities and would have a follow-up in some way or another. Some interviewees already mentioned certain activities that seemed to point in this direction, such as the formation of a new choir in which some of the members of the farmers' choir[3] participated, and a request for 'jewellery making' workshops to the woman who was responsible for the jewellery of some of the actors.

Lessons Learned

As this chapter stated at its start, working at the crossroads of science, arts, and society can be insightful for researchers interested in community engagement, place-based action, and community resilience. In this chapter, the Grutte Pier trilogy is extensively discussed, as researching such artistic practices can provide valuable insights and inspiration to researchers. Community arts or participatory arts projects such as this project are often embraced as a form of soft social engineering, with the idea that they can be useful to effect positive changes in society (Bishop, 2012). However, as discussed, evaluating such artistic practices is not a straightforward task. There are several tensions and concerns around evaluating the impact of the arts and there is no consensus on what the best evaluation methods are (Belfiore, 2006; Reeves, 2002). In drawing this chapter to a conclusion, what lessons can be drawn from the explorative research project into the impact of the Grutte Pier project on the village of Kimswerd?

The reflections on the research project support the need to adopt a critical perspective with regard to the value of artistic practices. In order to obtain both a general impression, as well as a more in-depth understanding of people's experiences with the arts, a mix of methods proved to be supportive. Moreover, interviewing people who were involved in the arts project to various degrees, and including those who were not involved, was instrumental in gaining a more nuanced understanding of the project's impact on the village. In the end, the findings from the explorative research project demonstrate that researching such artistic practices can provide insights and inspiration for researchers interested in community resilience.

[3] See footnote 2.

To start with, community arts projects seem to be an evocative way of engaging a community and can result in a variety of effects. Altogether, staying close to PeerGrouP's (2019) own terminology, the Grutte Pier project appears to have been a successful community arts project *in* Kimswerd, being accomplished together *with* the inhabitants, and with positive effects *for* the community. The interviewees pointed to several ingredients for its success: the popularity and authenticity of the story of Grutte Pier was regarded as a strength, and both the project's size and long duration were mentioned as reasons why the project achieved its effects to the extent it did. In addition, the fact that the project connected people with one another and involved the community to such a great extent was also mentioned as part of the project's strengths. Interestingly, the unfamiliarity of PeerGrouP's employees also appeared to play a role in the project's success, as this contributed to inhabitants' great enthusiasm to volunteer.

Inferred from the above, and in light of strengthening communities, it appears to be a successful formula to have artists, coming to a community as 'outsiders', actively engaging inhabitants in a large community arts project that is both locally grounded and offers the inhabitants various ways of participating themselves. This finding is different from what Rogers and Spokes (2003) concluded in their study on a community development project in small rural communities in Australia. They regarded the involvement of *local* artists as an essential element for the engagement of community members and community building objectives, noting that:

> local artists were already connected to the community, with a strong desire to improve their own profile and value to the community. (p. 7)

In Kimswerd however, PeerGrouP's 'integrate and infiltrate' approach seemed to play a considerable role in helping to enthuse inhabitants to become involved in the project. As noted, PeerGrouP strives to achieve lasting effects on the communities in and with which they work, and actually regards it as its civil mission to "challenge the community [...] to actively contribute to a transition to a sustainable society" (PeerGrouP,

2016, translated). This approach aligns with the earlier noted characterization of artists as potential 'agents of change', that can help to change people's mindsets and activate them to become involved (Horlings, 2017; Neal, 2015).

A final important remark needs to be made here, partly linking back to the earlier discussed concerns of the arts not being a panacea for communities. In light of community resilience it is important for community arts projects to achieve *sustainable* effects in order to have a lasting impact on communities (Askins & Pain, 2011; Carey & Sutton, 2004). The opening lines of the last part of the Grutte Pier trilogy spoke of Grutte Pier's story developing strong roots, and reminded the audience that "a tree with strong roots blossoms every year". While at the start of the Grutte Pier project, some people were quite sceptical about its overall feasibility, the project, eventually, showed the inhabitants what they can accomplish when they work together. Such effects can go a long way and can be a great boost for a community's 'organizational capacity' (McCarthy et al., 2004). The interviewees also raised the fact that the trilogy might serve as an inspirational influence for future activities. In this way, the community could reap the benefits of the blossoming 'Grutte Pier tree' for a long time.

References

Anwar McHenry, J. (2011). Rural empowerment through the arts: The role of the arts in civic and social participation in the mid west region of Western Australia. *Journal of Rural Studies, 27*, 245–253.

Askins, K., & Pain, R. (2011). Contact zones: Participation, materiality, and the messiness of interaction. *Environment and Planning d: Society and Space, 29*, 803–821.

Balfour, B., Fortunato, M.W.-P., & Alter, T. R. (2018). The creative fire: An interactional framework for rural arts-based development. *Journal of Rural Studies, 63*, 229–239.

Belfiore, E. (2006). The social impacts of the arts—Myth or reality? In M. Mirza (Ed.), *Culture vultures: Is UK arts policy damaging the arts?* (pp. 20–37). Policy Exchange Limited.

Belfiore, E. (2009). On bullshit in cultural policy practice and research: Notes from the British case. *International Journal of Cultural Policy, 15*(3), 343–359.

Belfiore, E., & Bennett, O. (2010). Beyond the 'toolkit approach': Arts impact evaluation research and the realities of cultural policy-making. *Journal for Cultural Research, 14*(2), 121–142.

Bishop, C. (2006). The social turn: Collaboration and its discontents. *Artforum, 44*(6), 178–183.

Bishop, C. (2012). *Artificial hells: Participatory art and the politics of spectatorship*. Verso.

Brice, S., & Fernández Arconada, S. (2018). Riding the tide: Socially-engaged art and resilience in an uncertain future. In E.-M. Trell, B. Restemeyer, M. M. Bakema, & B. van Hoven (Eds.), *Governing for resilience in vulnerable places* (pp. 224–243). Routledge.

Burnell, J. (2012). Small change: Understanding cultural action as a resource for unlocking assets and building resilience in communities. *Community Development Journal, 48*(1), 134–150.

Carey, P., & Sutton, S. (2004). Community development through participatory arts: Lessons learned from a community arts and regeneration project in South Liverpool. *Community Development Journal, 39*(2), 123–134.

Christopherson, S., Michie, J., & Tyler, P. (2010). Regional resilience: Theoretical and empirical perspectives. *Cambridge Journal of Regions, Economy and Society, 3*(1), 3–10.

Coaffee, J. (2013). Towards new-generation urban resilience in planning practice: From securitization to integrated place making. *Planning Practice & Research, 28*(3), 323–339.

Coemans, S., & Hannes, K. (2017). Researchers under the spell of the arts: Two decades of using arts-based methods in community-based inquiry with vulnerable populations. *Educational Research Review, 22*, 34–49.

Davoudi, S. (2012). Resilience: A bridging concept or a dead end? *Planning Theory & Practice, 13*(2), 299–307.

Derrett, R. (2008). *Regional festival: nourishing community resilience: The nature and role of cultural festivals in Northern Rivers NSW communities*. Southern Cross University.

Ferreira, I., & Duxbury, N. (2017). Cultural projects, public participation, and small city sustainability. In S. Asikainen, C. Brites, K. Plebańczyk, L. Rogač Mijatović, & K. Soini (Eds.), *Culture in sustainability: Towards a transdisciplinary approach* (pp. 45–60). University of Jyväskylä, Department of Social Sciences and Philosophy.

Guetzkow, J. (2002). *How the arts impact COMMUNITIES: An introduction to the literature on arts impact studies* (Working Paper Series, 20, Princeton University, Center for Arts and Cultural Policy Studies).

Hawkins, H. (2011). Dialogues and doings: Sketching the relationships between geography and art. *Geography Compass, 5*(7), 464–478.

Horlings, L. (2017). The role of artists and researchers in sustainable place-shaping. In S. Asikainen, C. Brites, K. Plebańczyk, L. Rogač Mijatović, & K. Soini (Eds.), *Culture in sustainability: Towards a transdisciplinary approach* (pp. 130–143). University of Jyväskylä, Department of Social Sciences and Philosophy.

Hutter, G., & Kuhlicke, C. (2013). Resilience, talk and action: Exploring the meanings of resilience in the context of planning and institutions. *Planning, Practice & Research, 28*(3), 294–306.

Jermyn, H. (2001). *The arts and social exclusion: A review prepared for the arts council of England*. Arts Council of England.

Kaika, M. (2017). 'Don't call me resilient again!': The new urban agenda as immunology ... or ... what happens when communities refuse to be vaccinated with 'smart cities' and indicators. *Environment & Urbanization, 29*(1), 89–102.

Khan, R. (2010). Going 'mainstream': Evaluating the instrumentalisation of multicultural arts. *International Journal of Cultural Policy, 16*(2), 184–199.

Larsen, L., Harlan, S. L., Bolin, B., Hackett, E. J., Hope, D., Kirby, A., Nelson, A., Rex, T. R., & Wolf, S. (2004). Bonding and bridging: Understanding the relationship between social capital and civic action. *Journal of Planning Education and Research, 24*(64), 64–77.

Matarasso, F. (1997). *Use or ornament? The social impact of participation in the arts*. Comedia.

Matarasso, F. (2007). Common ground: Cultural action as a route to community development. *Community Development Journal, 42*(4), 449–458.

McCarthy, K. F., & Jinnett, K. (2001). *A new framework for building participation in the arts*. RAND Corporation.

McCarthy, K. F., Ondaatje, E., Zakaras, L., & Brooks, A. (2004). *Gifts of the muse: Reframing the debate about the benefits of the arts*. RAND Corporation.

Merli, P. (2002). Evaluating the social impact of participation in arts activities. *International Journal of Cultural Policy, 8*(1), 107–118.

Mulligan, M., Humphery, K., James, Scanlon, C. P., Smith, P., & Welch, N. (2006). *Creating community: Celebrations, arts and wellbeing within and across local communities*. The Globalism Institute.

Neal, L. (2015). *Playing for time*. Oberon Books.

Newman, T., Curtis, K., & Stephens, J. (2003). Do community-based arts projects result in social gains? A review of the literature. *Community Development Journal, 38*(4), 310–322.

O'Hare, P., & White, I. (2013). Deconstructing resilience: Lessons from planning practice. *Planning Practice & Research, 28*(3), 275–279.

Ontario Arts Council. (1998). *Community arts workbook*. Ontario Arts Council.

PeerGrouP. (2016). *PeerGrouP 2017–2020 radius*. PeerGrouP.

PeerGrouP. (2019). *Bestuursverslag*. PeerGrouP.

PeerGrouP. (2020). *PeerGrouP—Site specific theatre*. Last accessed on 15 June 2020, via: https://www.peergroup.nl/english/.

Pendall, R., Foster, K. A., & Cowell, M. (2010). Resilience and regions: Building understanding of the metaphor. *Cambridge Journal of Regions, Economy and Society, 3*(1), 71–84.

Porter, L., & Davoudi, S. (2012). The politics of resilience for planning: A cautionary note. *Planning Theory and Practice, 13*(2), 329–333.

Reeves, M. (2002). *Measuring the economic and social impact of the arts: A review*. Arts Council of England.

Revell, P., & Dinnie, E. (2020). Community resilience and narratives of community empowerment in Scotland. *Community Development Journal, 55*(2), 218–236.

Rogers, M., & Spokes, J. (2003). Does cultural activity make a difference to community capacity? A key question addressed by the small towns: Big picture project. *Community Quarterly, 1*(4), 1–8.

Simoniti, V. (2018). Assessing socially engaged art. *The Journal of Aesthetics and Art Criticism, 76*(1), 71–82.

Steiner, A., & Markantoni, M. (2013). Unpacking community resilience through capacity for change. *Community Development Journal, 48*(3), 1–19.

Stocker, L., & Kennedy, D. (2011). Artistic representations of the sea and coast: Implications for sustainability. *Landscapes, 4*(2), 97–123.

van der Vaart, G. (2018). *Arts & resilience in a rural community: The value of arts-based community activities in resilience-building in Pingjum, northern Netherlands*. Netzodruk.

van der Vaart, G., van Hoven, B., & Huigen, P. P. P. (2017). 'It is not only an artist village, it is much more than that'. The binding and dividing effects of the arts on a community. *Community Development Journal*. https://doi.org/10.1093/cdj/bsx055.

Wang, Q., Coemans, S., Siegesmund, R., & Hannes, K. (2017). Arts-based methods in socially engaged research practice: A classification framework. *Art/research International: A Transdisciplinary Journal, 2*(2), 5–39.

White, I., & O'Hare, P. (2014). From rhetoric to reality: Which resilience, why resilience, and whose resilience in spatial planning? *Environment and Planning c: Government and Policy, 32*(5), 834–950.

8

How to Nurture Ground for Arts-Based Co-Creative Practice in an Invited Space: Reflections on a Community in North Netherlands

Scott Davis, Yanthe van Nek, and Lummina G. Horlings

Introduction

Place-based cultural projects are increasingly considered by governments as capable of fostering a greater 'sense of local identity' and strengthening social cohesiveness within local communities. The launch of such a project is often initiated by an external authority, such as a government, opening up and overseeing an 'invited space'. Creative practitioners and community members are then invited to create an artistic or creative output of public value as a cultural expression of the community. Situations in which the practitioner is assigned by the governmental authority can, however, be challenging from a power relations perspective. A lack of local agency over the project design and its direction can potentially result in scepticism and dis-trust of participants and, consequently, low levels of community engagement.

S. Davis (✉) · Y. van Nek · L. G. Horlings
University of Groningen, Groningen, The Netherlands
e-mail: s.davis@rug.nl

A. Franklin (ed.), *Co-Creativity and Engaged Scholarship*,
https://doi.org/10.1007/978-3-030-84248-2_8

In western-European countries there has been a generational shift in public governance from community engagement and empowerment principles as generalized rhetoric, to a situation whereby concepts such as participatory governance and bottom-up approaches can routinely be found embedded within public policy strategies (de Andrade, 2016; MacKinnon, 2002). Policies that aspire for more participative and engaged communities can be considered as place-based; working within local contexts to build on the characteristics and culture of a place and community. This is grounded in an appreciation that much of the knowledge necessary for local development is not held by public and private institutions but embodied by people on the ground (Barca, 2009). The rationale behind governments commissioning place-based cultural projects in order to shape community futures, resides in the understanding that 'bottom-up' practical projects are more effective in influencing the attitudes and behaviours of communities than the traditional delivery of 'top-down' rational, ideological doctrines (Svensson, 2012). The 'culture' of a place is therefore increasingly viewed by governments as a valuable asset for sustainable development, to strengthen societal resilience and improve people-centred social outcomes (Duxbury & Jeanotte, 2007; Dessein et al., 2015; UCLC, 2008; Hawkes, 2001).

As a conceptual term, culture has an array of definitions attributed to it, encompassing all the ways we make sense of our lives together, referred to as '*the social production of meaning*' (Hawkes, 2001). Culture encapsulates all of our values, practices and interactions involving both human and non-human forms, including socio-technical systems and technologies (Williams, 1980). The term can therefore be applied to almost any social context. Some critics posit 'culture' as an overused term as they consider it much too broad to hold significance within sustainability discourse; it can mean anything from a network of meaning, to a way of life, to high culture and arts (Throsby, 2008).

In this chapter we narrow the focus specifically to place-based culture. Places contain a vast amount of human history that informs cultural norms; the historic perceptions people have of their places are often connected to their attitudes, policies and political/economic consequences that result from these perceptions (Shortridge, 2005). Williams (1980, p. 67) articulates how our landscapes can also be considered a part of our place-based culture: '*The environment around us—our plants, gardens, wildlife, living things should be considered as an important part of this "culture" as 'the idea of nature contains an extraordinary amount of human history*'. The growing calls for place-based culture to be utilized as an asset for local sustainability planning, however, also enhance the threat of its mis-use. Place-based culture risks being utilized in an instrumentalist manner by those in authority to achieve their own pre-defined outcomes that fail to address, or are even in conflict with, local needs (see Van der Vaart, this book). Projects and initiatives that apply arts-based co-creative practice with communities are not immune from this charge and are in danger of being co-opted by managerialist governance approaches that can use such methods as cover to mask neo-liberal orthodoxies. This covert practice is described by MacKinnon (2002) as governing through community, operating as an instrumental governance strategy that functions to obscure existing power inequalities and therefore limits the potential for transformation and social change (Noorani et al., 2013).

In the Netherlands, but also in other countries (Grenni et al., 2020; Neal, 2015), we witness situations where community creative practitioners (i.e., those who utilize arts-based methods to engage with communities for social change) are commissioned by external authorities—commonly local or regional governments—to facilitate arts-based co-creative practice to provide a platform for local knowledge and values to be brought to the surface (Horlings, 2017). Community creative practitioners are often inspired by, or demonstrate a strong overlap with, participative action research (PAR) approaches. They work by applying their creative and facilitative skill sets to tap into the power of cultural activities (e.g., community music and theatre initiatives) that can provide opportunity for the construction of new forms of subjectivity and reach people on '*the affective*' level, promoting their capacity to perceive new possibilities (Mouffe, 2013). This is based on the assumption that engagement through cultural projects can unlock communities' '*transformative potential and thus challenge dominant representations and ways of knowing, facilitate dialogue across ideological and epistemological boundaries and change hearts and minds through building intellectual and affective understanding*' (Nunn, 2020, p. 4).

One might expect such a process to be inherently inductive, grassroots or 'bottom-up', operating as it does at the community level. However, this is not always the case and cannot be assumed when local, cultural projects are assigned by an external state authority. In this chapter we argue that a characteristic of paternalism may be present and, if so, must be considered whenever a creative practitioner is invited to work on a local, place-based project. While we encourage the state sector to embrace local, place-based approaches, we also advocate for heightened awareness regarding how state commissioned projects may serve or prioritize external agendas—partially or fully divorced from community needs and therefore not fully reflective of the bottom-up, grassroots values one would expect from projects engaging with place-based culture. We therefore suggest those who are embarking on an arts-based co-creative journey to become aware of, and take steps to address the challenges of working in participative spaces that are opened up within such projects. In doing so they are encouraged to regularly reflect upon and take seriously such questions as: who is initiating the invitation, on whose authority, and what interests are motivating it.

The questions guiding this chapter are therefore: How can creative practitioners deal with the constraints of being invited into a participative space when first engaging with a community? How then to respond to the wishes of the community on their own terms, rather than respond to external interests? How can this inform those who are considering using arts-based co-creative practice to engage with communities in the future? The three questions posed are critically explored by viewing the case through the conceptual lens of '*invited spaces*' (Gaventa, 2006) (see section 8.2 below).

This chapter is primarily an interpretative analysis of the case detailed by a researcher, the first author. The data for this chapter were derived from the researcher's personal observations of the creative practitioners' strategy in the field, and a series of open and semi-structured interviews with the second author, the creative practitioner Yanthe van Nek, who reflected on her time working in the village. The analysis is informed by the co-creative community engagement strategy that she undertook and documented in the described village. The third author, the book editor and an external reviewer provided valuable contributions to the chapter.

The creative practitioner was invited by the provincial government to work with village residents in order to facilitate a cultural expression of the place—in essence '*art-making as a way of knowing*' (Leavy, 2018, p. 4). The local project was one part of a wider programme of government-funded cultural projects across the province of Groningen in the Netherlands; commissioned to strengthen cultural infrastructure and social connectedness of residents to their places. This was in light of the provincial government's decision to centralize existing municipalities into larger but fewer entities. The residents within this particular village were also in the midst of dealing with the impact of energy transition developments as a result of government policy decisions, most notably the approved construction of a windpark within their village and immediate landscape. Throughout the time frame of the study, the residents were experiencing significant transformations externally imposed upon them that affected their daily lives and their immediate environment. As a result of these policy decisions, a local village protest group was formed in opposition to the windpark site construction. The decision of the regional government to commission and invite the village residents to work with a creative practitioner on a place-based cultural project of their village therefore re-raised questions noted earlier—who is doing the inviting and what are the motivations that underlie the invitation to participate?

While the specifics of this case are unique and we do not claim full generalizability, we aim to draw lessons that can support those participating in projects operating under similar conditions. The case described in this chapter illustrates what Neal (2015) describes as '*art of invitation*' principles in a community engagement strategy. We will argue that trust-building and fostering community agency are important in such a strategy, especially when working within a project under invited space conditions. Implementing an engagement strategy that addresses power imbalances and re-centres a project around community needs, can encourage project legacies that last beyond the timespan of a project and alleviate the criticism that practitioners and/or researchers parachute in and out of communities (Bastida et al., 2010).

The chapter continues as follows: the next section introduces the theoretical lens of invited spaces and the constraining consequences of an invitational space. We then explain the philosophy applied by the creative practitioner, Yanthe van Henk, addressing these constraints, and how her way of working reflects elements of participatory action research, notably the 'art of invitation' principles. The chapter then outlines the case and the research methodology. The results section describes and provides an analysis of how Yanthe engaged with the community in the initial stages of the project, nurturing the ground for creative art-based practice. The chapter closes with reflections on the findings and conclusions on the implications of the role of researchers and practitioners in externally commissioned place-based cultural projects.

The Concept of Invited Spaces

Invited Spaces as a Situational Constraint

In 2006, the political sociologist John Gaventa published a concept known as the power cube. The power cube is a conceptual framework designed to support the analysis of how communities interrelate with different levels, spaces and forms of power. This chapter focuses specifically on one component of the cube called the invited space.

Invited spaces are created where there is a request for community participation, involvement or consultation, usually from a particular governmental authority—typically a local, regional or national government (Gaventa, 2006). Spaces of this nature are designed as a governmental strategy with the view of strengthening the individual and collective agency of a population, by providing a participative opportunity for people to express their views, to potentially affect or influence future cultural, social and political discourses within their locality (Gaventa, 2006). The concept of an invited space helps us to understand the challenges that emerge for a creative practitioner when they are asked to facilitate a cultural process with a community by an external authority.

When a space is opened up by an authority and citizens are invited to participate, this can potentially present a channel where citizens can challenge dominant discourses, decisions, policies and relationships that affect their lives and interests (Gaventa, 2006). Within such an invited space however, as Gaventa critically explains, those in authority who govern the space also often shape the boundaries of these spaces of participation, what is possible within them, and which or what discourses and interests may enter the space. This results in restrictions of freedom for those participating, including how issues are framed and the limits over what can take place within these spaces (Gaventa, 2006). The notion of freedom within Gaventa's conceptual understanding of power is drawn from the work of Hayward, who defines freedom in a participative context as '*the capacity to participate effectively in shaping the social limits that define what is possible*' (Hayward, 1998, p. 2). Upon this interpretation free participation can be seen as not only the right to participate within a given space, but also the right to define and shape that space (Gaventa, 2006). This is closely aligned to the degree of agency a community is permitted within the invited space.

Typically, invited spaces are set up in more formal deliberative processes (e.g., community consultations) where citizens are consulted about specific projects or decisions. We would argue that place-based cultural projects commissioned by governments to achieve certain desired social outcomes, can also be viewed through the lens of invited spaces. These projects run the risk of mis-using culture whereby 'sense of place' interventions are applied in an instrumental, predetermined manner within externally set boundaries; curtailing what can be deliberated and achieved within the space rather than the framing of the space being co-designed with the community.

Regardless of whether the invited space is shaped in a more traditional community consultation format or through participative place-based projects, within both examples there is a risk that an invited space is created as a form of top-down steering, sponsored by authorities. As Aiyar (2010) argues, these initiatives inevitably contain power asymmetries whereby the invited space is somewhat bound by the norms of the state. This means that the purpose, mandate and remit of such spaces are circumscribed by the agendas of the implementing agencies, framing the perspective of the issues that surface and the grounds on which these issues can be debated (Cornwall, 2002). Furthermore, the authority monitors and holds a power of veto over project activities; therefore, the final say from any expressed desire still resides with the authoritative power (Gaventa, 2009). It can be said that the notion of an invited space—in this case where a place-based cultural project was commissioned—cannot be considered bottom-up and autonomous; but rather top-down designed, resulting in the emergence of paternalist characteristics. The concept of invited spaces and the skewed power dynamics that result from this type of community engagement can therefore be considered as a key constraint for practitioners working with communities co-creatively under these conditions.

To address these power imbalances inherent within projects commissioned under 'invited space' conditions, PAR approaches have been developed in the last decades to support the democratization of community engagement. PAR approaches have demonstrated the potential to disrupt existing power orthodoxies, by building alternative power bases from the bottom up, creating alliances between those involved in the process, resulting in more horizontal relationships (Anderson, 2017).

Participatory Action Research: An Overview

PAR was developed as a critical response to traditional, positivistic research approaches whereby the role and positionality of the researcher was one of a passive observer (Oakley & Marsden, 1985). Instead, PAR situates both the researcher and the community as active agents of the research process—conducting co-operative enquiry (Heron & Reason, 2006) with the objective of enabling structured transformation, often with the social justice aim(s) of changing the living conditions of people in a specific place (Kelly & van der Riet, 2001). PAR seeks to achieve this objective by integrating three basic principles: Participation (life in society and democracy), Action (engagement with experience and history), and Research (soundness in thought and the growth of knowledge) (Chevalier & Buckles, 2013).

PAR approaches within communities have become increasingly popular as a social science research method (Wilson et al., 2017) due, in part, to the potential of PAR to democratize the research process and empower communities for social change (Chevalier & Buckles,

2013; Hacker, 2013). PAR promotes democratizing elements that focus on a researcher's responsibility to the participants (Datta et al., 2014) rather than the relationship existing as '*hierarchical, vertical, dominating, and [potentially] exploitative*' (Kesby, 2005, p. 2051). In this vein, indigenous scholar Battiste (2008) suggests that the research should transfer power through the researcher's respect and accountability to the participants. The researcher should not only learn about the residents but also learn from them, exchanging experiences, understanding and empathizing with the value of their subjective experiences and ensuring active involvement and/or leadership roles in the conceptualization, design and implementation of the co-creative process.

This sphere of engagement created between a researcher and the community can be understood as a '*field of relationships built on mutual trust, and interests that, if not identical, converge around a certain set of activities where researchers' and participants' "respective paths cross or commingle"*' (Ingold, 2000, p. 145), highlighting the necessity for reciprocal engagement in co-creative practice to be effective. Co-creative practice within PAR can be defined as a joint or partnership-oriented creative approach between two or more parties, especially between an institution and constituents, towards achieving a desired outcome (van Westen & van Dijk, 2015). These co-creative processes with communities often involve emotionally labour-intensive relationship building (Facer & Enright, 2016), whereby researchers and practitioners are viscerally engaged in the 'messy realities' of other people's lives (Carter et al., 2013; Thomas-Hughes, 2018). While the term is sometimes used interchangeably with 'collaboration', co-creation is said to place a greater emphasis on process (van Westen & van Dijk, 2015; see also Franklin, this book). This process can assist in the democratization and empowerment process of PAR by valuing local knowledge, brokering connections, building trust and facilitating the emergence of collaborative problem solving and community leadership (Metz et al., 2019). However, to reach the point where effective arts-based co-creative practice is possible, it is necessary to first implement an appropriate community engagement strategy that is sensitive to the local context and addresses the constraint of being externally invited to co-create with the community. In the next section we will describe how these conditions can be created via the 'art of invitation'.

The Art of Invitation: Becoming a Voluntary Instrument of a Community

The 'art of invitation' can be seen as a PAR inspired approach. The term has been coined by Ruth Ben Tovim, Lucy Neal and Anne Marie Culhane, all artists in their own right, but also all members of Encounters, an organization established in 2003. Encounters was born from a community project whereby co-founders Ruth Ben-Tovim and Trish O'Shea took over a disused shop in Sharrow, Sheffield (UK), to open a creative dialogue about their rapidly changing local neighbourhood:

> *We see ourselves as bridge builders and space holders for an exchange to happen between a civic institution and its participants – many of whom might never have experienced or participated in a more formal art context before.* (Encounters A).

Encounters describes the art of invitation as follows: '*The Art of Invitation demonstrates how to engage different communities in becoming involved in creative projects; bringing people together to make social, cultural and ecological change happen*' (Encounters B). We draw here mainly on the work of Lucy Neal (2015), whose book: *Playing for time, making art as if the world mattered*, provides a rich overview of art-based practices of a wide group of creative practitioners. It is a handbook with narratives on collaboration and co-creation with communities. A key assumption is that '*art is in service of lives*', which enables holding one's gaze on the challenge of re-imagining the future (p. 14). Art-based engagement with communities can be considered as a place-based strategy, applied, for instance, in the transition town of Totnes, England (Hopkins, 2015), as a strategy of activism (Khan, 2015) and as a strategy to reclaim the (food) commons (Gordon-Fairleigh, 2015). Without making the claim to outline a formal methodology, key principles that outline conditions for effective community engagement for art-based activities have been described by Neal (2015) and are summarized and listed in Table 8.1.

Table 8.1 Ten principles of 'the art of invitation'

Principle	Description
Intention	The intention is to create conditions for change. This can feel open-ended, with no precise plan, but engaging in intent gives life to the choices we make about change. Intentions may have an impulse that is self-directed but contain a generosity towards others within this
Frame	A boundary of structure within which the freedom to play exists, creating the context or narrative for people to explore. Too large a frame can lack focus; too tight a one will not be expandable or open to change. It is a lens through which we can look at daily life anew. The lens can be changed as the project proceeds, opening choices up or closing them down
Work with community	A practice of community fosters self-awareness, empathy, vulnerability and realism to help us evolve. Art-based practices can bring people together that are not like-minded, and have different backgrounds and experiences
Facilitate	*Facil* means easy. Facilitation involves getting things to flow, overcoming obstacles such as scepticism. Some aspects of facilitation can be predicted and organized, but much will need improvizing and intuiting with a willingness to respond to what's needed, often without prior specialized knowledge
Hold space	Holding space focuses the energy of a group. The artists' presence can hold the space of a project when the setting is an unconventional one and especially at the arts, when there are lots of unknowns
Connect	Connecting makes space for coherent and holistic narratives of where we are now. When people feel a sense of interconnection, they transcend the limits of the individual worlds of 'I' and build a shared story that has meaning and can inspire the wider community with a sense of what can happen in between. Serendipities spring from such connections between ourselves, the visible and invisible, the past, present and future, etc

(continued)

Table 8.1 (continued)

Principle	Description
Work from commonality	This requires paying attention to what people have in common; their humanity and common values. It involves asking good questions and taking care to listen actively. It recognizes and includes people and makes multiple ways from them to enter and re-enter a project. Empathy lies at the heart of working with commonality
Collaborate	Collaboration needs humour, open-heartedness and negotiation, all of which are part of its creative dynamic. It is a transformative, complex human process and requires a surrender of some control, but not of rigor and care. It requires an openness to working with others to allow creative journeys to be co-created
Change	People's perspectives and sense of capabilities can change. Doors to new possibilities open that might have stayed closed which people jump through into new stories

Source Neal 2015, pp. 81–93)

We consider the work of the creative practitioner Yanthe Van Nek within our own case as an example of a PAR-inspired approach, where she interpreted and applied these principles in her own way.

Yanthe is a community artist based in the North of the Netherlands and describes her creative interest in place-based cultural projects as one that resides within chaos—in the space between perfection and destruction. Her philosophy views participatory art as more than just having pleasant conversations. Art, research and the social context must coincide. Therefore an essential part of all of Yanthe's projects is entering and engaging deeply into each other's worlds, building relationships with local people and understanding the dynamics of the place she is working within. This can result in an emergent collective wish or desire for change that then invites her to continue her work with the community on more tangible, visible arts-based activities, having nurtured a degree of trust and consent through engagement with the community. Yanthe views cultural practice as a method to expose structural conflicts, mobilize communities and shine light on alternative pathways rather than merely community consultation through art (Pritchard, 2017).

Yanthe addresses the challenge that comes with working within invited spaces through her strategy of becoming '*a voluntary instrument of the community*'. To explain this we can use a musical instrument as a metaphor. Yanthe invites the community to 'play' her as if she is an instrument in whichever way they choose. The terms of engagement as to how the community participates in the project are therefore open-ended rather than prescribed. Yanthe is capable of offering a rich repertoire of arts-based skills that they can choose from, but the community—acting as the musician—decides how the instrument is played. Only the community can play Yanthe to a tune of their choosing, external actors (e.g., project administrators) existing outside the co-creative space are not permitted to 'play' Yanthe.

Yanthe's philosophy of co-creation posits the initial engagement process as crucial, immersing herself in both the social and ecological environment of the community. If appropriate and safe to do so, it begins by physically moving into and living within the community space and thereafter conducting an investigation of the desires of community members through the cultivation of meaningful relationships with the residents. This immersive investigation can be guided by community members, moving organically from person to person, picking up information about residents that informed her decisions as to who she should speak to next, including who is marginalized or not spoken about. This provides valuable opportunities to understand local power dynamics—including the dynamics within the community and between the community and the governing authority. The wishes that exist within a community can be explored and can be creatively conceptualized into a collective wish or shared story. This is done through holding the space with individuals for significant periods of time; initiating conversations, listening actively and probing with questions that demonstrate authentic interest. To assist in this investigative process, Yanthe carries around a large book during her conversations, known as '*The Wish Book*'—a private book—not shared with any external actor, in which she collects the local residents' inner wishes for the future of the community. She makes it clear that whatever collective desire/wish emerges and can be agreed upon, she will utilize her artistic expertise to co-create a cultural expression of their place.

Methodology

Research Methods

Empirical evidence for the illustrative case was collected using a mixed-methods approach to qualitative social research. The focus of the data discussed in this chapter was on how Yanthe applied PAR principles, when first engaging with a community in an invited space setting. A combination of six semi-structured and reflective interviews, unstructured walking conversations and a phone interview were conducted with Yanthe over a four month period during Spring 2020. The choice to use observations and interviews as methods of data collection followed from the aim to get a deeper insight into, and understanding of, the practitioner's experiences within the village; these would be much harder to access through more structured data collection methods such as a series of questionnaires (Gillham, 2000; Ritchie & Lewis, 2003).

While spending time with Yanthe in the village, observations were gathered on how she implemented the first stage of her engagement strategy with the community to nurture the ground for arts-based co-creative practice. The data centred around Yanthe's time setting-up a space for co-creation in a public venue within the centre of the village. The interviews were conducted at the beginning, throughout, and the end of this initial engagement process. The semi-structured interviews and the phone interview were designed to explore Yanthe's strategy of community engagement, with a focus on power dynamics between those who administer the project, her own role and the community. The walking conversations were unstructured in order to encourage Yanthe to freely reflect on her experiences within the village.

The original research plan was to conduct multiple rounds of interviews with the village residents to ensure triangulation and incorporate perceptions of community members on the village developments and co-creative practice. Unfortunately, due to the advent of the COVID-19 pandemic and subsequent access restrictions implemented by the Dutch government, data collection on-site ceased in May 2020. Instead, the opportunity was taken to deepen the collaboration *with* Yanthe. Specifically, Yanthe prepared a reflective transcript that summarized her time in the village, and reflected on our conversations and interviews. As well as offering further rich insight, this transcript also helped to guide the analysis. Quotations from this transcript have been included within the results section alongside relevant quotes from interviews to provide context to the analysis. While the analysis of the results remained the responsibility of Davis (lead author) throughout, Yanthe provided valuable contributions to various drafts of the chapter and was active in the editing process to ensure that her philosophy and strategy were accurately described.

Context of the Case

The chapter employs a single case study approach. This section first provides a description of the illustrative case followed by justification of its suitability for this chapter. The place-based cultural project was part of a wider government-sponsored programme to support the cultural infrastructure and social cohesiveness of the province. The situation in the village was one of additional importance to the provincial authorities, the village having accrued a recent reputation as a place with strong local opposition to their decision to construct a windpark within their surrounding landscape. The windpark was planned as part of a broader national energy transition agenda delivered across the northern Netherlands, and resulted in areas of local resistance in various parts of the province opposing the implementation of energy transition policies. In this village a local action group protested for over seven years in opposition to the windpark. The resistance was triggered by the top-down decision-making process of the provincial government, most notably regarding the large scale of the park, the height of the windmills, and the close distance to the residential area. Community resistance movements like this can be unfairly characterized as NIMBY responses by government actors (Not-In-My-Backyard), but can be alternatively understood as rational concern for the future of their place, stemming from a strong local sense of place and serious concerns about local identity (Devine-Wright & Clayton, 2010). In this village, the windpark enacted feelings of distrust and volatility towards the provincial government. The protest group acted not just against the provincial deputy in charge, but also against farmers willing to sell their land to the government, therefore enabling the park and further affecting social cohesion within the community.

Illustrative cases are employed to shed light on a particular situation or set of circumstances where social relationships and processes are embedded within them. The overarching purpose of an illustrative case is to address an audience that may not yet be greatly informed about the topic and can offer understandable insights without oversimplification. Therefore, this approach is congruent with our goal of communicating lessons that can be drawn from the creative practitioner's experience to help inform those interested in embarking on an arts-based creative practice journey for the first time.

We regard this as a suitable illustrative case because it is a solid representation of how place-based cultural projects are often externally commissioned in the Netherlands. This project was assigned by a provincial government who opened up the participative space. The same authority supported a series of top-down sustainability and resilience policy decisions that resulted in dis-trust and local opposition. This makes it an interesting and unique case that deserves investigation as to how a creative practitioner can engage with a village from a power relations and trust-building perspective. While we aknowledge that results may not be generalized to other contexts, lessons can nevertheless be derived from this case, for future place-based cultural projects.

Outline of the Engagement Process

Yanthe's roots lie in the same province as the community village, which was advantageous in respect to having a familiarity and connection to the local context. While she resided in the same general locality, she was not 'local' to the specific community. That is to say that she was initially unfamiliar with the community and the dynamics that existed within it upon being invited to work within the space. However, this was also advantageous in a number of ways. It allowed her freedom to enter into the community not bound by existing prejudices or unwritten laws, power dynamics and social norms that consciously or subconsciously determine the actions/behaviours of individuals within the community. Instead, with being a relative outsider, a valuable, anthropological neutrality was present.

As stated, the goal of the initial engagement phase was to nurture the ground for effective arts-based co-creative practice to occur. Yanthe did this by embedding herself within the village and building meaningful relationships with village residents so as to understand and conceptualize the wishes of the community. Although the creative practitioner's strategy was not consciously or explicitly pre-designed based on PAR in a formal sense, she applied methods that closely aligned with the PAR principles that exist within Neal's (2015) 'art of invitation'. She entered the community with an open-ended intention, inductively facilitated the framing of the direction of the project around community desires, and did so by spending time fostering connections that could support transformational change. This helped to democratize the process, resulting in the direction and content of the project resting largely in the hands of the community. Yanthe and the residents engaged co-creatively as explorers, rather than by determining specific outputs beforehand (Reason & Bradbury, 2008). These explorations took place through living with the community, holding the space in the community for conversations by setting up open-surgeries for residents in the village centre and engaging in local community practices in order to nurture fertile ground for co-creative practice to take place. Unfortunately, due to the onset of the COVID-19 pandemic and subsequent social distancing restrictions, the project was temporarily suspended shortly after this engagement phase, as Yanthe was no longer able to physically hold space within the village due to national COVID social distancing/household regulations.

Results and Reflections

What lessons can be learnt from the initial engagement process with a community to nurture ground for co-creative practice? The next sub-section reflects on how the creative practitioner dealt with the constraints of working within an invited space. The following sub-sections describe how this constraint was reversed via practicing 'art of invitation' principles and by the practitioner's method of becoming a voluntary instrument of the community. This resulted in the community inviting the creative practitioner to continue the project on their own terms. The last sub-section briefly reflects on time, as a particular constraint that Yanthe could not overcome through the art of invitation.

Power Dynamics and Community Agency Within the Invited Space

Upon accepting the invitation and entering the village, Yanthe experienced multiple instances of suspicion and cynicism towards the project from a proportion of the village residents. This was predominantly due to opposition within the community to the government's approval for the construction of the windpark, with residents recognizing that the cultural project was administered by the same governmental authority. This resulted in some residents initially rejecting the invitation into the participative space and by consequence rejecting the practitioner. Yanthe articulated:

> *It was not only the arrival of the windpark, but especially the way in which this change was delivered; it brought an accumulation of unrest within this old village. Feelings of skepticism and a lack of trust from the community were therefore also initially directed towards the project.*

This quote highlights how distrust towards external governmental authorities was directed towards the project, not only because of the windpark decision in isolation, but because the community felt they had no real influence or say on the decision-making process itself. By understanding this context, the creative practitioner was able to empathize with the initial suspicion and cynicism towards a government administered project to facilitate a cultural expression of their place when they had thus far felt unheard in other participative arenas.

Lessons can be learnt as to why, when entering a community, a creative practitioner or researcher should understand the existing relationship between relevant governmental authorities and the community when the project is funded or sponsored by the state. Such knowledge can inform the community engagement strategy and help to understand resident behaviours that may then be exhibited towards the creative practitioner or researcher upon arrival into the space. If the practitioner is viewed as a representative of the state, this will likely affect the level of trust and time needed to build relationships within the project. It is therefore necessary to fully grasp and empathize with the nature of this distrust by putting into practice the 'art of invitation' principle of working with the community. This requires continuous reflexivity on the practitioner's positionality as a facilitator within the broader social, cultural and political context of the project (see also Horlings et al., 2020).

Having moved into and then begun living within the space, it was necessary for Yanthe to quickly become acquainted with the power dynamics at play and it became necessary for her to impress on the village residents that although her work was funded by an external authority, she was there with the purpose of becoming a voluntary instrument of the community rather than as a commissioned instrument of the state.

Yanthe noted that the governmental authorities were greatly aware of the significant unrest that existed regarding the recent spatial transformations implemented in the village, most notably the windpark construction. She explained that those running the cultural project had expressed concerns to her about the potential for disruptive consequences to occur throughout its duration, due to the current situation. While it could be said that the project was broadly commissioned in the spirit of fostering social cohesiveness and building a cultural infrastructure, the governmental authorities were still concerned with how it would affect the local resistance towards the windpark construction and therefore sought to monitor the co-creative process. This speaks to Gaventa's Hayward-inspired critique of invited spaces whereby the framing and limits of what is possible may be at risk of being decided and/or influenced by the external authority in charge of the project rather than by the community. It was therefore important for Yanthe to assert her position as the community engagement specialist to the authorities.

Yanthe made it clear to the government officials overseeing the project that she had been commissioned on the basis that she could implement her methodology of becoming a voluntary instrument of the community, and that it was not possible for her to carry out her methodological process and the subsequent arts-based activities if there was significant external interference. From the beginning of the project, governmental officials had gently probed Yanthe in order to learn about 'what was going on' within the village. This was further demonstrated when one of the public officials suggested they should be present at meetings between the community and Yanthe. Yanthe checked with the residents whether this was appropriate and then communicated that they would not consent to this. This guaranteed that the space being held and facilitated by Yanthe was a private, safe one, creating potential for deeper connections through a demonstration of loyalty to the community, supporting the process of trust-building and community engagement. Participation in practice is '*rarely a seamless process; rather, it constitutes a terrain of contestation, in which relations of power between different actors, each with their own "projects", shape and reshape the boundaries of action*' (Cornwall, 2008, p. 276). Yanthe navigated the terrain of contestation and mitigated external interference, securing greater community agency over the project process. She also made it clear to those administering the project from the outset that if residents wanted to use her as an instrument to protest against the arrival of the windmills, that she would facilitate this:

> *By communicating my chosen position clearly and the position of the participants from the first stages of the project to the funder, it helped in dealing with possible resistance and change of expectations of my client when the project starts naturally evolving in response to the wishes of the community… Due to me outlining clearly the terms of engagement, the local government promised to take the backseat and trust the process. It was very important to emphasise to them the importance of co-ownership and creative freedom of the creators. In my work it is always too easy for me to just push the funders'*

wishes through. After all, in my vocation, I build a relationship of trust [with communities] that is easy to abuse.

The quote illustrates that when a practitioner or researcher aims to apply co-creative arts-based methods it is important to ensure that their preferred community engagement methods are both understood and accepted from the outset by those responsible for initiating an invitational space within a community.

Through building connections with the residents, a shared feeling began to emerge that the residents would rather use the space to create an arts-based symbol of togetherness rather than to further any resistance movements to the windpark construction, disproving concerns from the authorities that the project would be utilized '*disruptively*'. Yanthe describes the benefits of a community being given the opportunity to frame the content and project direction:

The community becomes co-creators and co-owners of the project… there are benefits for the funder in relinquishing control within these spaces as the funder gains genuine insights into the wishes, questions, wants and foremost the power and wisdom of the village itself. Instead of seeing a deficit and using culture and 'a creative' to fix it, they can instead trust the process of letting me submerge within a community.

This quote underpins our plea towards governmental authorities, to acknowledge the benefits of relinquishing control over (the framing of) a project to the community.

Building Connections via the Art of Invitation

Yanthe applied a series of 'art of invitation' principles through her engagement process.

She described the beginning of her process as conducting an investigation by allowing herself to be led by the community:

I literally let myself be taken away and carried away by the residents, it was they who let me know what they wanted from me as an artist and what any artwork that emerged from this process should bring about. I was a voluntary tool for the inhabitants of village… The residents of the village took me from meeting to meeting. Out on the street and into the houses. I immersed myself in the village and started an intense investigation with all my senses on edge. My days were full of encounters, from the billiard club, the knitting club, boxing, singing, the library, the village school, the village corporation, the party committee, the Youth Box… I spoke with the oldest generation, the parents and the youth. Wherever they took me, I followed every reference.

The essential element of all my projects is to be a part of each other's world. Entering into a relationship, I want to get to know the dynamics of the place where I am going. I found a place to stay at one of the farmers, "wind farmers" as fellow village people call them. I found my workplace at a local car garage, offered to me by one of the village people I had met…From fruitful conversations a widely supported wish emerged. Central themes came to the table such as a feeling from the villagers of having to choose between supporting and opposing the windmills, lamenting the loss of togetherness and a need to reaffirm and perpetuate the feeling of belonging with the help of the existing structures around the rich community life in the village.

Yanthe chose to live and share her evening meals with a local resident, illustrating the art of invitation principle of connecting. This was a microcosm of what Yanthe was trying to achieve across the community as whole. She created a series of connections across the village where a shared story or wish would eventually emerge, building interconnections that transcend the limits of the individual world of 'I' and result in a shared story (Neal, 2015). Yanthe helped this process by ensuring she stayed within the public gaze, 'holding space' in the centre of the village at a local automotive garage. If she was not found behind a table in the garage foyer, she would be seen walking the streets, at people's doors or engaging in local community practices. Always available to actively listen, collaborate and work with the community in their own time.

The Emergence of Shared Stories and Wishes via the Art of Invitation

A shared narrative that emerged from individual community desires and was conceptualized into a communal wish, was one of music and festivals. Through her investigation and connections with individual residents, a shared story emerged of the village as how it was once an important part of the blues music scene in the Netherlands. This previous identity of the village had now been largely forgotten and replaced as a village with a reputation for windpark protests. The communal wish was therefore to create an arts-based symbol of the community that paid homage to this musical and festival spirit of the past. Doing so, it was hoped, would foster a renewed sense of community, belonging and togetherness.

Yanthe noted that while there were people passionately for and against the windpark construction, many residents were also broadly neutral on the issue and had more interest in protecting the community from fracturing permanently across this ideological fault line. She reflected:

> *In all my history, I've never felt such a strong desire for a sense of community from a group of village residents.*

While solidarity and resilience within the village was clearly demonstrated through the protests, the deep wishes of the community also highlighted the pressures that the windpark construction decision and resultant protests were placing on the community, exposing potential fragilities. Hence the desire emerged to rekindle a longing for reunification and togetherness and alter the reputation of the village with a symbol to demonstrate that they are not only a village of protest.

Through nurturing connections with the residents and noting their wishes, Yanthe therefore conceptualized a broader shared story that could frame the arts-based co-creative practice—the residents creating their own boundary or structure with which freedom to play can exist (Neal, 2015) while protecting the freedom of participation (Hayward, 1998). In this instance the communal wish was to create a new symbol of togetherness for the village through their shared place-based history of music and festival. The process of arts-based co-creative practice began by a local resident expressing this wish at a community gathering through the recital of a poem in the local dialect (see below).

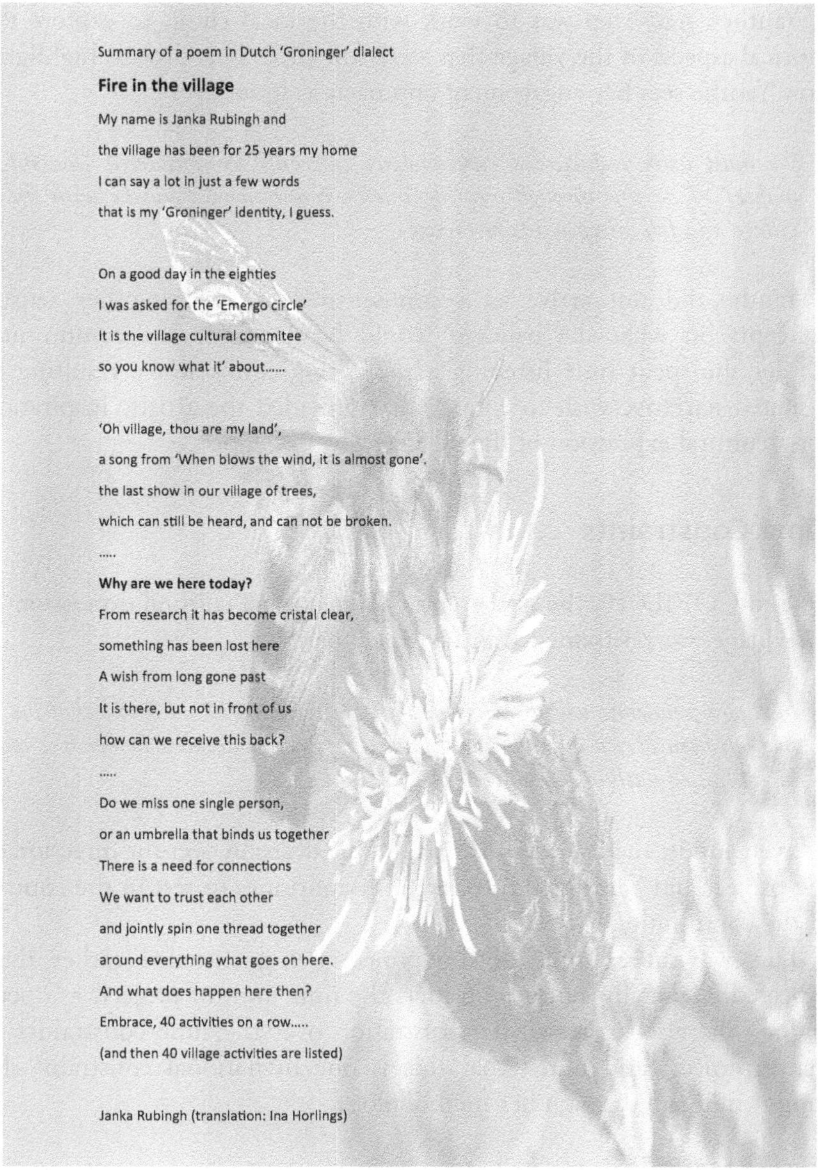

Summary of a poem in Dutch 'Groninger' dialect

Fire in the village

My name is Janka Rubingh and
the village has been for 25 years my home
I can say a lot in just a few words
that is my 'Groninger' identity, I guess.

On a good day in the eighties
I was asked for the 'Emergo circle'
It is the village cultural commitee
so you know what it' about......

'Oh village, thou are my land',
a song from 'When blows the wind, it is almost gone'.
the last show in our village of trees,
which can still be heard, and can not be broken.
.....

Why are we here today?

From research it has become cristal clear,
something has been lost here
A wish from long gone past
It is there, but not in front of us
how can we receive this back?
.....

Do we miss one single person,
or an umbrella that binds us together
There is a need for connections
We want to trust each other
and jointly spin one thread together
around everything what goes on here.
And what does happen here then?
Embrace, 40 activities on a row.....
(and then 40 village activities are listed)

Janka Rubingh (translation: Ina Horlings)

Yanthe's next step was to work with the local choir, to explore the musical aspects of the village that were still alive. This process highlights how Yanthe sees her engagement approach as inductive:

We must work inductively, otherwise we can only be considered "content makers". I do this through creating intense connections. I awake with the village and fall asleep with the village

Yanthe did not arrive as a content maker, imposing her artistic concepts on what she believed would be good for the community. Rather she spent time listening and creating connections, resulting in a shared narrative/wish to emerge that provided the artistic inspiration for a cultural expression of the village to be developed.

Time Constraints

Due to COVID-19, the next step of co-creating a cultural expression of the village was postponed. Yanthe explained that:

In a post-pandemic world it may be that the residents let me know that my role has been played out. They will have the power to decide whether it is to be continued with or without me

Even amidst the project's suspension, the decision over the direction of the project and Yanthe's future role in it continues to rest in the control of the community.

This was an example of the project being curtailed earlier than expected. Generally Yanthe considers the time limit of projects as open-ended. She also expressed her frustration over the time constraints of many projects and noted that this is one invitational constraint she cannot mitigate through her methodology:

The funders expect that Friday is my last day and I say goodbye. You raise hopes for the wish, you set things into motion and then due to the constraints of the project you have to leave. I give them a platform to share their wishes... so I will not leave until I know they can do it on their own.

In this case, Yanthe stayed in the village by searching and then securing external funding from other sources in order to continue her work with the community after the social distancing measures of COVID-19 were lifted.

Time constraints and juggling pressures are familiar issues with which participatory action researchers also routinely contend. Researchers may not be able to stay beyond an initial agreed project time-scale due to their commitments and demands placed upon them as employees of university institutions. This means that the risk of a '*parachute*' effect remains somewhat for social researchers entering communities, resulting in the build-up of relationships and then leaving before the community is ready for the project to end.

In order to enable mutual learning and to learn from the voices and experiences of community members, projects require sufficient time in a community to maintain an open-ended, exploratory engagement with community members. However, there is rarely the political will to provide significant funding for longer term community projects that can facilitate this.

Palmer et al. (2017) advocates the necessity of time, specifically the virtue of a practitioner or researcher 'waiting' within these types of projects, describing this as an important component of community-based and ethnographic research approaches. This waiting can involve days of drifting and 'nondirective discovery' (Okely, 2012) with discoveries from these open explorations requiring protracted periods of time (Atkinson et al., 2001; Hammersley & Atkinson, 2007).

Summing Up

A major theme that emerged from the creative practitioner's engagement strategy was one of relinquishment of control over the direction of the project to the community members. This strategy promoted an authentically inductive approach and coincided with one of the key beneficial tenets of PAR—the democratization of the research process. The creative practitioner adjusted the pre-existing paternalistic power dynamic by devolving power bestowed on her by the external authority directly to the community; offering herself voluntarily as an instrument to be used by the community within the invited space. By relinquishing power within the space to the community, this allowed the project objectives to emerge inductively from the residents' wishes. The creative practitioner recognized that being invited into the space by the governmental authority rather than by the community was the first major challenge that needed to be overcome in order to nurture the ground for a tangible cultural expression to take place through arts-based activities.

The artist countered the paternalistic nature of the invitational space by entering into a deep inductive process, spending time living within the village to develop unconditional relationships with the residents. This process reduced suspicion and cynicism of the community members towards the space and instead increased trust towards her motives.

Relinquishing power over the framing and the direction of the project to the community resulted in an emergent wish to use the project to foster togetherness rather than as a vehicle to further protest against the windpark's construction. While still holding strong views on this issue, the project was treated as a reprieve from the continual protests.

Conclusions

In this chapter we reflected on an illustrative case in the Netherlands to answer the question how a creative practitioner can apply co-creative practice in a community when entering into an invited space, while dealing with situational constraints. We also discussed how best to respond to the wishes of communities on their own terms, rather than to external interests.

Through the conceptual len of an invited space we identified constraints that contributed to the paternalistic manner in which place-based projects are run by governmental authorities, such as the prescribed framing of the project and attempts of external influence over what can be deliberated within the space.

When embarking on arts-based co-creative practice, it is therefore imperative to recognize the power dynamics often inherent within such spaces. 'Bottom-up' cultural projects ideally facilitate spaces where communities are free to navigate not only consensus-driven themes but also conflicts and existing asymmetrical social relationships:

> *...where in times of political turmoil, social unrest, chronic housing crises and public sector decimation, culture should be utilised as a way to forge new alliances, social movements and collectives to push against these ills. Culture is not something that should be utilised to maintain the status quo, rather it is an instrument for change* (Mould, 2017).

The case discussed in this chapter illustrates how a creative practitioner utilized 'art of invitation' PAR principles to reverse the notion of being invited into the village by an external authority, and instead received consent or an 'invitation' from the community to continue the project on their own terms. The creative practitioner became a 'voluntary instrument of the community', by engaging in an open-ended, inductive process, therefore building trust, connections and collaboration, and encouraging a shared story or communal wish to emerge. As a result, it can be said that the constraints of the invited space were 'flipped on their head' into a community invitation, where the village residents asked her to co-creatively construct a joint arts-based symbol for the village building on the past memories of music festivals that had, in the past, been facilitated by a prominent community member.

The described case provides lessons for those interested in embarking upon an arts-based creative practice journey within a community. These lessons include, first an awareness that entering a village with preconceived assumptions of what is best for the community can create resistance towards connecting and building the relationships needed for an effective project with a lasting legacy. Second, in order to explore the wishes of a community instead of externally pre-described outputs or impacts, the principles of the 'art of invitation' offer an added value to place-based cultural projects. These principles should not be applied as an instrumental tool box but should be used to guide an inductive process. Third, the case showed that through the practitioner becoming a voluntary instrument of the community, an authentically inductive approach can be established that can build trust and secure increased community agency that help to mitigate external interference within the participatory space. Fourth, time constraints remain a challenge for government-administered projects.

This chapter argues that when place-based cultural projects take place under invited space conditions, the relationship between researcher/practitioner, the community and the external authority can be viewed as a ménage à trois, with the external authority learning to 'take a back seat' as the project progresses. This is especially relevant for governments in situations where communities might have different ideas or interests about the future of their cultural, physical or environment assets; therefore the projects are not compelled to be framed within these dominant discourses and policy interests. We suggest that practices of '*dissensus*' should be welcomed within participatory governance initiatives (Anderson et al., 2016; Kaika, 2017) and that those who work on arts-based creative practice and PAR projects continue to encourage governmental institutions to further democratize participatory spaces. Even when faced with navigating invitational spaces, communities can still then represent and frame their views within co-creative projects, with tangible opportunities to materially influence the future of their place.

Acknowledgements This research received funding from the European Union's Horizon 2020 research and innovation programme under the Marie Sklowdoska-Curie Grant Agreement No. 765389.

References

Aiyar, Y. (2010). Invited spaces, invited participation: Effects of greater participation on accountability in service delivery. *India Review, 9*(2), 204–229.

Anderson, G. (2017). Participatory action research (PAR) as democratic disruption: New public management and educational research in schools and universities. *International Journal of Qualitative Studies in Education, 30*(5), 432–449.

Anderson, M., Hall, D., McEvoy, J., Gilbertz, S., Ward, L., & Rode, A. (2016). Defending dissensus: Participatory governance and the politics of water measurement in Montana's Yellowstone River Basin. *Environmental Politics, 26*(5).

Atkinson, P., Coffey, A., Delamont, S., Lofland, J., & Lofland, L., et al. (2001). Handbook of ethnography: Editorial introduction. In P. Atkinson, A. Coffey, & S. Delamont (Eds.), *Handbook of ethnography* (pp. 1–8). Sage.

Barca, F. (2009). *An agenda for reformed cohesion policy: A place-based approach to meeting European Union challenges and expectations*. Independent report prepared at the request of D. Hübner, Commissioner for Regional Policy, Brussels.

Bastida, E., Tseng, T. S., McKeever, C., & Leonard, J. (2010). Ethics and community-based participatory research: Perspectives from the field. *Health Promotion Practice, 11*, 16–20.

Battiste, M. (2008). Research ethics for protecting Indigenous knowledge and heritage: Institutional and researcher responsibilities. In N. K. Denzin, Y. S. Lincoln, & L. Tuhiwai Smith (Eds.), *Handbook of critical and Indigenous methodologies* (pp. 497–509). Sage.

Carter, K., Banks, S., Armstrong, A., Kindon, S., & Burkett, I. (2013). Issues of disclosure and intrusion: Ethical challenges for a community researcher. *Ethics and Social Welfare, 7*, 92–100.

Chevalier, J. M., & Buckles, D. (2013). *Participatory action research: Theory and methods for engaged inquiry*. Abingdon-on-Thames.

Cornwall, A. (2002). *Making spaces, changing places: Situating participation in development* (Working paper series, 170). Brighton: IDS.

Cornwall, A. (2008). Unpacking 'participation': Models, meanings and practices. *Community Development Journal, 43*(3, 269–283.

Datta, R., Khyang, N. U., Khyang, H. K. P., Kheyang, H. A. P., Khyang, M. C., & M. C. K., & Chapola, J. . (2014). Participatory action research and researcher's responsibilities: An experience with an Indigenous community. *International Journal of Social Research Methodology, 18*(6), 581–599.

de Andrade, M. (2016). Tackling health inequalities through asset-based approaches, co-production and empowerment: Ticking consultation boxes or meaningful engagement with diverse, disadvantaged communities? *Journal of Poverty and Social Justice, 24*(2), 127–141.

Denzin, N. K., & Lincoln, Y. S. (Eds.). (2005). *The Sage handbook of qualitative research* (3rd ed.). Sage.

Dessein, J., Soini, K., Fairclough, G., & Horlings, L. (2015). Culture in, for and as sustainable development. *Conclusions from the COST ACTION IS1007*.

Devine-Wright, P., & Clayton, S. (2010). Introduction to the special issue: Place, identity and environmental behaviour. *Journal of Environmental Psychology, 30*, 267–270.

Duxbury, N., & Jeannotte, E. (2007). *Culture as a key dimension of sustainability: Exploring concepts, themes, and models* (Working Paper 1). Creative City network of Canada; Centre of Expertise on Culture and Communities: Vancouver.

Encounters A. https://civicroleartsinquiry.gulbenkian.org.uk/resources/encounters.

Encounters B. https://civicroleartsinquiry.gulbenkian.org.uk/resources/encounters.

Facer, K., & Enright, B. (2016). Creating living knowledge: The connected communities programme, community-university partnerships and the participatory turn in the production of knowledge. *Arts and Humanities Research Council*. https://connectedcommunities.org/index.php/creating-living-knowledge-report/.

Gaventa, J. (2006). Finding the spaces for change: A power analysis. *IDS Bulletin, 37*(6), 23–33.

Gaventa, J. (2009). Power pack: Understanding power for social change. http://www.powercube.net/wpcontent/uploads/2010/01/PowerPack_web_version.pdf.

Gillham, B. (2000). *Case study research methods*. Continuum.

Gordon-Fairleigh, J. (2015). Reclaiming the commons. In L. Neal (Ed.), *Playing for time: Making art as if the world mattered*. Oberon Books Ltd.

Grenni, S., Horlings, L. G., & Soini, K. (2020). Linking spatial planning and place branding strategies through cultural narratives in places. *European Planning Studies, 28*(7), 1355–1374.

Hacker, K. (2013). *Community-based participatory research*. Sage.

Hammersley M., & Atkinson P. (2007). *Ethnography: Principles in practice*. Routledge.

Hawkes, J. (2001). *The fourth pillar of sustainability: Culture's essential role in public planning*. Common Ground Publishing Pty Ltd.

Hayward, C. R. (1998). De-facing power. *Polity, 31*(1), 1–22.

Heron, J., & Reason, P. (2006). The practice of co-operative inquiry: Research "with" rather than "on" people. In P. Reason & H. Bradbury (Eds.), *Handbook of action research* (pp. 144–154). Sage.

Hopkins, R. (2015). Relocalisation and the transition movement. In L. Neal (Ed.), *Playing for time: Making art as if the world mattered*. Oberon Books Ltd.

Horlings, L. (2017). The role of artists and researchers in sustainable place-shaping. In S. Asikainen, C. Brites, K. Plebańczyk, L. R. Mijatović & K. Soini (Eds.), *Culture in sustainability: Towards a transdisciplinary approach* (Vol. 139, pp. 130–143). University of Jyväskylä.

Horlings, L. G., Nieto-Romero, M., & Pisters, S. (2020). "Operationalising transformative sustainability science through place-based research: The role of researchers. *Sustainability Science, 15*, 467–484.

Ingold, T. (2000). *The perception of the environment: Essays on livelihood, dwellong and skill*. Routledge.

Kaika, M. (2017). 'Don't call me resilient again!': The new urban agenda as immunology … or … what happens when communities refuse to be vaccinated with 'smart cities' and indicators. *Environment and Urbanization, 29*, 89–102.

Kelly, K., & van der Riet, M. (2001). Participatory research in community settings: Processes, methods and challenges. In M. Seedat, N. Duncan, & S. Lazarus (Eds.), *Community psychology: Theory, method and practice. South African and other perspectives* (pp. 159–188). Oxford University Press.

Kesby, M. (2005). Re-theorising empowerment-through-participation as a performance in space: Beyond tyranny to transformation. *Signs: Journal of Women in Culture and Society, 30*, 2037–2065.

Khan, R. (2015). *Art in community: The provisional citizen*. Palgrave McMillan.

Leavy, P. (2018). Introduction to arts-based research. In P. Leavy (Ed.), *Handbook of arts-based research* (pp. 3–21). Guilford Press.

MacKinnon, D. (2002). Rural governance and local involvement: Assessing state—Community relations in the Scottish highlands. *Journal of Rural Studies, 18*(3), 307–324.

Metz, A., Boaz, A., & Robert, G. (2019). Co-creative approaches to knowledge production: What next for bridging the research to practice gap? *Evidence & Policy, 1*(7).

Mouffe, C. (2013). *Agonistics: Thinking the world politically*. Verso.

Mould, O. (2017). *Why culture competitions and 'artwashing' drive urban inequality*. Open Democracy. https://www.opendemocracy.net/en/opendemocracyuk/why-culture-competitions-and-artwashing-drive-urban-inequality/.

Neal, L. (2015). *Playing for time: Making art as if the world mattered*. Oberon Books Ltd.

Noorani, T., Blencowe, C., & Brigstocke, J. (2013). *Problems of participation: Reflections on democracy, authority, and the struggle for common life*. ARN Press.

Nunn, C. (2020). The participatory arts-based research project as an exceptional sphere of belonging. *Qualitative Research.*

Oakley, P., & Marsden, D. (1985). *Approaches to participation in rural development*. ILO.

Okely, J. (2012). *Anthropological practice: Fieldwork and the ethnographic method*. Berg.

Palmer, J., Burton, L., & Walsh, A. (2020). Emerging spheres of engagement: The role of trust and care in community–university research. *Qualitative Research, 20*, 749–766.

Palmer, J., Pocock, C., & Burton, L. (2017). Waiting, power and time in ethnographic and community based research. *Qualitative Research, 18*(4), 416–432.

Pritchard, S. (2017). *Artwashing: Social capital & anti-gentrification activism.* Colouring in Culture. https://colouringinculture.org/uncategorized/artwashingsocialcapitalantigentrification.

Reason, P., & Bradbury, H. (2008). *The Sage handbook of action research* (2nd ed.). Sage.

Ritchie, J., & Lewis, J. (2003). *Qualitative research practice: A guide for social science students and researchers.* Sage

Shortridge, J. R. (2005). Regional image and sense of place in Kansas. *Kansas History: A Journal of the Central Plains, 28*(Autumn), 202–219.

Svensson, E. (2012). Achieving sustainable lifestyles? socio-cultural dispositions, collective action and material culture as problems and possibilities. *Local Environment, 17*(3), 269–286.

Thomas-Hughes, H (2018). Ethical 'mess' in co-produced research: reflections from a U.K.-based case study. *International Journal of Social Research Methodology, 21*(2), 231–242.

Throsby, D. (2008). *Culture in sustainable development: Insights for the future implementation of art. Convention on the protection and promotion of the diversity of cultural expressions.* Unesco.

United Cities and Local Governments (UCLG). (2008). *Agenda 21 for culture*. http://www.agenda21culture.net/sites/default/files/files/documents/multi/ag21_en.pdf.

van Westen, R., & van Dijk, D. (2015). *European Union: RICHES—Good practices and methods for co-creation.* https://resources.riches-project.eu/wp-

content/uploads/2015/12/RICHES-D4-2-Good-practices-and-methods-for-co-creation_public.pdf.
Williams, R. (1980). *Problems in materialism and culture*. Verso.
Wilson, E., Kenny, A., & Dickson-Swift, V. (2017). Ethical challenges in community-based participatory research: A scoping review. *Qualitative Health Research, 28*(2), 189.
Yamini, A. (2010). Invited spaces, invited participation: Effects of greater participation on accountability in service delivery. *India Review, 9*(2), 204–229.
Yanthe Van Nek. (2020). https://www.yanthe.com/.

9

Reflections on Doing Cross-Cultural Research Through and with Visual Methods

Kei Yan Leung

Introduction

Language is the dominant medium on which social scientists rely in their research practices; they use language both to create knowledge and in their choice of interpretative methods to communicate this knowledge (Davies & Dwyer, 2007). However, our daily lives are composed of many different dimensions, and not all knowledge is reducible to language. Instead of seeking singularity and certainty to make sense of that one reality through language, there are multiple realities, and we need new ways of knowing in order to navigate through the diffuse and messy world (Law, 2004). When we aim to understand the mindsets and practices of interviewees, focusing solely on spoken words may be limiting. In the context of my own research, farmers do not necessarily engage in

K. Y. Leung (✉)
University of Natural Resources and Life Sciences, Vienna, Austria
e-mail: natalie.leung@boku.ac.at

A. Franklin (ed.), *Co-Creativity and Engaged Scholarship*,
https://doi.org/10.1007/978-3-030-84248-2_9

word-based cognitive reflection when they interact with plants, animals, soils or tools.

Instead of focusing solely on how we communicate our thoughts and experiences through language, scholars have been shifting their focus to more-than-representational experiences such as emotions, affects and sensuous experiences[1] (Law, 2004; Lorimer, 2005). The relational, emotional and affective aspects in research practices regarding the interactions between humans and non-humans have been increasingly acknowledged (e.g., Campbell et al., 2019; Hitchings, 2003; Krzywoszynska, 2016). In addition to increasing explorations on methods that invoke emotions, different attempts have also been made to theorize and interpret emotions and affect (Anderson, 2006; Harrison, 2000; Pile, 2010; Thien, 2005; Tolia-Kelly, 2006). In these works, the world is conceived as full of sensibilities, and researchers seek ways of knowing beyond words and languages through openness and reflexivity. With the assumption that knowing is more-than-representational by considering both representations and affective, sensuous experiences, this chapter uses the context of cross-cultural research to explore the limitations of knowing solely through language.

Cross-cultural research is a fertile ground used to explore the role that meanings beyond language play in our understanding and engagement of the world. When conducting research in one's own native language, the researcher might also struggle with language, but they focus more on making sense of how discourses and practices work together (Krzywoszynska, 2015). In the context of cross-cultural research, the researcher does not instinctively know all the experiences that are associated with words; knowing also involves the process of understanding the emotive and embodied relationships that are specific to the language, place and cultural practices (Krzywoszynska, 2015). When language cannot give a full account of experiences, it unfolds the opportunity for cross-cultural researchers to seek alternative understandings of the interviewees than a local researcher might do. In this sense, a cross-cultural

[1] There is an extensive body of literature on non-representational and more-than-representational approaches that do not prioritize the role of representation and reasons, they also take into account the role of practices, affects, emotions to account for the interactions between humans and non-humans (e.g., Anderson & Harrison, 2016; Lorimer, 2005).

researcher is perhaps similar to a blind person: they might not be able to see through direct vision like a local researcher, but they can see hidden meanings through a range of sensitivities and sensibilities that a sighted researcher may otherwise neglect. Indeed, it will take longer for a cross-cultural researcher to understand their interviewees because of the cultural and language differences. They will also learn less about certain things because the cultural and language gaps just could not take them there. However, they can potentially learn a wider range of realities, or even participate in the making of those realities, because they are outsiders who always seek more thorough explanations from the interviewees.

Visual methods have the potential to supplement the limitations of verbal research methods in both cross-cultural and same cultural settings. Our vision is not only limited to an objective process that is associated with discourses, meanings and judgements. When we see, we also develop subjective experiences such as sensibilities, and embodiment (Rose & Tolia-Kelly, 2012). Undertaking interviews with images allows people to go beyond the verbal thinking mode and include a wider aspect of their experiences at the emotional level, or layers of experiences that cannot be easily put into words (Bagnoli, 2009).

In addition to data collection, visual methods also have the potential to improve scientific explanation and understanding of scientific knowledge to both scientific and non-scientific audiences (Rodríguez Estrada & Davis, 2015). Scientists have been using graphs and figures to communicate scientific results visually for centuries (Tufte, 1997, cited by Darnhofer, 2018). More recently, visual communications are also increasingly used to connect non-scientific audiences. For instance, Bartlett (2013) used cartoons to communicate her research findings about issues related to misconceptions of dementia. She noted that cartooning helped to present serious topics in a more playful way, making it easier to engage audiences (Bartlett, 2013). Darnhofer (2018) found that using comic-style posters to share preliminary findings with her research participants was effective in engaging the participants to share their feedback and facilitate more in-depth discussions of the research topic.

In this chapter, I will reflect on my experiences as a cross-cultural researcher during my field work in Japan, where I conducted interviews with farmers using photo-elicitation to understand how they build relationships with artworks and their farming. As a Hongkonger, I had previously worked at the field site for three months on an art project about farming. However, I am not able to speak Japanese, so I worked with a local interpreter to conduct interviews with farmers with Japanese–English translation. Through the experiences of working with photo elicitation to collect data, and the attempt to convey research results through illustrations, I argue that visual methods help to uncover different realities that are beyond the scope of linguistic relevance, but nevertheless, fundamental to understanding the mindsets and practices of farmers.

In the following, I start by providing a brief account of how I applied photo elicitation in my doctoral research. I then first discuss how experiencing the challenges of communication brought by cultural differences pushed me to rearrange the interview questions. Second, I elaborate on the limitations of translation in communicating experiences, thoughts and emotions that are tied to cultural practices. Third, I illustrate how photo elicitation helped to unfold different layers of experiences by the farmers during data collection. Last but not least, I discuss the potential challenge of conveying results from research in Japan to non-Asian audiences, and why I combined visual illustrations with verbal quotes to give a more in-depth portrayal of the experiences of Japanese farmers to Western audiences.

The Study: Talking About Art Through Photo Elicitation

In my doctoral research, I used photo elicitation in interviews to explore how Japanese farmers build relationships between their farming and artworks. My aim was to understand how farmers perceive artworks and the potential impacts of art on their farming. The study was conducted in Tokamachi (a remote mountainous area in Northeastern Japan), in two respective field works in the winter of 2019 and 2020. I selected this particular area because it is where the Echigo-Tsumari Art Triennale (ETAT) takes place. Similar to many rural areas in Japan, Tokamachi is facing the problem of dwindling population and increasing numbers of abandoned houses and fields (Kitagawa et al., 2015). ETAT was the first art project designed to address the issue of rural revitalization in Japan; it is the world's largest rural art festival. By placing art installations in abandoned landscapes and rice fields, ETAT uses art as the means of reconnecting with traditional farming practices, and the farming landscape they produce (Kitagawa et al., 2015). Through these, ETAT aims to invite urban visitors and local residents to rediscover existing but neglected local cultural and natural resources, and reflect on their values.

In total, I interviewed 25 farmers and an artist who has been working with local farmers to market their products. The farmers were selected based on their engagement in more agro-ecological farming practices than mainstream farmers. These agro-ecological practices are broadly defined by the way they substitute environmentally sound inputs and practices for industrial ones, and reconnect with traditional practices that are more well-suited to local agro-ecosystems (Gliessman, 2017). In Japan, farming is dominated by small-scale, mechanized rice farming under the co-ordination of state-supported Japanese Agricultural Cooperatives (JA). JAs manage government subsidies, provide advice and input, and govern standardized, industrial rice production and sales (Esham et al., 2012; Mulgan, 2005). The state-initiated agricultural liberalization since 1990 enabled the emergence of this group of agro-ecological farmers who intentionally engage in more sustainable farming practices. This group of farmers were selected because their farming practices resonate with ETAT in re-signifying dying, traditional villages through innovation and re-interpretation of traditions. Such similarity suggested that these farmers might be more able to relate to ETAT's artworks. However, this is not a point that my research can prove because I did not interview mainstream farmers.

The interviewed farmers were selected through snowball sampling. Based on the information of the respondents, it is believed that these 25 farmers are most, if not all, of the agro-ecological farmers known in the area. Among these farmers, five of them work for ETAT on a part-time basis to take care of fields that host artworks; the rest of the farmers work on fields that do not have artworks and they were not involved in the selection and management of the artworks. This reflects the fact that that most agro-ecological farmers are not directly involved in the artworks or the art festival. After the interviews in Winter 2019, I subsequently carried out three group discussions with them to deepen my understanding of how they make sense of selected artworks that are related to agriculture and landscape in Winter 2020. Similar to many of my research respondents, the interpreter moved to the area from Tokyo to seek a traditional lifestyle in the countryside.

During the first field work, in Winter 2019, I interviewed the farmers with nine pictures of seven selected artworks. The artworks were selected mainly based on their high publicity in ETAT, conspicuous locations on rice fields, and the link to traditional agricultural practices in Japan. In the pre-designed interview guide, I started by asking briefly about their farming background, i.e., how they started, and about their farming characteristics and challenges. The rest of the questions were guided by photo elicitation: I showed the farmer photos of artworks, invited them to choose one that they wanted to talk about, and asked them to share how the artwork is relevant to their farming.

Adaptation of Interview Questions

The first few trial interviews were a mess, and I very quickly realized there is a gap that limited my understandings of what the interviewees said. The questions about the artwork were too difficult and abstract for some of the farmers. They were anxious about not being able to give a 'correct' answer about what the artworks convey. Responses like 'I know nothing about art', 'I am just a farmer, I am not interested in art' were common. Although Tokamachi is dotted with artworks, art is still something unfamiliar and distant to some of the farmers, because they consider aesthetic experiences as something separated from ordinary experiences. Originally, I thought that the multivocality and abstraction of art could invite the farmers to relate to it in a diverse way, but it turned out that the indirectness of art made some of them feel anxious about not being able to give a model answer. I tried to understand why they are not interested in art; we were exchanging words through the interpreter but we were not able to communicate. They were trying to understand what my strange questions were trying to capture, and I was trying hard to understand the meaning of their answers. Gradually, I realized that what was hindering the communication was my lack of understanding about their ways of being a farmer in the area; I was not able to relate to their mentalities, and to their emotional struggles about being an outsider in the group-oriented village culture in Japan.

For instance, in the following conversation, I originally intended to find out why the farmer was not interested in the artworks and what kind of qualities the artworks lack. At the beginning when I was not aware of the social pressures they were referring to, I had difficulty to logically relate to the distinction they made between artificial things and cultural events/traditions:

Q: Why are you not interested in the artworks?

A: I don't understand the meaning of the works… I am not really interested in those art, so I never thought of anything about them. I have seen some of them, but it is so awkward to put artificial things in nature I don't like the idea, it feels so unnatural. I don't get anything about them, I love nature more than artificial things.

Q: What about Matsuri? It is more of a cultural event.

A: Those ceremonies with fire, all the traditional events, that maybe seem like art to foreigners, I think they have meanings in their cultural background, so I can understand them.

Q: How do you understand them? Does it mean you do not like the artworks because they have no cultural background?

A: From those cultural events, I don't know if I like or dislike them or not, I just accept them. I accept them because the meanings are inherited from the past, I just accept the ideas. But for those artificial things, they are just so awkward, so I don't understand any of them.

Q: You just mentioned that the cultural events are meaningful because they are from the past…what do you mean by just accepting the ideas because they are from the past?

A: (thinking about it hard) Hmm… I think to follow those traditional events, rules and life they have here… it's about respecting the life here… I am following local rules and habits from the past, just like the locals are following their lifestyle from the past. I started growing

> rice and veggies because my neighbours and the elderly also do it in this way, it is really important for me to follow the kind of life they have had for long time, to me this is the meaning of living a local life… in this village we prepare rice shooting collectively and they would use chemicals, I cannot avoid it. It is part of the cooperation. If I do not follow, I will be ditched and complained about by the community. To assimilate in the community and receive support from others, I have to accept something that I don't agree with. For me, it is really important to live in this village peacefully and happily without having any troubles with villagers.

It turned out that the context behind this farmer's rejection of the artworks and acceptance of traditions related to their situation as an incomer in the village and the emotional pressures to assimilate into the community and local traditions. I came to understand that the relationship between art and farming is not just about how the art installation itself is related to the meaning of their farming practices. It involves far more diverse relationships that are connected to their mode of life as a farmer who engages in agro-ecological farming practices and a newcomer in the village, and the emotions and feelings that are tied to it.

As a cross-cultural researcher, I needed to know more about their form of life in order to understand the possible relationships they would develop with the artworks. Therefore, I changed the arrangement of the interview questions, I allocated more time to talking about their farming values and practices before moving to the questions of art. Not only did this help me to make better sense of their farming life, talking about things they are more familiar also helped to empower the farmers and reduce the anxiety of talking about art.

The Difficulty of Translating Life and Practices Under Cultural Differences

Although rearranging the questions inspired more diverse discussions with the farmers over the artworks, I noticed that the limitations of translating emotions and local experiences through language were still present.

After all, I still depended mainly on words to communicate with the farmers and understand how they saw the photos. I noticed that even a perfect word-for-word translation did not help me to proceed smoothly with the interview. As a foreign researcher, I simply lacked the cultural sensitivity and fluency to ask the question in a culturally relevant way to delve into topics in which I was interested.

This is illustrated in the following extract of a group discussion, where apparently the farmers and the interpreter have a different understanding of 'nature' than my original question intended. They do not see nature as separated from humans; they also see humans as part of nature in their cultural concept of Satoyama. When I framed my question based on the western concept of the human/nature split and tried to understand how they make sense of such distinction, it did not lead me anywhere. My question got lost in translation because the interpreter and the farmers were not able to relate to my question:

(Q: me; I: interpreter; A: interviewee)

Q: In your opinion, what is landscape? The artificial rice terrace or the wild forest?

I: It is a hard question, it's not difficult to translate but the question itself is difficult to understand...

Q: Do you make a distinction between artificial rice terraces or natural/wild forests?

A: We called it *satoyama*, there is no translation. It means the middle part of wild nature and where people live. The landscape here or the future of here is the *satoyama*, that's why in the mountain there are rice terraces here, we call it *satoyama* view.

Q: Earlier some of them mentioned being captured by some beautiful moments in life; is that because the nature/landscape you see involves human participants It involves villagers, community, and you also live in this environment...

I: I think it depends; it is different for everyone. What do you want to ask?

Q: I just want to know if they find the nature here beautiful because they are part of it? Like the example of going to the wild nature: countryside people go there and they still find it beautiful, but they cannot live there, so those kinds of moments that they think as beautiful won't last. But for people here, they can observe the changes in different seasons; is that because they live a life here/they share a life with nature?

A: When you go to the deep mountain here, if there is a path, it means there was life there many decades ago. I see the beauty when I feel the traces of humans, just like I also feel the beauty from the piece of farmland that elders just weeded by hand, I feel the beauty of their work.

These moments were clumsy and awkward, because I just could not find words to communicate my thoughts. It is not because we cannot find the 'right words', but more that I do not have the cultural context to understand their connection with nature in their everyday lives, and therefore their experiences associated with these connections. Meanwhile, the farmers also did not understand what I was asking because the nature/human split simply does not exist in their cultural understandings of landscape. Through these experiences, I realized that there are limitations in language that I simply cannot transcend as a cross-cultural researcher.

Cross-cultural researchers in social sciences tend to focus on solving 'problems' in translation and making communication effective (Turner, 2010). For instance, how to get precise translation, a correct version of an interview transcription to minimize the discrepancy caused by language differences (Turner, 2010). Many social scientists are often preoccupied with words because they tend to believe that the world is static, definite and predictable, and through precise words they can discover these definite states that exist out there in reality (Davies & Dwyer, 2007; Law, 2004). However, my fieldwork experiences clearly showed that the world is more complex and textured, so that the challenge is less in finding an exact translation, than our capacity to understand what farmers try to convey.

Instead of getting frustrated by not being able to maximize accuracy in language, I chose to be reflexive about the role and limitation language plays in cross-cultural research. The language differences made explicit the cultural dimension and demonstrated that meaning is made outside of literal translation. Pereira et al., (2009, p. 5, cited in Krzywoszynska, 2015) believed that 'a lot of insight can be found, and a lot of knowledge can be produced, through explicit and critical reflection on the challenges and incommensurabilities of language difference'. For instance, there is no such concept of '*Satoyama*' in English. It resembles the concept of 'countryside' but is more than that. If it is translated from Japanese to English literally, it means 'the area between mountains and human settlements'; it is covered with managed woodlands and terraced rice fields (Brown & Yokohari, 2003). However, the translation neglects the values of nature implicit in the concept, that human communities (*sato*) and non-human nature (*yama*) coexist side-by-side in harmony (Yokohari & Bolthouse, 2011). The non-nature/human split does not only shape traditional agricultural practices in Japan, it also shapes the spiritual connections and respect farmers have with nature. There are also *Satoyama* landscapes in various places in other Asian countries like China. The distinctiveness of the Japanese one is the spiritual ties to nature, in which the Japanese believe that there are eight million deities present in nature (Iwatsuki, 2008). As a buffer zone between human settlements and deep mountain areas (*okuyama*), the *Satoyama* area is where they set up shrines to worship the deities so as to ensure co-existence and their guardian in daily life (Iwatsuki, 2008). If I just adhere to the mainstream western approach to strive for a precise translation and omit the cultural understanding that farmers have of nature, the possibility of exploring different approaches of nature/human relationship in another culture could be easily overlooked.

In addition, I also found that the language differences highlighted affective experiences that are tied to cultural practices. Feelings and affects do not just come with words; they are entangled with the specific social and cultural life of the interviewees. Although I went to the field with a local interpreter who acted as a vital cultural broker, it was not easy to convey feelings across cultural and language barriers. For instance, in the conversation about landscape, when I was trying to understand how farmers categorize and distinguish human and nature, the farmers were relating it to how they feel the beauty from the collaboration between humans and nature in an affective way. However, a perfect linguistic translation did not make me experience the feeling of beauty they have experienced. In the following conversation, I was still trying hard to understand from the interpreter the kind of beauty they were referring to:

> *Q*: What are the things that they found beautiful? What is it that captured them?
> *I*: There were moments that nature and the view just overwhelmed us, those are moments that we are not able to tell what captured us... I just asked them about it, it is not about any specific thing but the whole atmosphere at that moment just captured us in our daily life.

It is therefore important for the interpreter to be sensitive about cultural differences; they have to understand peoples' feelings, reframing them and making them 'reasonable' to researchers from a different cultural context (Turner, 2010).[2] However, in the process of making feelings 'reasonable', the quality of the feelings described is incomparable with how it was experienced (Harrison, 2007). One of the respondents of Giustini (2019, p. 195) illustrated the limitations of language when it comes to the expression of emotions: 'sometimes we can't find the linguistic or cultural expression that would match the same level of emotion, but we try to do as much as we can to impact the audience'.

Opening up New Ways of Seeing Through Photo Elicitation

The use of photography in research is not something new; it first appeared in 1957 to study how different ethnic groups adapt to residence and new forms of work in urban factories (Collier, 1957, cited in Harper, 2002). Since then, photography has been increasingly used in various social sciences disciplines. There are different methods, as photos can be provided by the researcher or taken by research participants through

[2] The roles and influences of interpreters in cross-cultural research is a subject in its own right, and this book chapter is too limited in scope for more in-depth discussions. When one can speak a particular language, it does not automatically mean that you can represent a culture. The sensitivity to cultural differences is not only shaped by one's cultural background, it is also shaped by one's social background, positionality, personality traits, language proficiency, and so on (Turner, 2010; see also: Caretta, 2015; Temple, 2002; Temple & Young, 2004).

a camera handed to them. Also, different terms, such as photo voice, participatory photography and reflexive photography, have emerged to denote the varying use of photography in research.

Photo elicitation is broadly defined as a qualitative research method where photography is used to enrich and complement research data (Harper, 2002; see also Axinte, this book). The advantages of using photography in research are well documented, including research about farmers, or those in natural resources settings (e.g., Beilin, 2005; Sherren & Verstraten, 2013; Sherren et al., 2012). For instance, visual materials prompt respondents to reflect on things that they did not get to discuss in talk-only interviews (Rose, 2014). By putting farmers in control of the conversations that emerged, photo elicitation helps researchers to study complex issues that can be very personal and deeply held by farmers (e.g., farm landscape management), in a manageable and sensitive way (Sherren et al., 2010, 2012).

In photo elicitation interviews, pictures are used to invoke comment and discussions in the course of interviews, and therefore to make various realities visible in data collection (Banks, 2001; Rose, 2014). This approach uses the structure of showing, and then telling what is shown; the image is simply used by the researcher as an inscription device to visualize a certain reality to research respondents (Rose, 2014). Even if the pictures are taken by the respondents, researchers tend to focus more on what is pictured and making meanings by working with what the image shows. For example, photography is also used to study farmers, but it is used in a similar way to trace knowledge and experiences that the researchers are looking for. Harper (2001) used aerial photos of farmlands and historical photos to interview elderly farmers about their memory and interpretations of farm life in the 1940s. By using the photos to make the old way of farming visible again, Harper (2001) elicited rich details from the interviewees about technological transition to industrial farming in the US, what social relationships were like before the transition and, more importantly, their feelings about those old days.

However, Rose (2014, p. 31) finds that in visual research methods there is an almost total neglect of the 'symbolic and communicative components that are specific to the culture'; she refers to these components as visuality. Visuality is the cultural construction of visual experience; it means what the respondents see, how they are able to see, and how this seeing and unseeing are governed by their cultural understandings of the messages inherent in the images (Foster, 1988, cited in Rose, 2014). In my research, my unseeing of the seeing of the farmers has, in many ways, highlighted how their seeing is specific to the social and cultural contexts of Japan.

Initially, I aimed to use the photos to elicit more information and make it easier for the farmers to talk about relationships between art and their farming. In other words, I also used the photos as an inscription device to confirm the reality that I was looking for, without taking into account how the farmers see. Indeed, the photos helped a lot in relating the artworks to the daily life of the farmers. The combination of artwork and landscape in the photos encouraged them to look at the artworks as pictures or sceneries and share their perspectives freely. It turned out that the photos encouraged different imaginations among the farmers, and they shared more insights than what messages the photos conveyed to them. This was especially so in the case of those who said they knew nothing about art at the beginning of the interview. For instance, they related the photos to the past farming scenes in the area, the resemblance with their current practices in terms of values and actual things they have created like farmhouses, and their vision of the future of the local area. Indeed, their seeing was not just restricted by the art piece itself, but also the background of the artwork and the combination of the artwork with the background. Some of them also built relationships between their seeing and farming life, in which they projected themselves to certain artwork and saw how it signified the rhythms and characteristics of their farming life.

Through these experiences I found that the cultural differences of how some of the farmers and I see visual materials in the photos can serve as a good opportunity for me to understand the social life of the farmers and highlight their associated affective and sensuous experiences. When they talk about the images in the interview, to me it is more than just about what the images show, but also about how they see the images in certain ways. The fissure between the seeing and unseeing among us provided rich material to uncover different realities; this might well remain implicit and thus hidden in research conducted in the same cultural context.

For example, when I saw artistic elements from the photos of traditional festivals where people are playing traditional Japanese musical instruments, to some of the farmers those are cultural events that worship the spirits of dead people and convey the meanings of assimilation. Some of them saw the projection of traditional cultural landscapes in Japan from the photos, I saw nothing but just some forests with abandoned fields. It is these fissures that inspired me to ask more questions to seek clarification from the taken for granted but unobservable thoughts of the farmers. As a foreign researcher who lacked understanding about how Japanese farmers make sense of farming, this has opened up different realities of local farming context to me. These are realities very different from the one I encountered from doing a literature research, in which the literature mainly highlights the challenges farmers face under different structural forces in Japan.

By sharing how they see the pictures and what the pictures visualize to them, the farmers illustrated how they make sense of their farming life through a different form of knowing, one that does not separate reasoning, feelings, affective and sensuous experiences. These more-than-representational understandings are embodied sensations, such as the sight of rice terraces, or the physical difficulties and embodied feelings of working on the land.

I also found that the point of using visual method is not at all about filling the difference between language and experiences resulting from the gap in cultural understanding in cross-cultural/same-cultural research to make communication 'adequate'. There is never a translation where language is 'adequate', especially if we also take more-than-representational experiences into account (Harrison, 2007). What is significant about using visual methods in both cross-cultural and same cultural research is that it traces the limits and possibilities of mainstream social analysis, and inspires researchers to stop preoccupying themselves with language, and start considering how meanings beyond language can enrich our understanding of our research subject. The following are some examples of how the use of photos in interviews helped to uncover different layers of meanings through the seeing of the farmers and my unseeing.

Ikeda was amazed by the photo of the 'Rice House' in Fig. 9.1. He saw the life of a household in the frame, he saw a kitchen, a living room and a family living there. He related the picture to the concept of *Tanaka*, which literally means a 'house surrounded by rice terraces', and he saw the traditional view of the local village hundreds of years ago from the photo. As a non-Japanese farmer, I was not able to see the rich cultural connotation that the 'Rice House' carries. Yet, such unseeing captured my attention towards the nostalgia that Ikeda had towards such traditional landscapes, and the disappointment he had when he gave up some rice terraces last year, as cultivating them was too physically demanding. His seeing from the photo did not allow me to experience the same emotional connections he has with the landscape, but it highlighted the importance of considering the emotional aspects that are keeping farmers motivated to persist traditional, labour-intensive practices.

Fig. 9.1 'Rice House'—the metal frame merges with the landscape and rice terraces to form an harmonious picture in different seasons (*Source* Photo by author)

Fig. 9.2 'Scarecrow Project'—the red silhouettes represent the past scene of the family of the landlord working on their ancestral land (*Source* Calvin Wong)

Morita takes care of the rice fields hosting the 'Scarecrow Project' (Fig. 9.2, above). He was working on the fields every day; he did not think much about the artwork because it was too connected to his daily life. Instead, he focused on the uneven growing conditions of the rice field in the picture. The rice field has experienced mudslides, resulting from rice fields in higher terraces being abandoned. They damaged the field, making it difficult to work on because some parts are deep and muddy. He related the artwork to these embodied experiences of working the field, and this reminded him of the need to keep rice fields active to preserve the landscape. This is in contrast to all I could see from the picture: the red silhouettes. Yet these red silhouettes did not matter to Morita at all. All the photo reminded him of were the uneven growing conditions of the rice plants. From his words, I was able to know the hardship he had endured when working with this field, but I was not able to experience his embodied experiences of physical fatigue and the emotional connections with landscape through language. However, his seeing and embodied feelings uncovered his sentiment towards preserving the *Satoyama* landscape, which is getting lost because in this mountainous area rice fields are increasingly abandoned. As mentioned earlier about the *Satoyama* landscape, to local farmers the loss of the landscape is not just about its physical disappearance, it is also the breaking up of human–nature collaboration and the spiritual connections. To him, the physical fatigue he was enduring was the bridge to rebuild this collaboration, it was associated with the emotional connections to the culturally meaningful landscape.

The strength of how visual materials are more capable in stressing emotional experiences became obvious when Shibata and Keiko talked about their feelings towards snow, respectively through words and picture. The lack of emplaced experiences made it too abstract for me to relate to Shibata's feelings about snow. Also, he believed that as a visitor, I would not be able to relate to the beauty they find from the snow in the area.

> It is the light and darkness about the beauty. If you don't live here for a whole year, you don't know how tough it is to live through the winter. When you just come and see the beauty of snow in winter, the joy is different, and you also would not feel our thankfulness to spring. (Shibata)

Although I am not able to share the same feeling towards the beauty of snow with the farmers, the emotion related to seasonal changes was made explicit when another farmer, Keiko, relieved the feelings she experienced from snow, triggered by a photo from the artwork 'Human re-entering nature' (Fig. 9.3) in winter. Her experiences signified the emotional changes and embodied experiences of living a rhythm of life that is dependent on seasonal changes. For instance, the relief from working on terraced rice fields that are not accessible by machinery in summer, and therefore the calm and stillness she associated with snow in winter.

> I wonder what he is thinking, I can imagine many things by looking at him… I wonder why he is standing like this, nowadays people don't stop, they are all moving around. But when snow comes, we feel relieved because winter has finally come. Maybe he feels relieved to see the snow and he can finally take a rest. It is such a good picture. (Keiko)

Conveying Research Results Through Visual Illustrations

The cross-cultural position as a non-Japanese person has helped me to unfold different layers of meanings that might remain hidden to a Japanese researcher working in their own cultural context, as they might be too obvious to merit mentioning. However, as an Asian I also faced the challenge of adequately conveying the research results to non-Asian audiences through words.[3] It was already difficult for me to 'fully' understand the farmers given the cultural differences between Hong Kong and Japan. As a result, the words of the farmers had already gone through a first layer of cultural translation by the time the interpreter translated the words to me. There would then be a second layer of cultural translation when I communicate my research results through words to western audiences, who are even more culturally distant to Japanese farmers than I am as a Hongkonger. The idea of visualizing interview quotes first emerged

[3] Communicating research results to non-Asian audiences is necessary as the research is intended for publication and doctoral examination in a western context.

Fig. 9.3 'Human re-entering nature', Winter—a 4-metre high human figure that changes with the season (*Source* Anna Mak)

when I checked if the interview transcription is 'correct' with the interpreter. She told me that some of the quotes sound right literally, but the spirit and the emotion of the speaker is gone in the English sentences. For example, when Yoshida spoke in the following quote, the emphasis was on the positive and thankful feelings for being able to work together with the beautiful landscape to produce food. However, the interpreter felt that these emotions were lost, as the English translation sounded more like he was frustrated by the physical workload, which he uses to justify the use of chemical pesticides. She thought that it was important to emphasize the positive attitude through a supplementary note.

> I always enjoy farming in the mountains, I feel proud that my rice and wheat came from these beautiful mountains, and I preserve the Satoyama landscape... I am taking care of the abandoned farmland, but it is taking too much labour and time to do the organic way, I cannot do all the work by myself, I have to use chemicals to weed. I want to reduce the chemical usage on wheat, I am looking for a good way to use them. (Yoshida)

Therefore, I thought of working with a local Japanese artist, whom I interviewed during the first field work in Winter 2019, paying her to produce graphic illustrations of selected quotes. I could use the illustrations to supplement the farmer's verbal quote and depict the scene of how particular experiences were understood by the farmers. Similarly, Dahl et al. (2012) used a graphic novel to retell the life stories of five homeless people they interviewed in academic research. Through portraying the life events preceding their homelessness, and how they experienced these events emotionally, the graphic illustrations draw the attention of a wider group of audiences to the complicated social issue of homelessness.

Together with the interpreter, we therefore identified three interview quotes in which the emotional and sensuous experiences of the farmers are especially absent in the English translation. Afterwards, I arranged a meeting with the local artist, in which the interpreter and I discussed these quotes with the artist and shared with her how the emotions and spirits of the farmers were lost in translation during the interviews. I invited the artist to draw three illustrations based on the three quotes; I also invited her to convey the quotes in her way. Shortly after the meeting, the artist presented me a first draft with five illustrations. I picked three illustrations that I thought fitted most with the quotes, and I gave her some comments and suggestions for aspects that I thought could be further highlighted in the illustrations. After that, she incorporated my comments, revised the first draft and delivered the final illustrations to me.

As a rare artist working with local farmers in this remote rural area, the artist has been using artistic design to help local farmers to market their products to both local and urban customers The artist therefore provided highly valuable thoughts and ideas because she understood the language and experiences of local farmers, and she is also experienced in converting the language of farmers to non-farming audiences through visualization. Rodríguez Estrada and Davis (2015) point out that not all visualizations are effective communications, and visualization is also not a blind process of integrating written discourses with visual illustrations. It is important for the communicator to not just understand the written discourses, but she should also be able to connect with non-specialist audiences through their language. In the design approach of the artist, she first understands the farming approaches of farmers, then she helps them identify the distinctiveness of their products. Therefore, she is familiar with local farming landscapes and the settings of the farms of local farmers, and knowledgeable about the farming approaches and practices of local farmers. In her design work for local farmers, she communicates these aspects of local farmers to both local and urban customers. Through these experiences and background, she can utilize the techniques in graphic design, e.g., composition, colour, layers, as highlighted by Rodríguez Estrada and Davis (2015), to convert the selected quotes of the farmers to paintings that are easier for non-Japanese audiences to capture emphasis of the quotes.

Using a visual illustration has the potential of supplementing the verbal quote to draw audiences' attention to the more comprehensive experiences of the farmers. Visual illustrations can help convey the emotions and sensations that farmers were trying to convey in the interviews in an imaginative way. To address the common misconceptions related to dementia, Bartlett (2013) used cartoon-style drawings to portray the lifeworld of dementia patients in a playful way with imaginative scenes in an exhibition. The exhibition received positive feedback, and the cartoons were able to generate fluid and open-ended interpretations. This example shows that illustrations can relate non-verbal connotations to readers of one's research in a metaphorical way. In striving for this, the hope is to inspire the reader to take these aspects into account in understanding the mindsets and practices of farmers. For instance, by just reading the following quote, the embodied aspect of learning mentioned by the farmer can easily be overlooked. The quote can be seen as being merely about cognitive understanding of farming knowledge. In fact, during the conversation the farmer did not separate the cognitive aspect of learning from the embodied one in the quote: how he feels hopeful and is looking forward to reaching the embodied level, more than the fact that he needs more confidence and experiences. It is difficult to capture his emphasis just by reading the quote:

> I need more experiences, many things can change the condition... soil, water, weather and everything, I need more confidence in what I am actually doing, then I can move on after 2-3 year. What I have learnt from books and my mentor will be more in my body. I will feel more confident and have an actual sense of what I was taught. (Yoshida)

The following is a short account of how the artist interpreted the quotes, how she incorporated aspects that I wanted to highlight and visualize the quote.

Figure 9.4 shows a simple and subtle scene of a farmer touching and checking the soil and seedling with two hands: the artist intended to use this ordinary moment to convey different meanings expressed in the quote. She focused on the word 'soil' and used it as an anchor to link different aspects of learning Yoshida mentioned in the quote. In the opinion of the artist, soil shows the impact of weather, humidity and the environment; soil also shapes the condition of plants. Since farmers accumulate knowledge of soil through experience and practices, the artist uses the act of touching and caring of soil in Fig. 9.4 to symbolize the accumulation of knowledge as indicated in the quote. As I wanted to highlight the aspect of learning from an embodied level in the illustration, the artist presented the farmer without a head. The

Fig. 9.4 Illustration of a farmer caring for the plant through his body to accumulate knowledge and experience (*Source* Megumi Hirose)

blue background has two implications: first, it shows a blue sky and aims to bring a positive and hopeful emotion that the quote expressed; second, the background meanwhile can also be the chest of the farmer, the combination of different colours signifies the fluidity of affect and feelings he has for the plant in his heart.

Illustrations can also visualize in a more realistic way the scene of how farmers interact with non-humans through various sensuous experiences in their farming practices. In the following quote, Yanaga shared how she used her body to feel and care for the plants to understand their condition:

> The scent, you can smell it… you can feel it, touch it and you can taste it. If the crop is not doing well, I can see the failure from their shapes. If you look at a tomato closely, you feel the hair…. (Yanaga)

When she expressed the quote, she was excited and showed how these interactions with the plants and attention to small details motivated her in her everyday farming life. However, these emotions are also not explicit in the quote. In Fig. 9.5, the artist wanted to highlight the love and affection that a farmer develops towards her plants and tomatoes in the process of nurturing and caring for them. The facial expression of the farmer in Fig. 9.5 communicates such excitement when she is examining the tomato through the touch of her hand and her smelling the tomato. The farmer is surrounded and embraced by her flourishing plants in the composition of the illustration; it conveys the reciprocity of care and happiness between the two, in which the plants flourish because of the care of the farmer, and the farmer is happy when the plants grow well.

Similarly, Shibata illustrated how he plants rice by hand. In the quote, it is difficult to imagine at the cognitive level how human hands can transfer energy to the plants, within a western scientific understanding. The 'energy' he is referring to is related to the concept of 'ki-energy'

Fig. 9.5 Illustration of a farmer appreciating her tomato like an artwork (*Source* Megumi Hirose)

in Japanese culture. The meanings of 'ki-energy' are very different from what the English word 'energy' conveys: 'ki-energy' signifies a vital force that flows between animated things and inanimate things in the world. In the context of this quote, it is a force that flows between the body of the farmer, the rice plants and the soil:

> We like planting rice by hands because it makes rice more delicious. When we use our hands to touch the plants, some good energy is transmitted to the plants and the Earth. I believe in the power of it, that's why I want more people to be involved in my farming, and I can get a lot of good power from a variety of people. (Shibata)

Both the artist and I identified 'ki-energy' as the key idea to be highlighted in the illustration. The first image that popped up in her mind from the quote is a farmer transplanting rice with their hands in a *Satoyama* landscape. Being an incomer from an urban area, the artist does not just experience *Satoyama* through the visual landscape but also the smell in the air. She believes that a lot of energy is also stored in the air in the landscape. In Fig. 9.6, the white path signifies the moving of 'ki-energy' in the air embracing the farmers, the villages, mountains and forests. The white path also visualizes how 'ki-energy' is generated from manual rice transplantation, an aspect that I wanted to highlight. As the farmer transplants the rice plants using his hands, the 'ki-energy' gathers and spreads to the villages between the rice fields and the mountains. It signifies the process of how the delicious rice produced by traditional rice planting method brings good energy to people. The illustration also highlights that rice transplantation by hand involves more than the physical movement and visible touch of the plants. It also includes the invisible connections that the farmer feels with the plants and soil in the process of transplanting them.

Fig. 9.6 Illustration of a farmer enjoying rice planting by hands to cultivate the circulation of 'ki-energy' between his body, soil and the plants (*Source* Megumi Hirose)

Conclusions: 'Seeing' Meanings Beyond Language

Visual methods can be a useful tool to unlock the sensibilities of a cross-cultural researcher, or in general a same-cultural researcher who is overly or insufficiently familiar with the cultural context. They allow the researcher to better understand emotions and affects that are neglected in the research practices seeking singularity and certainty in social sciences. Through doing cross-cultural research in Japan, I experienced the insufficiency of language in understanding the experiences, emotions and feelings that farmers associated with their practices and forms of life. This led to many awkward moments when words were exchanged, but did not convey meaning. Using the photos in the interviews allowed me to gain a deeper understanding of the everyday farming life of my respondents in light of their specific cultural context. The photos also triggered associations that allowed us to discuss aspects of experiences that are difficult to put into words. Reflecting on what farmers saw, but I could not see, enabled me to co-creatively identify various more-than-representational dimensions they associated with their farming practices, i.e., the emotions and feelings. Reflecting on the cultural constraints during data collection and the resulting difficulties of conveying meanings fully through using only words, I also explored the use of illustrations by a local artist to complement quotes, so as to more fully convey research results to western audiences.

It was through reflecting on the feelings of frustration and embarrassment associated with the language differences that I was able to notice how visual methods can help to uncover different layers of understanding. As noted by Law (2004), method is not just a set of techniques, every method is performative, it depends on what kind of social science we want to practice. Only when I stopped desperately looking for the 'right' words did I become more mindful of how the seeing and unseeing of the farmers reveal a different reality, one that I did not encounter in the literature on farmers in Japan. Indeed, 'method goes with work, and ways of working, and ways of being' (Law, 2004, p. 10). Being a foreign, clumsy and mute body in the field unavoidably influenced the interactions I had with the farmers. My questions often sounded strange to them, as I was not able to formulate my questions in a way that was meaningful in the cultural context of rural Japan. Fortunately, having a local interpreter who already knew most of the farmers helped to encourage the farmers to be more generous in sharing their thoughts and experiences. The farmers did not feel offended and they were willing to explain their seeing in a more in-depth way, making meanings and implications explicit that were a taken-for-granted part of their experiences.

I have found being a cross-cultural researcher, in the context of this field research, in some ways like being a blind person—I could not see directly how experiences and practices are tied to culture. However, through the incorporation of visual methods, this has not been a limitation; rather, the visual methods helped me to capture meanings hidden behind language. I suggest that cross-cultural research is a good opportunity to broaden, to subvert and to remake research methods. It clearly reveals that the world is so complex that we cannot fully grasp it. It also inspires us to transcend the habit of looking for security and the definite; to recognize the importance of opening up our sensibilities to uncover multiple realities through methods that might otherwise be dismissed as slow, vulnerable and imprecise.

Acknowledgements This research received funding from the European Union's Horizon 2020 research and innovation programme under the Marie Sklowdoska-Curie Grant Agreement No. 765389.

References

Anderson, B. (2006). Becoming and being hopeful: Towards a theory of affect. *Environment and Planning d: Society and Space, 24*(5), 733–752. https://doi.org/10.1068/d393t

Anderson, B., & Harrison, P. (2016). *Taking-place: Non-representational theories and geography*. Routledge.

Bagnoli, A. (2009). Beyond the standard interview: The use of graphic elicitation and arts-based methods. *Qualitative Research, 9*(5), 547–570. https://doi.org/10.1177/1468794109343625

Banks, M. (2001). *Visual methods in social research*. Sage.

Bartlett, R. (2013). Playing with meaning: Using cartoons to disseminate research findings. *Qualitative Research, 13*(2), 214–227. https://doi.org/10.1177/1468794112451037

Beilin, R. (2005). Photo-elicitation and the agricultural landscape: 'Seeing' and 'telling' about farming, community and place. *Visual Studies, 20*(1), 56–68. https://doi.org/10.1080/14725860500064904

Brown, R. D., & Yokohari, M. (2003). Ideological contributions of satoyamas. In K. Takeuchi, R. D. Brown, I. Washitani, A. Tsunekawa, & M. Yokohari (Eds.), *Satoyama: The traditional rural landscape of Japan* (pp. 1–6). Springer.

Campbell, L. K., Svendsen, E. S., Reynolds, R., & Marshall, V. (2019). Material and social relations in a coastal community garden assemblage. *Social & Cultural Geography*, 1–23. https://doi.org/10.1080/14649365.2019.1658800

Caretta, M. A. (2015). Situated knowledge in cross-cultural, cross-language research: A collaborative reflexive analysis of researcher, assistant and participant subjectivities. *Qualitative Research, 15*(4), 489–505. https://doi.org/10.1177/1468794114543404

Collier, J., Jr. (1957). Photography in anthropology: A report on two experiments. *American Anthropologist, 59*(5), 843–859. https://doi.org/10.1525/aa.1957.59.5.02a00100

Dahl, S., Morris, G., Brown, P., Scullion, L., & Somerville, P. (2012). *Somewhere nowhere. Lives without homes*. Salford Housing and Urban Studies Unit.

Darnhofer, I. (2018). Using comic-style posters for engaging participants and for promoting researcher reflexivity. *International Journal of Qualitative Methods, 17*(1), 1–12. https://doi.org/10.1177/1609406918804716

Davies, G., & Dwyer, C. (2007). Qualitative methods: Are you enchanted or are you alienated? *Progress in Human Geography, 31*(2), 257–266. https://doi.org/10.1177/0309132507076417

Esham, M., Kobayashi, H., Matsura, I., & Alam, A. (2012). Japanese agricultural cooperatives at crossroads: A review. *American-Eurasian Journal of Agricultural and Environmental Sciences, 12*(7), 943–953. https://doi.org/10.5829/idosi.aejaes.2012.12.07.175

Foster, H. (1988). Preface. In H. Foster (Ed.), *Vision and visuality* (pp. ix–xiv). Bay Press.

Giustini, D. (2019). "It's not just words, it's the feeling, the passion, the emotions": An ethnography of affect in interpreters' practices in contemporary Japan. *Asian Anthropology, 18*(3), 186–202. https://doi.org/10.1080/1683478X.2019.1632546

Gliessman, S. (2017). Agroecology: Building an ecological knowledge-base for food system sustainability. *Agroecology and Sustainable Food Systems, 41*(7), 695–696. https://doi.org/10.1080/21683565.2017.1335152

Harper, D. (2001). *Changing works: Visions of a lost agriculture*. University of Chicago Press.

Harper, D. (2002). Talking about pictures: A case for photo elicitation. *Visual Studies, 17*(1), 13–26. https://doi.org/10.1080/14725860220137345

Harrison, P. (2000). Making sense: Embodiment and the sensibilities of the everyday. *Environment and Planning d: Society and Space, 18*(4), 497–517. https://doi.org/10.1068/d195t

Harrison, P. (2007). "How shall I say it…?" Relating the nonrelational. *Environment and Planning A, 39*(3), 590–608. https://doi.org/10.1068/a3825

Hitchings, R. (2003). People, plants and performance: On actor network theory and the material pleasures of the private garden. *Social & Cultural Geography, 4*(1), 99–114. https://doi.org/10.1080/1464936032000049333

Iwatsuki, K. (2008). Sustainable use of biodiversity, with reference to the Japanese spirit of worshipping nature. In N. Furuta, K. Iwatsuki, H. Nishida, & M. Kawamichi (Eds.), *Conserving nature: A Japanese perspective* (pp. 4–11). Biodiversity Network Japan.

Kitagawa, F., Breslin, L., & Favell, A. (2015). *Art place Japan: The Echigo-Tsumari Art Triennale and the vision to reconnect art and nature*. Princeton Architectural Press.

Krzywoszynska, A. (2015). On being a foreign body in the field, or how reflexivity around translation can take us beyond language. *Area, 47*(3), 311–318. https://doi.org/10.1111/area.12202

Krzywoszynska, A. (2016). What farmers know: Experiential knowledge and care in vine growing. *Sociologia Ruralis, 56*(2), 289–310. https://doi.org/10.1111/soru.12084

Law, J. (2004). *After method: Mess in social science research*. London: Routledge.

Lorimer, H. (2005). Cultural geography: The busyness of being 'more-than-representational'. *Progress in Human Geography, 29*(1), 83–94. https://doi.org/10.1191/0309132505ph531pr

Mulgan, A. G. (2005). Where tradition meets change: Japan's agricultural politics in transition. *The Journal of Japanese Studies, 31*(2), 261–298. https://doi.org/10.1353/jjs.2005.0053

Pereira, M. D. M., Scharff, C., & Marhia, N. (2009). Interrogating language difference and translation in social science research: Towards a critical and interdisciplinary approach. *Graduate Journal of Social Science, 6*(3), 1–12.

Pile, S. (2010). Emotions and affect in recent human geography. *Transactions of the Institute of British Geographers, 35*(1), 5–20. https://doi.org/10.1111/j.1475-5661.2009.00368.x

Rodríguez Estrada, F. C., & Davis, L. S. (2015). Improving visual communication of science through the incorporation of graphic design theories and practices into science communication. *Science Communication, 37*(1), 140–148. https://doi.org/10.1177/1075547014562914

Rose, G. (2014). On the relation between 'visual research methods' and contemporary visual culture. *The Sociological Review, 62*(1), 24–46. https://doi.org/10.1111/1467-954X.12109

Rose, G., & Tolia-Kelly, D. P. (2012). Visuality/materiality: Introducing a manifesto for practice. In G. Rose & D. P. Tolia-Kelly (Eds.), *Visuality/materiality: Images, objects and practices* (pp. 1–11). Ashgate Publishing Ltd.

Sherren, K., Fischer, J., & Fazey, I. (2012). Managing the grazing landscape: Insights for agricultural adaptation from a mid-drought photo-elicitation study in the Australian sheep-wheat belt. *Agricultural Systems, 106*(1), 72–83. https://doi.org/10.1016/j.agsy.2011.11.001

Sherren, K., Fischer, J., & Price, R. (2010). Using photography to elicit grazier values and management practices relating to tree survival and recruitment. *Land Use Policy, 27*(4), 1056–1067. https://doi.org/10.1016/j.landusepol.2010.02.002

Sherren, K., & Verstraten, C. (2013). What can photo-elicitation tell us about how maritime farmers perceive wetlands as climate changes? *Wetlands, 33*(1), 65–81. https://doi.org/10.1007/s13157-012-0352-2

Temple, B. (2002). Crossed wires: Interpreters, translators, and bilingual workers in cross-language research. *Qualitative Health Research, 12*(6), 844–854. https://doi.org/10.1177/104973230201200610

Temple, B., & Young, A. (2004). Qualitative research and translation dilemmas. *Qualitative Research, 4*(2), 161–178. https://doi.org/10.1177/1468794104044430

Thien, D. (2005). After or beyond feeling? A consideration of affect and emotion in geography. *Area, 37*(4), 450–454. https://doi.org/10.1111/j.1475-4762.2005.00643a.x

Tolia-Kelly, D. P. (2006). Affect—An ethnocentric encounter? Exploring the 'universalist' imperative of emotional/affectual geographies. *Area, 38*(2), 213–217. https://doi.org/10.1111/j.1475-4762.2006.00682.x

Tufte, E. (1997). *Visual explanations. Images and quantities, evidence and narrative*. Graphics Press.

Turner, S. (2010). Research note: The silenced assistant. Reflections of invisible interpreters and research assistants. *Asia Pacific Viewpoint, 51*(2), 206–219. https://doi.org/10.1111/j.1467-8373.2010.01425.x

Yokohari, M., & Bolthouse, J. (2011). Keep it alive, don't freeze it: A conceptual perspective on the conservation of continuously evolving satoyama landscapes. *Landscape and Ecological Engineering, 7*(2), 207–216. https://doi.org/10.1007/s11355-010-0116-1

10

The Eye of the Beholder: Applying Visual Analysis in an Historical Study of Lynxes' Representations in the Bavarian Forest Region

Zhanna Baimukhamedova

> Yes. They will forget. That is our fate, nothing can be done. What appears to be serious, significant, so very important—when the time comes—will be forgotten or seem so trivial.
>
> —A. Chekhov, *Three Sisters*

Introduction

Wild animals occupy a special place in the human psyche: our oldest companions on the planet, their wilderness often served as a backdrop to highlight people's humanity. One of the earliest examples of art, cave paintings of game and hunters, is repeated century after century in different forms, media, and connotation. What has largely remained

Z. Baimukhamedova (✉)
Rachel Carson Center, Ludwig Maximilian University of Munich, Munich, Germany
e-mail: zhanna.bai@rcc.lmu.de

A. Franklin (ed.), *Co-Creativity and Engaged Scholarship*,
https://doi.org/10.1007/978-3-030-84248-2_10

the same is fascination with the wild, a hope to tame—when not in flesh, then in image. John Berger notes that looking at an animal, having it in one's proximity—even in the form of a still image—offers a person some sort of consolidation to the inescapable feeling of "loneliness of man as a species" (1991, p. 6). This "great divide" separating humanity from our natural environment commenced eons ago with the first instances of domestication and the beginnings of agriculture. This was exacerbated by the Industrial Revolution and aggravated to an extreme degree by capitalism's voracious greed for extraction and profit, marginalizing animals by reducing them to a mere commodity: a raw material, a fodder, a plaything, a spectacle. Never in the history of the planet were people so outnumbered by species that are raised specifically for consumption—and our appetites grow still.[1]

In qualitative academic research, employing visual analysis methods often evokes associations with a rather niche application. Rarely do they compose the cornerstone of an investigation in fields outside of, for example, media studies, art history, or *Bildwissenschaft*[2]—in short, outside of disciplines that *literally* characterize, situate, and interpret images. However, while by no means a universal tool to fill the gaps in all academic enquiry, employing visual analysis builds up on a rather prosaic act of seeing and making sense of the outside world, imaginatively. Visual analysis is necessarily a creative process, and as a research method it opens up the possibility of forming conclusions from one's personal interaction with what is seen. To borrow the words of Murray Bookchin, who wrote under the pseudonym Lewis Herber, a task of enriching academic research to make sense of phenomena, whose comprehension might stall

---

[1] For the progression of numbers in the past several decades, see, for instance, an article by Alex Thornton "This Is How Many Animals We Eat Each Year" at the World Economic Forum website.

[2] *Bildwissenschaft*, like visual studies broadly, is a discipline that tries to understand and interpret the significance of imagery. Originating from German-speaking academic circles, *Bildwissenschaft* employs a variety of methods (often interdisciplinary) to analyse a vast variety of visual sources. Moreover, it focuses predominantly on the visual properties of an image rather than its other characteristics (i.e., its history or the political, economic, or social conditions of its making). For a fuller explanation of differences between *Bildwissenschaft* and other visual studies, see Jason Gaiger, "The Idea of Universal Bildwissenschaft", *Estetika: The European Journal of Aesthetics*.

when traditional methods applied, "calls for imaginative departures from conventional approaches" (2018, p. 17).

In this chapter I reflect upon employing visual analysis to corroborate an historical narrative, tracing the development of human–wildlife relationships in the Bavarian Forest region. Using the example of Eurasian lynx, I closely study their visual representations in the span of the past 50 years. Unlike many other large carnivores, lynxes with their reticent, somehow shy appearance tend to be viewed favourably and are generally considered a charismatic species.[3] The local population has been progressively more accepting of lynxes' presence.[4] Therefore, the aim of this chapter is to answer the question: how these changes are visually represented in the local media? Critically analysing representations of lynxes helps us understand how this fluctuating dynamic has been visually captured over the years.

The task appears simple and complex at the same time. Its simplicity lies in the fact that, in principle, media reflects general public attitude, and since the trajectory of lynxes' dispersal in the area can be considered a success story, one expects the local media to have a generally positive coverage. At the same time, as famously noted by Edward S. Herman and Noam Chomsky (2002), media have the capacity to form popular opinion based on the assumption that they truthfully represent the opinion of the public. The capacity of media to affectively inform people's opinions makes it a powerful instrument to propagate a certain agenda. Moreover, local media sources commonly evoke a feeling of proximity: since physically they are often located in or very close to a place that they cover in their reports, they appear more accessible and trustworthy than regional, state, or international media. The way local media cover a sensitive topic, such as the return of large carnivores, can

---

[3] Charisma here, as Jamie Lorimer explains, is a set of characteristics that have a popular appeal for circulation in the media and, by extension, markets. Aesthetic charisma, he continues, "describes the visual appearance of a species in print, on film, or in the spectacular encounters of ecotourism" (2015, p. 40). An animal whose look people tend to view favourably can become the face of a campaign and thus lend its attractiveness to the campaign itself.

[4] It is, however, worth noting that the presence of large carnivores is still a contentious issue. For a more detailed account of the tortuous path of the resettlement of large carnivores in the Bavarian Forest region, see Ulrich Schraml (2019) *Wildtiermanagement für Menschen* (in German).

have a direct impact on the general acceptance of this species. And since photographs may tell a story of their own depending on their interpretation and intentionality, critically analysing their content uncovers a wider web of relationships, interests, and attitudes.

Regardless of where, discussions around the reintroduction or return of large carnivores are typically laden with conflicting opinions. In the Bavarian Forest region, the slow resettlement of lynxes has been amply documented by local news outlets: from fearful to tamed, from welcomed to despised—the articles have followed the furtive movement of the lynx through the forest and at times in proximity, often fatal, to human settlements. Through this coverage lynxes' imagery slowly started to make its way into people's homes: not necessarily welcome as creatures of flesh and blood, these large cats entered collective cultural memory (Kuhn, 2007), enriching the vernacular with ecological jargon and, for once, ushering the possibility of coexistence of humans and wild beasts.[5] While overall attitude towards lynxes' presence in the Bavarian Forest region has improved dramatically over the past several decades, analysing how lynxes have been represented could shed light on whether images illustrate this changing dynamic. Andrew Isenberg (2002, p. 60) once said that "[our] representations of wildlife are inescapably expressions of human values", and visual analysis allows looking behind a textual narrative to discern whether what we *see* of the wildlife corresponds to what we *understand.*

This chapter is organized in the following order: a general introduction to visual analysis that talks about why visuality is such an intrinsic part of human experience and what makes it conducive for research purposes; the following section deals specifically with wildlife photography, its affective quality that sets it apart from a visual representation of inanimate nature; the section on methodology tackles the procedural part of conducting visual analysis—the what, how, and why of applying this method; in the empirical section, two photographs of lynxes are analysed; the discussion section covers the applicability of the present findings in similar kinds of research on studying animals' visual representation in

[5] See, for example, a volume edited by Marco Heurich and Christof Mauch, *Urwald der Bayern: Geschichte, Politik und Natur im Nationalpark Bayerischer Wald* (2020, in German).

the media; and finally, the chapter ends with a conclusions section where merits and challenges of the method are discussed.

Seeing as a Creative Process

As a species, we largely make sense of the world through the act of seeing: we intake, comprehend, store, and represent reality imaginatively rather than linguistically. By no means a "truthful", or in any way precise, visual perception is a bridge between the outside world and ourselves (Barry, 1997). Walter Benjamin once said that "[d]uring long periods of history, the mode of human sense perception changes with humanity's entire mode of existence" (1969, p. 5), yet what stretches further both through time and geographical terrain is the fact that principally humans perceive in images. The unprecedented proliferation of social media and streaming platforms testifies to a predominance of visual media. The entire fabric of certain types of contemporary work, leisure, entertainment, and education, to mention but a few, is woven in glaring threads of alluring, ever-changing images. Imaginary thinking is a *lingua franca* that makes a conversation, however restricted, possible across different scales, contexts, and geographies.

Visual analysis, broadly defined as a collection of methods to interpret and understand imagery, has been employed widely in studies of art and art culture, and more recently in other fields of inquiry as well. However, as Stephen Spencer notes, despite the growing popularity of visual analysis methods, social sciences have been rather reluctant to recognize and apply them as valid information sources on a par with text. He emphasizes the potentiality of visual methods to "provide a deeper and more subtle exploration of social contexts and relationships" (2010, p. 1) while at the same time acknowledging that within some strands of social sciences visual analysis might have limited applicability. It is the ability of visual analysis to show the familiar from an unfamiliar angle, coupled with the recognition of actors—producers, subjects, and "consumers"—of visual material, which affords this sort of research a novel, creative twist. With that in mind, Spencer also warns that visual analysis should not be viewed as a scholarly silver bullet, a sort of innovative instrument that can singlehandedly substitute all other methods of inquiry.

In historical analysis, engaging with the visual might uncover how a certain subject has been portrayed and perceived throughout the years. By its nature, an historical study requires looking at a phenomenon as it unravelled over a period of time. Photographs are momentous, but a compiled body of several of them taken at different points could tell a story beyond a text, enrich textual description or give it more depth. Apart from a natural improvement of technology, looking at photographs could illustrate the development of a photographed subject: what is depicted in an image and how, which details are present and whether they change over time, what is brought to the foreground, what is obscured, and the reason for either method. Taking a photograph, especially with the proliferation of smartphones, has become a routine occurrence, and while the mechanics of taking a picture might, by and large, be the same, *what* we tend to photograph, which instances we deem worthy of preserving for future use and reflection, differ.

The affective faculty of imagery and its expository significance for academic research has been noted by many. For instance, in *Camera Lucida* (1981) Roland Barthes's describes images that impel shifts in people's psyche that can later be translated into actions for a cause. More recently, Susan Sontag analysed the participatory competence of photographs, both taking and seeing. Images here serve as a sort of language through which both the producer and the consumer of a photograph enter a dialogue, creating a narrative that spans over time and space. Still images, Sontag contends, allow a way of examining reality that is otherwise impossible since a human's focus tends to wander. While appreciating this gift of a fixed gaze, she acknowledges that perceiving something in complete immobility is unnatural to our senses: "Life is not about significant details, illuminated in a flash, fixed forever. Photographs are" (2005, pp. 63–64).

Visual analysis of large carnivores can demonstrate how they were perceived at a certain point in history. For instance, in early modern English heraldry, Kathryn Will attests (2016), animals represented a set of highly specified characteristics, attributed to each beast and, by extension, to a person whose property a said beast visually inhabited. More contemporarily, as in cases with the reintroduction of large carnivores across the globe, their images go through a rigorous process of selection in order to convey a specific, interest-laden message. Varying from proponents of their return, employing images of large carnivores playing and tending for one another, to photographs depicting an attack on livestock or a blood-stained muzzle below a pair of empty glowing eyes, used by those whom proximity to carnivores appears rather perilous, these pictures hit every point on the gradient of attitudes. Canons of photography change over time, but their main purpose—that of an affective illustration—withstand the caprices of fashion.

A question might arise: how much background knowledge is necessary to conduct visual analysis? A trained eye, it can be argued, might see something inaccessible to a layperson, yet the methodological and conceptual innocence of the latter might offer insights to which a trained eye could be blind. An image is such a medium whose interpretation easily differs from person to person. Often even the intentionality of a maker can be misinterpreted by a viewer, endowing an image with a whole new meaning. However, the specific positionality of an image-maker should not be overlooked: although what they try to convey might be taken in a million different ways, it is the unique moment of their engagement with a subject that then turns into a visual material in the examination. Susan Sontag confirms this influence of image production by saying that, "[in] deciding how a picture should look, in preferring one exposure to another, photographers are always imposing standards on their subjects" (2005, p. 4). A photographer snatches a moment in all its intricacy that afterwards starts living a life of its own.

Perceiving an image, intending to understand it, and take it in amounts to an act of profound retrospection. Visual analysis is a highly reflective exercise where a viewer must confront their situatedness to see something in an image that lays way beyond the apparent. By acknowledging their positionality, the viewer enters a dialogue with an inanimate representation of reality, getting a glimpse of the world that feels like it is more available, more accessible than it is in reality. Committing to visual analysis is not just confrontation of personal ethical and intellectual baggage, but also a manifestation of a certain degree of trust in one's ability to contemplate, discern, and appreciate "the immediacy of the moment" (Grady, 2008) and all its history, in making and being. As a tool, the applicability of visual analysis depends on the kind of research question one wishes to ask—and some answers might well be found in the visual.

Wildlife Photography

Wildlife photography takes a somewhat privileged position in nature photography. Not only does it try to depict living creatures who, at any point, might flee the frame, but also, as in the case with large predators, it attests to exceptional patience and physical ableness of a photographer. Juha Suonpää, a researcher and himself an avid photographer, speaks about the derivative value of a photograph that correlates to the amount of effort a photographer must put in to "snap" that one prized (sometimes literally) shot. He concludes that, "whatever animal is rare, timid or exceptionally spectacular is considered to be especially valuable when captured on film" (2000, p. 59). This statement speaks to what Robert Castel and Dominique Schnapper discuss when they describe how, for a photograph to attain deeper meaning, they must demonstrate "weighty, imposing subjects" (1990, p. 120).

Wildlife photography as an art form has its canons—tropes, repeated scenarios, preferred angles, and such—that in turn aim to represent an animal in its natural environment *authentically*. There are, of course, other forms of wildlife photography that take place in zoos and enclosures. In such cases, its mobility restrained and behaviour rather predictable, an animal going about its everyday life offers a wide array of photo opportunities that can be easily transformed into a collection of spectacular images. Tamed to a degree, an animal nonetheless appears unquestionably wild not only because its look corresponds to what people tend to have in mind when they think of wildlife. In a sense, wildlife photography engenders the source of its own proliferation: since very few people have unmediated access to wild animals in their most agitated states, the imagery comes chiefly from the creative representative work of wildlife enthusiasts whose choice of subjects and compositions becomes the subjects and compositions inhabiting people's imaginations.

For instance, Kelly Enright (2012) observes that in wilderness accounts of many scholars and explorers in the first half of the twentieth century in the United States, the idea of what one would encounter in the jungle often preceded an actual departure: the conceptualization of what wilderness entails was projected on what was yet to be seen. As a result, many such writings reinforced the mythical appearance of the wild—and this work of imagination translated to an array of ecological practices, both in conservation and representation. An eternal debate on which nature should be protected (and, by extension, which species, large and small) always reflects human values—and these values are often informed by imagery, manufactured on purpose or incidentally. When it comes to wildlife photography, Derek Bousè notes that, "animals in the wild rarely appear as majestic (or as *cuddly*) as they do on the screen" (2003, p. 124, original emphasis), yet the sheer volume of dramatic shots—especially of large predators—makes it hard to believe that those animals can be anything but stunning, always.

While looking, perceiving, and appreciating the look of an animal, a viewer bases their judgement, at least in part, on pure viscerality, on the feeling of interacting with a beast in a rather unmediated way. In the process, personal history, experiences, and professional occupation of an observer can surface without one's active will—after all, none of us can help knowing what we know. What triggers a certain association often bypasses consciousness, and that is what makes interpretation an act of art rather than simple calculation. Close analysis of an image could enrich a narrative in a way unattainable by the means of text alone: while inherently subjective and rather indiscriminate as to what kind of background knowledge a person possesses, visual analysis opens up an opportunity to create a story, compellingly different from what a photographer might have intended. Certainly, text might have just as many readings as an image, yet most text has some sort of a structure that makes its interpretation a more orderly affair.

Methodology

Unlike most other research methods, visual analysis is an activity that we all engage in constantly across our personal and professional lives. Saturated with images, everyday life offers innumerable opportunities for engaging with the visual. Employing visual analysis for research purposes does not differ drastically from observing the outside world by means of seeing—rather, the differences lay in the level of consciousness, as well as rigour, depth of critical reflection, and purposes towards which the act is targeted. Applying visual analysis in research necessitates the creation of a mental space where a person is invited to derive information using a regular yet often overlooked medium—the visual. Exploring possibilities of visual analysis makes an important contribution to our knowledge because, building on an everyday activity of seeing, it propels one to question their first visual impressions, to look more intently in order to really see.

Before attempting an historical study of photographs, one is inevitably confronted with several logistical issues: which images to pick, how to access them, which timeframe is sufficient to show an historical progression, how many photographs constitute a minimum necessary sample. There is no one universal answer and addressing these issues already constitutes the first step of the research process. Just like any other method of investigation, visual analysis aims to answer a specific research question. Therefore, first and foremost, one needs to think about what interests them in a matter, and depending on the answer, select a research method that is best suited to tackle the issue. John Grady (2008) suggests that a good starting point is to ask oneself whether the images, however many, produce sufficient information necessary to answer a research question. Since images often offer an innumerable number of interpretations, it is crucial to limit oneself with a clearly defined framework of a research question.

When analysing a photograph, Annette Kuhn (2007) outlines a four-step procedure, each addressing a specific part of the production, reception, and study of an image: (1) subject/s; (2) context of production; (3) technology and canon; (4) audience. These parts are not clearly separated and often blend into one another, allowing the forming of a coherent narrative rather than a check-list. To conduct a visual analysis of animal imagery, Matthew Brower (2011) expands this process by pointing to a question of whether the animal portrayed is, in fact, a living animal or a diligently reconstructed stuffed replica. While in present times wildlife photography predominantly showcases living creatures, in its inception this genre of visual representation did not shy away from depicting corpses propped to look alive[6] or to paint zoo animals as if they roamed free.[7] Keeping these components in mind when attempting visual analysis helps to navigate an image and produce insights that might be overlooked from cursory contemplation.

In principle, photographs of wild animals—including lynxes—fall into one of two broad categories: the authentic ones, acquired by luck or through the medium of a camera trap, and the "staged" ones, where the animals look majestic and wild, although snapped in the (dis)comfort of an enclosure or another kind of confinement. The arranged characteristics of the second kind of photographs are rarely recognized, for the underlying idea of an image is to convey the wildness of a beast—or rather to project what the wildness of a beast constitutes in human imagination (Fudge, 2002). To borrow Roderick Nash's apt observation, wilderness and its inhabitants are currently gaining popularity, yet in people's struggle to see the wild, the wild is under constant human encroachment and runs the risk of "being loved to death" (2001, p. x): to capture an animal behaving naturally, it must be under continuous, incessant surveillance.

At its core, the wildlife genre has an underlying assumption that the scenes portrayed are unmediated representations of animals as they are in their natural environment—even when in reality each snapshot comes as a result of a lengthy production process (Chris, 2006). To help situate an image within a broader historical background, titles and accompanying

[6] Contemporary hunting photography inherited some of this lore whereby fallen animals, when photographed, are portrayed as if still alive and even at times in motion. See, for example, Kalof and Fitzgerald "Reading the Trophy: Exploring the Display of Dead Animals in Hunting Magazines" (*Visual Studies, 18*(2), 112–122, 2003).

[7] One of notable exceptions from this rather commonplace practice was a German artist, Wilhelm Kuhnert, who habitually painted wild animals in situ. His elaborate drawings of animals informed not only the imaginative perception of African wildlife, but also filled pages of many scholarly publications of the time.

captions might serve as guides. Admittedly, there is no clear consensus on whether they should be considered when analysing a visual representation of an animal. Titles are more universal when considering a work of visual art and captions almost always accompany a publication in a periodical. However, it is important to keep in mind that not all visual materials come with a textual description (including, for example, a person's own photography). They can provide important insights into the context of an image, but at the same time there is a possibility of involuntarily adopting certain judgements passed on by whoever created a caption. The literature on the matter in the majority of cases appears to leave it to the discretion of a researcher.

Wildlife imagery and the context of its production offer insights into broader social and historical changes. For this chapter, I selected two photographs of lynxes from a local newspaper, *Grafenauer Anzeiger*. The newspaper primarily covers local affairs of the districts *Freyung*, *Grafenau*, and *Waldkirchen*, and its headquarters are located in the same town as the administration of the Bavarian Forest National Park. For years, *Grafenauer Anzeiger* has been reporting on matters of the national park as well as at times serving as a means of communication between the locals and the national park staff. This proximity has naturally allowed the newspaper to take a close look at the national park's affairs and report on them sooner and in more detail than other newspapers.

The first of the selected photographs (Fig. 10.1) dates back to 1973, shortly after the reintroduction of lynxes to the area. Vanished from Bavaria for more than a hundred years, lynxes were secretly brought from the mountains of Slovakia, soon after the establishment of the first national park in Germany. The second photograph (Fig. 10.2) appeared in an online publication of *Grafenauer Anzeiger* from April 2016 to illustrate a report on a traffic collision with the lynx portrayed. While more visual material could be employed for the purposes of this chapter, after careful consideration a number of similar photographs were disregarded to keep the chapter succinct. These two photographs seem ample enough to illustrate the applicability of visual analysis in observing changing attitudes towards lynxes in the area. When analysing the photographs, captions proved to be an excellent complementary source of information.

Fig. 10.1 One of the first photographs of lynxes in an article covering plans for their return (Source: *Grafenauer Anzeiger*, 6 October 1973)

Seeing the Lynx

The October edition of *Grafenauer Anzeiger* contained one of the first photographs of lynxes in the Bavarian Forest National Park (Fig. 10.1). The title reads simply "Soon to be released in freedom: lynxes in the National Park".[8] In the photograph, two animals are depicted sitting very close to each other on what appears to be a large boulder. The one on the right, whose fur is ornate with more dramatic shadowing, is located closer to the camera; it squints. The one on the left sits slightly to the

[8] "Bald in Freiheit enlassen: Luchse im Nationalpark"—my translation.

side: we see more of its elongated, so unmistakably feline body. Advanced understanding of lynxes' morphology would be required to guess the gender of the two individuals, but since the animals form an apparent pair, it could indicate the duty of proliferation bestowed upon them by their status of harbingers of the large carnivore's return. Conversely, the lynxes could also be siblings transported in a pair for company and comfort. The camera catches a human figure behind a wooden fence—in the background and out of focus. The person nonetheless is a part of the composition, which indicates that the photograph was taken amidst people. The lynxes appear calm, human presence does not seem to bother them much. Both attentively look at one thing on a far side behind the camera; the camera angle takes the shot from a lower angle. The lynxes look imposing, yet one cannot help but see them as enclosure lynxes.

This photograph is interesting to look at because, as described previously, it corresponds to the strand of wildlife photography that tries to depict the wilderness by showing a captive subject. There are no cage bars that are normally associated with enclosed animals, yet the visible part of the fence and a significantly more elevated position of a person behind leave no doubt that lynxes are photographed while under a vigilant eye and, subsequently, control of humans. With many a debate that the establishment of the Bavarian Forest National Park and the following reintroduction of lynxes instigated, the photograph serves as a sort of reassurance that the animals would not cause much trouble—they sit, docile, free yet tamed. The photograph has a certain visual appeal with animals glancing to the right as if following the flow familiar to a western viewer.[9] This direction is described by some as the "gaze in the future"—a fitting visual accompaniment for a report on further plans to bring in new species to the national park.

The photograph also exemplifies a sort of commodification of wild beasts: living in an enclosure, lynxes have relatively sufficient room to prowl, play, and hide. However, an enclosure serves as a spacious yet restricted stage where visitors of the national park can marvel at lynxes, behaving *as if* in their natural element. To be fair, enclosures in national

[9] For more detail on the validity of culturally specific positioning in visual media, see Bode et al., "Left-Right Position in Moving Images: An Analysis of Face Orientation, Face Position, and Movement Direction in Eight Action Films" (*Art & Perception, 4*(3), 241–263, 2016).

park management practices across the globe often represent a necessary trade-off—sensitizing people to the presence of often problematic species and attending to educational purposes of teaching about animals' characteristics[10]: they are frequently one the few possible ways of reintroducing/providing more roaming range for certain animals. Based on the caption, the two lynxes in the photograph were not destined to stay in an enclosure forever; however, an enclosure as a manifestation of wider human–wildlife relationships speaks to the fact that even the perceived wilderness of an animal can become a commodity, a spectacle. Not solely by the means of visitation and wonder, people consume animal objects through imagery, both making and as postcards, photographs, promotional material, etc.

Based on the analysis and literature review, it is possible to think of visual representations of large carnivores in three complementary ways: (1) awesome and fearful; (2) tamed and disciplined; (3) problem species. Throughout the years, all three have been employed by *Grafenuaer Anzeiger* to depict how general predisposition towards lynxes' presence in the area has been progressing. For instance, in the beginning, there was much uncertainty, for lynxes had been absent from the forest for more than a century and, as often lamented in reports dealing with clashes between people and wild beasts, knowledge of how to coexist did not withstand the test of time. In the meantime, farmlands and road networks had expanded dramatically, and locals were not necessarily convinced that the animal could survive in the altered cultural landscape. Figure 10.1 might be interpreted as putting forward an image of lynxes as both fearful and tamed—their posture invites admiration yet the surroundings indicate that, as impressive as the animals appear, they have been put safely under human management.

Much in the same vein, more recent photographs of lynxes portray animals that are free to wander but at the same time are subtly domesticated. An online issue of *Grafenauer Anzeiger* from April 2016 contains a photograph of the lynx Patrick (Fig. 10.2). The caption under the

10 "Problematic species" is an interesting concept because it speaks to broader human attitudes towards animals and nature in general: there is no intrinsic "problem" with a certain species—rather, it is the degree of its living and predation habits negatively affecting people's interests that renders an animal a problem.

Fig. 10.2 Lynx Patrick (Source: *Grafenauer Anzeiger*, 13 April 2016)

photograph reads: "Lynx Patrick was photographed with a tracking collar in the winter of 2015/2016 in Farrenberg bei Finsterau (administrative district Freyung-Grafenau). On March 24, he was run over on B12 close to Philippsreut (administrative district Freyung-Grafenau) and then died".[11] Unlike the lynxes in Fig. 10.1, this one had a name, which lends a feeling of proximity and familiarity; he was not just a lynx, but lynx Patrick. The frame includes a boulder on which Patrick is resting and bare, snow-covered trees. Set against the backdrop of the overcast winter sky, the colour palette of the photograph appears rather monotonous with the only colourful element being Patrick himself and the tracking collar adorning his neck. Patrick is looking to the viewer's right; he seems to be unaware of or ignoring the camera. As in the previous photograph, the angle here is from a slightly lower position. Patrick is figuratively and thematically in the centre of the image.

[11] "Luchs Patrick, der mit einem Sender versehen war, im Winter 2015/2016 aufgenommen am Farrenberg bei Finsterau (Landkreis Freuyng-Grafenau). Er wurde am 24.März bei Philippsreut (Landkreis Freyung-Grafenau) auf der B12 überfahren und kam dabei ums Leben"—my translation.

A barren, almost featureless background makes it easy to concentrate on a smallish, yet central animal figure. All of the elements of the photograph that have any motion to them (the many angles formed by conjunction of sprawling twigs, the curvature of rocks, the thick framing line of a naked tree, the overall movement from the top left of the photograph where the highest point trespasses the limits of the frame to the bottom-middle right where some other vegetation, out of focus, softly encloses the field of view) are at the fore, and Patrick is crowning the composition. The undeniable centrepiece, whose posture and diverted face give a sense of disinterest, almost palpable feeling of separation, Patrick looks to me as if informed of his soon fatal encounter with a human artefact, a car. He is free in his element yet is also clearly tamed—the collar betrays Patrick's otherwise supposed wilderness. Animal collars are inextricably objects of domination: they control, signify belonging, identify, and locate. Collaring wild animals has a number of practical and scientific reasons, yet *visually* a collar turns an animal into a pet. What a name does in a text, so a collar does in an image: Patrick stands for wilderness made a domestic matter.

In the forty years separating the photographs lynxes have established themselves as quite a regular species in the Bavarian forest. Still observed by very few outside of enclosures, they are monitored and—like Patrick—some tagged with a tracking collar. There are sightings of lynxes provided by camera traps that at times make it to the pages of local and regional newspapers. Progressively, as technology improves and more people engage in wildlife photography, the depictions of lynxes appear more "heroic": whereas in the beginning, it seemed like the purpose of a photograph was to familiarize a viewer with the look of lynxes, more recently an image tells a story. To reiterate the point, Patrick is not just a lynx, but a named lynx whose life was followed and whose death was featured. In the photograph, Patrick does not look in the camera[12]; he is present yet detached. As we learn from the caption, Patrick was run over. The photograph transmits the feeling of separation quite well.

[12] Eye contact is a common occurrence in wildlife photography and documentaries, serving to make a viewer feel a sense of connection and intimacy with an animal portrayed. See, for example, Bousè (2003).

Discussion

Representation is never neutral and, as such, visual representation of animals has often been employed as a sort of leverage point, an emotional trigger aimed to deliver a certain message. J. Keri Cronin notes that, "images of nature are always already bound up in political, social, cultural, and environmental processes" (2011, p. 19). The history of humanity is intertwined with the life paths of our animal companions: enchanted by their unpredictable and, therefore, mysterious behaviour, people have been watching wildlife in the hope of glimpsing behind animals' impervious gaze (Peterson, 2013). Following the overarching momentous nature of photography, visual representation of wild animals affords a close-up look that is impossible for the vast majority of people (Bousè, 2003). It is this closeness, this manufactured intimacy[13] of contact that makes wildlife photographs so powerful—when wild animals enter our mental space, it takes less strain to feel affected by and affectionate towards them (Serpell, 2004).

Historically, visual representations of charismatic animals have often been employed by different organizations to evoke an emotional response in viewers. These images in the context of the *return* of the animals, i.e., in an allusive acknowledgement of their nativity to the land and human-afflicted temporal absence, imply something similar to what Bernhard Gissibl described as remorse over an implied loss of peaceful coexistence between animals and humans. Framing this process in terms of "return" (unlike, for instance, invasion or colonization), signifies that for a while animals inhabited this area—and if humans did too, then at least for a time they managed to live there together. However, Gissibl continues, "[h]umans and wild animals coexisted in dynamic adaptation, and hunting was the main way of human interaction with them" (2016, p. 36). The story of large carnivores in Bavaria, as with many other stories from all around the globe, went along similar lines: predators embodied

[13] Matthew Brower tackles this issue in his book *Developing Animals: Wildlife and Early American Photography* (2011). There he talks about how wildlife photography necessarily erases humans and the visible results of their activity from the frame so as to heighten the "wild" nature of a subject exposed. The authenticity of an animal is therefore a result of embellishment and manipulation on the part of a photographer or composition.

peril for humans, and humans retaliated—often, to the complete decimation of a competing species. Paradise was never on offer, yet as a mental image or a conservation strategy, it stands for a fine destination indeed.

As demonstrated by the examples in this chapter, the progression of an historical narrative can be traced in visual material. By closely observing the differences in photographs of lynxes in the media, we can engage in the process of trying to understand what messages are being put forward by employing certain portrayal canons and the choice of photographed subject/s. While it is possible that more photographs could paint a fuller picture of the changing attitude towards the presence of lynxes in the area, the two photographs in Figs. 10.1 and 10.2 offer an insight into the applicability of visual analysis as a research method in an historical analysis. A close reading of a photograph can engender material that an accompanying text might not necessarily transmit. At the same time, having background knowledge and analysing captions helps contextualize the photograph itself. Combining the two strategies could open access to a rich source of data, although, as mentioned previously, one should study imageries without relying too much on the seeming factuality of a written text: being informed should help one see details that might otherwise go amiss and not create a tunnel vision of what one expects to see.

Conclusions

Visual analysis has indisputable merits for a wide range of scholarly examination, beyond those dealing specifically with aesthetics. As a method of inquiry, it invites a researcher to engage with imagery to piece together information via one of the most common practices of all—seeing. Images often serve to accompany or illustrate a written narrative, but just as often they represent their own source of research data. As discussed previously, many decisions need to be made to produce a visual artefact: photographs, for instance, snap a photographer's momentous interaction with an object in a frame, yet they are also a product of all the thought work, editing, and narrating that come before, during, and after the click of the camera. Comprehending an image requires an equal amount of decision-making—to truly see, a viewer must recognize what they are looking for in the image and try to solve this sort of puzzle, where each piece carries its own significance. In the case of scholarly research, this translates to being conscious of the kind of question one is trying to answer and consider whether imagery suffices for the task.

In an historical study, visual analysis can create a space for more direct involvement with a slice of the past. Engaging with a long-gone moment by means of seeing is one of the closest experiences to an unmediated interaction; one observes as if being present in a moment, yet the still nature of imagery allows us to take time to scrutinize every little detail to form a conclusion, to narrate a scene. Again, people derive a great deal of understanding of the outside world by taking in and processing imageries, and when one comprehends an historical image, the past comes to life. Text of any sort necessarily conveys the predispositions of an author that, in many cases, is precisely what makes them an excellent source of information. In other instances, however, images can aid overcoming this innate subjectivity for they might have a higher degree of independence from the author's intentions. Certainly, like any other medium, images are far from objective, yet, depending on a research question, they might offer more possibilities to form one's own opinion on a matter.

At the same time, there are certain challenges and limitations to employing visual analysis as a research method in an historical study. As demonstrated in this chapter, certain insights require reading captions to grasp the idea of a photograph more accurately: in many cases an accompanying text could help to situate an image and have a fuller understanding and appreciation of it. While it is always beneficial to have a theoretical understanding of canons of photography and their affective faculties, images do not always fall neatly into one category. A wider spectrum of photographs of the same event might mitigate this issue, yet it is equally important to avoid selection bias by intentionally choosing images that fit a historical narrative. Additionally, depending on the timeline of research, finding suitable imagery can itself become a treasure hunt: while in the past century, and specifically after the advent of smartphones, photography as a means of recording a moment has been dramatically increasing both in volume and subjects photographed, some events and processes of earlier history might have a more modest visual output.

Visual analysis highlights an inevitable degree of creative constructedness in a research process. While interpreting a photograph, a viewer might possess sufficient background knowledge to situate themselves in very close approximation to the inhabitants of an image, yet as Rasmussen (1962) points out, the atmosphere *inside* of a photograph—its sounds, smells, almost imperceptible movements, the play of light—is not easily, if at all, translatable through visual representation alone. While many conclusions derive from trying to interpret the intentionality of a visualized message, one can *imagine* what it could have been like to be in the frame—the sensual experiences that Rasmussen mentions—and corroborate or accentuate their impressions based on their knowledge of the image, its making, or its other properties. As a research method in an historical study, visual analysis could give depth to a narrative, illuminating details overlooked or mentioned in passing in a text. The interpretive richness of visual analysis broadens research material by inviting creativity in the process of contemplation. At the same time, it is crucial to always be mindful of what end visual analysis serves: a research method is like a tool, and the success of its application depends on the intended task and one's willingness to learn.

Acknowledgements This research received funding from the European Union's Horizon 2020 research and innovation programme under the Marie Sklowdoska-Curie Grant Agreement No. 765389.

References

Barry, A. M. S. (1997). *Visual intelligence: Perception, image, and manipulation in visual communication.* State University of New York Press.

Barthes, R. (1981). *Camera lucida: Reflections on photography*. Macmillan.

Benjamin, W. (1969). The work of art in the age of mechanical reproduction. In H. Arendt (Ed.), *Illuminations* (1st ed., pp. 1–26). Schocken Books.

Berger, J. (1991). *About looking*. Vintage Books.
Bergman, C. (2013). Hunger makes the wolf. In K. Nagy & P. D. Johnson (Eds.), *Trash animals: How we live with nature's filthy and unwanted species* (pp. 39–67). University of Minnesota Press.
Bode, C., Bertamini, M., & Helmy, M. (2016). Left-right position in moving images: An analysis of face orientation, face position, and movement direction in eight action films. *Art & Perception, 4*(3), 241–263.
Bousè, D. (2003). False intimacy: Close-ups and viewer involvement in wildlife films. *Visual Studies, 18*(2), 123–132.
Brower, M. (2011). *Developing animals: Wildlife and early American photography*. University of Minnesota Press.
Castel, R., & Schnapper, D. (1990). Aesthetic ambitions and social aspirations: The camera club as a secondary group. In P. Bourdieu (Ed.), *Photography: A middle-brow art* (pp. 103–128). Polity Press.
Chris, C. (2006). *Watching wildlife*. University of Minnesota Press.
Cronin, J. K. (2011). *Manufacturing national park nature: Photography, ecology, and the wilderness industry of Jasper*. UBC Press.
Enright, K. (2012). *The maximum of wilderness: The jungle in the American imagination*. University of Virginia Press.
Fudge, E. (2002). A left-handed blow: Writing the history of animals. In N. Rothfels (Ed.), *Representing animals* (pp. 3–18). Indiana University Press.
Gaiger, J. (2014). The idea of universal Bildwissenschaft. *Estetika: The European Journal of Aesthetics, 2*, 208–229.
Gissibl, B. (2016). *The nature of German imperialism: Conservation and the politics of wildlife in colonial East Africa*. Berghahn Books.
Grady, J. (2008). Visual research at the crossroads. *Forum: Qualitative Sozialforschung/Forum: Qualitative Social Research, 9*(3), Art. 38.
Herber, L. (2018). *Our synthetic environment*. Martino Fine Books.
Herman, E. S., & Chomsky, N. (2002). *Manufacturing consent: The political economy of the mass media*. Pantheon Books.
Heurich, M., & Mauch, C. (2020). *Urwald der Bayern: Geschichte, Politik und Natur im Nationalpark Bayerischer Wald*. Vandenhoeck & Ruprecht Verlag.
Isenberg, A. C. (2002). The moral ecology of wildlife. In N. Rothfels (Ed.), *Representing animals* (pp. 48–64). Indiana University Press.
Kalof, L., & Fitzgerald, A. (2003). Reading the trophy: Exploring the display of dead animals in hunting magazines. *Visual Studies, 18*(2), 112–122.
Kuhn, A. (2007). Photography and cultural memory: A methodological exploration. *Visual Studies, 22*(3), 283–292.

Lorimer, J. (2015). *Wildlife in the anthropocene: Conservation after nature*. University of Minnesota Press.
Nash, R. (2001). *Wilderness and the American mind*. Yale University Press.
Peterson, A. L. (2013). *Being animal: Beasts & boundaries in nature ethics*. Columbia University Press.
Rasmussen, S. E. (1962). *Experiencing architecture*. The MIT Press.
Serpell, J. A. (2004). Factors influencing human attitudes to animals and their welfare. *Animal Welfare, 13*, 145–151.
Schraml, U. (2019). Wildtiermanagement für Menschen. In M. Heurich (Ed.), *Wolf, Luchs und Bär in der Kulturlandschaft: Konflikte, Chancen, Lösungen im Umgang mit Großen Beutegreifern*. Eugen Elmer KG.
Sontag, S. (2005). *On photography*. Rosetta Books LLC.
Spencer, S. (2010). *Visual research methods in the social sciences: Awakening visions*. Routledge.
Suonpää, J. (2000). Taming predators through photography. *Visual Studies, 15*(1), 51–64.
Thornton, A. (2019). *This is how many animals we eat each year*. World Economic Forum. https://www.weforum.org/agenda/2019/02/chart-of-the-day-this-is-how-many-animals-we-eat-each-year/. Accessed January 8, 2021.
Will, K. (2016). When is a panther not a panther? Representing animals in early modern English heraldry. *Early Modern Culture, 11*, Art. 6.

11

Back to the Drawing Board: Creative Mapping Methods for Inclusion and Connection

Talitta Reitz

Introduction

Amid the vast array of precious items in The British Museum, a small, broken clay tablet might appear rather unremarkable: a plain object, to inattentive eyes. However, this seemingly ordinary artefact contains a star-shaped engraving known to be the very first world map.[1] In the *Imago Mundi* (Fig. 11.1), triangular shapes mark mysterious islands that the Babylonians thought were the edges of the planet (Unger, 1937, p. 1). Rectangles and lines represent the city of Babylon and the Euphrates River, both circumscribed by water: the *Earthly Ocean* separates the city from unexplored territories, and the *Heavenly Ocean* bridges constellations and the Earth. Drawings and cuneiform writings narrate ancient

[1]Object, *The Map of the World*, The British Museum, London, UK.

T. Reitz (✉)
Rachel Carson Center, Ludwig Maximilian University of Munich, Munich, Germany

A. Franklin (ed.), *Co-Creativity and Engaged Scholarship*,
https://doi.org/10.1007/978-3-030-84248-2_11

Fig. 11.1 Babylonian map. The *Imago Mundi* conveys the origins and configuration of the world, according to Mesopotamic cosmology. Maps express culture and spirituality, combining knowledge, belief, and imagination (*Source* The British Museum, 600 BC)

Mesopotamic beliefs in gods, beasts, and other mystical creatures (Smith, 1996, p. 209). The map carries an all-too-human mixture of symbolism, knowledge, and belief. In other words, this invaluable relic epitomizes humankind's long-standing reliance on maps for understanding the world and telling stories.

Beyond its historic importance, the *Imago Mundi* articulates language, spatial information, and narrative. Extremely important to maps, these elements were retained across many cultures through centuries of mapping practices. But what are, in fact, the core attributes of maps? Some understand them as strictly graphic objects, but perhaps maps could also use non-visual languages (Harley & Woodward, 1987, pp. 1–2).[2] Perhaps there are maps that do not relate to tangible spaces, but rather to imaginary scenarios and abstract relationships. And is it really possible to create a "neutral" map—an object with no story, no bias, no point of view? The concern for the inextricable principles of maps has produced interminable debates with impossible-to-answer questions such as these.

Until the advent of cartography as a discipline, different mapmaking practices unsystematically explored these and other reflections with no substantial attempts to achieve a consensus. Nevertheless, with the alliance of mapmaking and science (Cramptom & Krygier, 2010, p. 11) in the nineteenth century, maps were regarded by academics as precise, visual, and impartial objects. They were definite and their purpose was to accurately depict space. On the one hand, this ambition expanded the scope of representation, mapping methods, technology, and survey subjects. On the other hand, it narrowed the range of experimentation, excluded narratives, and created a problematic divide: the artistic (subjective) versus the technical (objective) map. Predilection for scientific mapping methods produced an illusion of objectivity; the sacrifice of art, aesthetics, and opinions favoured precision. However, in previous traditions of skill and knowledge, this binary model would have been considered irrelevant and absurd, because the differentiation of art and technique—concepts of indissociable origins—is modern (Williams, 1985, pp. 42, 315–316).[3] Nevertheless, this segregated discourse reigned in cartography for almost two centuries, perpetuated by mapmaking elites (map houses, academics, and the state) until the eye-opening defiance of post-modern ideas (Cramptom & Krygier, 2010, p. 12). *Critical*

---

[2] The authors acknowledge as valid the definition of maps as graphic texts.

[3] About the origins of *art* and *science*, Williams explains: "Until (the eighteenth Century) most sciences were arts; the modern distinction between science and art, as contrasted areas of human skill and effort, with fundamentally different methods and purposes, dates effectively from (the middle of nineteenth Century), though the words themselves are sometimes contrasted, much earlier, in the sense of 'theory' and 'practice'". This complex set of historical distinctions between various kinds of human skill (…) is evidently related both to changes in the practical division of labour (…). It can be primarily related to the changes inherent in capitalist commodity production, with its specialization and reduction of use values to exchange values. (…) This is the formal basis of the distinction between art and industry, and between fine arts and useful arts (the latter eventually acquiring a new specialized term, in technology)" (p. 42).

cartography sustains the intrinsic impossibility of objective or impartial maps[4] and condemns the ethical consequences of such segregating viewpoints. To destroy this divide, the critique proposes new creative mapping methods for academic researchers and practitioners, which is the central topic of this chapter.

The surgency of new cartographic methods was progressive, rather than eventual. It was through a gradual discipline reform that cartography managed to take a hard look at itself and restructure mapmaking practices and power systems. Inceptive debates on the agency of maps and discontentment with cartographic "scientification" received attention during the 1990s (Cramptom & Krygier, 2010, p. 19). One notable effort was carried by cartographers John Brian Harley and David Woodward. Together, they traced a global history of cartography (Harley & Woodward, 1987),[5] exploring maps as worldwide cultural artefacts. This all-encompassing work uncovered overlooked maps and mapping traditions since pre-history, including their impacts and geopolitical repercussions. Their research raised reflections about the realities and actors that maps can either favour or harm.[6] Thanks to Harley, Woodward, and others,[7] the critique slowly grew in debate circles, where scholars exposed and rejected the positivist "crimes" of traditional cartography. This acknowledgement opened the academic space for the new practices and theoretical approaches under the umbrella of critical cartography.

Amidst the critique, both inside and outside academia, aspirations of objectivity started to become irrelevant. Expanding notions and agents of maps have legitimized other forms of sensing, representing, and relating to space. As a consequence, the mapping *process* became the focus of cartography. Geographers Jeremy Cramptom and John Krygier define critical cartography as "a one-two punch of new mapping practices and theoretical critique" (Cramptom & Krygier, 2010, p. 11). Therefore,

---

[4] Maps produced with a claim to objectivity or impartiality are addressed in this chapter as "scientific", "positivist", or "traditional".

[5] Complemented by a third volume in 1998, by Woodward and Lewis.

[6] In *The New Nature of Maps*, Harley discusses these inquiries more deeply, through concepts such as *cartographic silence* or (i.e., the omission, falsification, or manipulation of maps for the benefit and dominance of a specific group or authority). He outlined multiple principles of cartographic misuse that enable or reinforce abuse of power and hegemonic perspectives.

[7] Other notable critical cartographers are geographers Denis Cosgrove and Jeremy Cramptom.

its mapping methods are necessarily untraditional and constructivist. They support uncertainty. They discover new stories and inclusive meanings of space. Rather than the outcome (the map as an object), other mapmaking aspects prevail, such as transparency and participation. More relevant questions arise: which ideas, senses, and values are included or excluded in the mapping process? Who is heard and who is silenced? What purpose does the map serve and which transformations can it unravel?

This chapter discusses these questions through the framework of two creative mapping methods: social cartography and deep mapping. These approaches do not attempt to reject, but rather reclaim the partial, subjective nature of maps. Since cartography is inevitably limited, these methods recognize that creativity is vital to attend neglected necessities. Each section introduces one of the two methods with their origins, theoretical frameworks, reception, and applications. More importantly, they will show how these methods solve the challenges of traditional cartography, with hopes of demonstrating their incredible potential for participatory and environmental research.

One of the major problems denounced by critical cartography is the reinforcement of hegemonic perspectives. To avoid it, cartographers must explore the unseen, the concealed. Stories of the less privileged. Stories of the environment. In sum, maps need *to include*. All individuals and groups should have the opportunity to reflect, express, and opinionate about their territories and spatial practices; to have their stories told. In response, the first part of this chapter introduces social cartography, a method of acknowledging and legitimizing underrepresented standpoints. The section shows how the method was born within the social sciences and gained visibility in participatory planning. It also analyses its effectiveness against scientific maps, exposing its reformative principles. Finally, the chapter brings some examples of how the method gained momentum among artists, academics, and activists. Social cartography opened up new, free, bottom-up, hands-on forms of mapping.

Another problem relates to perception. As a visual language, maps have limitations in depicting personal and sensorial experiences. Abstract qualities of space challenge representation: the passing of time, flows and rhythms, memories, values, and interests. Maps could certainly encompass more. In addition, traditional maps fail to recognize the importance of subjective spatial experiences—precisely those that foster connection and place attachment. Sometimes, understanding a space requires it *to be felt*. Deep maps focus on such qualities. They aim to surpass the objective, the pictorial; a goal helpful not only to those who cannot see, but to all who do not feel particularly engaged through visuality. With literature, storytelling, artistic performances, and other creative practices, deep mapping bridges these sensorial and imaginative gaps. This method generates a richer and more nuanced conception of space, a reason why it is so necessary. The second section shows how the book *PrairyErth* pioneered deep mapping in the form of literary cartography, opening the floor to other creative approaches. Selected examples illustrate the method's emphasis on process and immersive experiences, with both informative and transformative benefits (for further extended discussion and application of deep maps as a co-creative form of research practice see Humphris et al., this book).

These two approaches are transdisciplinary and conceptually boundless. Oftentimes they overlap with different methods; an openness that can be received with discredit. However, their creators and users aim not at formality or definition but take advantage of this theoretical amplitude to generate new debates and push the boundaries of spatial representation.

Traditional cartography may be dead (Wood, 2003, p. 4), in some sense, but mapmaking still lives. There is not only space but demand for methods that challenge power relationships, both inside and outside academia. This chapter proposes that environmental activists and researchers of all disciplines incorporate and/or develop new cartographic approaches in their work. They can contribute to critical cartography by including unheard perspectives and inviting others to imagine how spaces can be different, better. As participatory practices within communities, these methods promote dialogue, empowerment, and transformation. Mapmaking is not exclusive to the elites and maps are not unidimensional objects. This is the great lesson of critical cartography. One must simply gather creative tools to overpower the old "Age of Cartography" (Wood, 2003, p. 4).

Social Cartography: A Method for Inclusion

"I'm surprised my body isn't in the Tyne. Before Elliott gave me a bed, I was homeless, I wanted to die". These are the difficult memories of Sam, a man without housing living in the United Kingdom—one of the many to have his needs and experiences daily brushed off by politicians, city planners, and the regular passer-by. His impactful words were written in the composite map (Fig. 11.2) created during the workshop *Imaging Homelessness in a City of Care* (Irving & Moss, 2018, pp. 270–275).[8] To spark meaningful debates and defy preconceptions of homelessness, the project authors collected the marginalized voices of 30 people living in the streets of Newcastle-upon-Tyne. They discussed how social cartography can offer perspectives beyond conventional views of urban space. "Not allowed" areas, or notes such as "slept here" and "I was married there!" mark the places of homelessness' struggles. An innocent "where I live—nice area, family around me" revealed, in fact, a surprisingly difficult condition: the label was placed on a cemetery. The concealed, "undesirable" practices of the homeless—regularly addressed with pity or judgement—are openly discussed in the project, provoking a reaction, calling for attention and care. The map presents one of the many possible approaches to social cartography and its great potential to oppose exclusionary narratives of space, open new perspectives, and inspire social change.

But exactly how does social cartography create democratic mapping practices? To explain maps as socially constructed forms of knowledge, geographer J. B. Harley once identified two sets of rules guiding dominant mapping discourses (Harley, 2001, pp. 153–158). One set concerns technological systems that dictate *how to map*. These systems originate from the false assumptions:

> (...) that the objects in the world to be mapped are real and objective, and that they enjoy an existence independent of the cartographer; that their

[8] The 2014 project, by social policy researcher Adele Irving and social/cultural geographer Oliver Moss, was financed by the Economic and Social Research Council (ESRC), UK (kollektiv orangotango+, 2018). More information on: https://esrcimaginghomelessness.wordpress.com/.

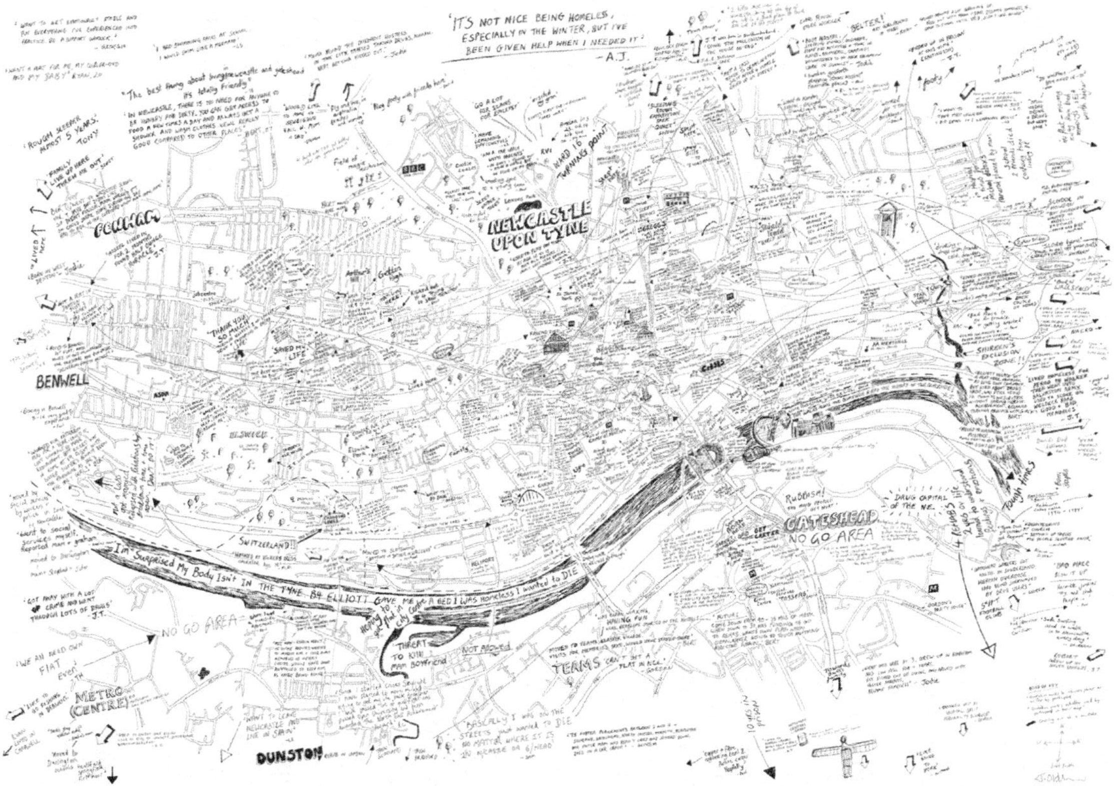

Fig. 11.2 Spaces of homelessness. Counter-map for the project *Imaging Homelessness in a City of Care*, in Newcastle-upon-Tyne, UK. With legends and quotes, the map situates the daily homelessness experiences of thirty participants (*Source* Adele Irving and Oliver Moss, 2018. Creative Commons License: https://creativecommons.org/licenses/by/4.0/legalcode. URL: http://nrl.northumbria.ac.uk/id/eprint/39919)

> reality can be expressed in mathematical terms; that systematic observation and measurement offer the only route to cartographic truth; and that this truth can be independently verified. (Harley, 2001, p. 154)

In other words, exclusionary systems of technology, skills, and standards categorize (and homogenize) the cartographic knowledge. Social cartography questions these rules by disputing the authority of selected parameters. It defends the legitimization of bottom-up, creative, and participatory maps. The homelessness map, for example, was drawn with the assistance of a visual artist but conceptualized by the workshop contributors—those who have in-depth experiences of homelessness but lack visibility and mapmaking authority.

Another group of rules, elusive but equally powerful, comprises cultural values. Related to ethnicity, politics, religion, and social class (Harley, 2001, p. 156), these influences guide *what to map*. Here, social cartography defies the presumed objectivity of maps by inquiring which realities are portrayed; by whom, and for whom. To exercise these examinations is to reveal the inexorable subjectivity of maps. Furthermore, as narratives, maps will always conceal one side of the story. Harley calls it *the silence of maps*. He argues that maps "exert a social influence by their omissions as much as by the features they depict and emphasize" (Harley, 2001, p. 67). Therefore, social maps are subversive for their focus on systemically neglected perspectives—another point exemplified by the homelessness map, which uncovers "inconvenient" topics of social inequality.

Those are the defiant principles of contemporary social maps. In sum, the method's contribution to the post-modern critique comes from the rejection of elitist cartographies, in favour of bottom-up and horizontal decision-making processes. As a result, social maps can generate dialogue or resistance against social and environmental injustice.

Origins, Academic Use, and Criticism

Social cartography holds no single definition or methodological delimitation. Different methods across disciplines have been designated as such, and their first appearances have neither been attributed to specific scholars or works nor necessarily sustained subversive intentions. Urban researcher Laura Vaughan understands social maps as those "whose purpose is to represent specific aspects of society at a given time and place" (Vaughan, 2018, p. 1). Tracing them back to the 1790s, the author links the emergence of social cartography to industrialization and urbanization. In this context, she emphasizes the work of social researcher Charles Booth and his role in "the phenomenon of the social reformer as urban investigator" (Vaughan, 2018, p. 2). At the time, and mostly in the UK, social scientists started to use maps for locating and visualizing patterns of social dynamics in cities. One contemporary example of similar use would be the racial dot map (Fig. 11.3), which represents, with colourful dots, demographic data on race or ethnicity. Produced in many different countries, such maps reveal logics of spatial segregation, especially when contrasted to urban infrastructure or poverty maps

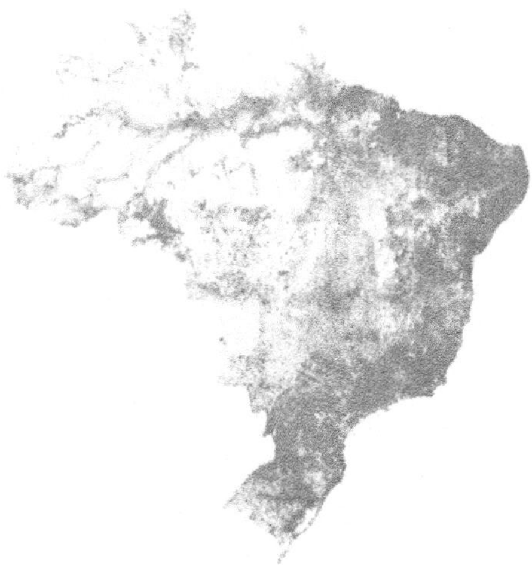

Fig. 11.3 Racial dot map of Brazil. Each dot represents a person of declared race/ethnicity as follows: blue for white, green for brown (mixed race), red for black, yellow for Asian, and brown for indigenous. The map reveals a predominance of black and mixed populations in the North and Northeast, coinciding with the lowest income areas, as indicated by the *Population in Poverty Map*, Fig. 11.4 (*Source* Post Advertising Technology Agency (PATA), 2015. Creative Commons License: https://creativecommons.org/licenses/by/4.0/legalcode. URL: https://patadata.org/maparacial/en.html)

(Fig. 11.4). In Vaughan's definition, the subversive trait of social maps is not intrinsic—they merely reflect existing conditions of society and space. More often than not, however, they have defied biased positions simply by exposing the reality.

In another methodological direction, social scientists Rolland Paulston and Martin Liebman adopted social cartography for studies in educational policy (Liebman & Paulston, 1994) in the field of comparative education, in the 1990s. They applied the method in an entirely different way from the social reformers, illustrating its interdisciplinary diversity. With a rather loose notion of social maps (which, nevertheless, fits within Vaughan's definition), their research focused on social relationships instead of spatial dynamics. Their maps were similar to concept,

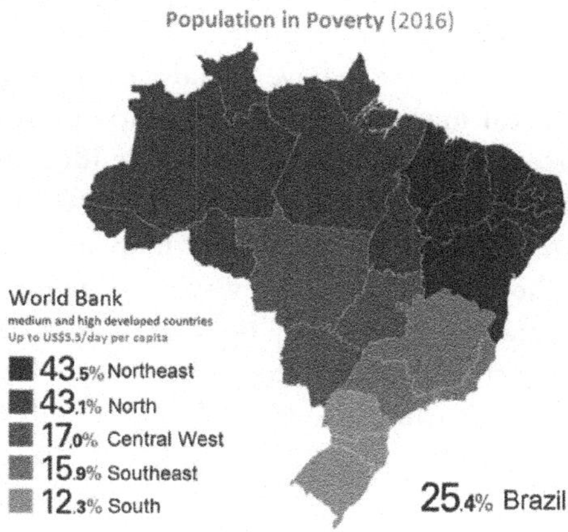

Fig. 11.4 Brazilian population in poverty map. The map shows that North and Northeast areas of Brazil had almost half of the population in poverty. The comparison with the *Racial Dot Map* (Fig. 11.3) suggests a correlation between these areas and people of black or mixed race. The two maps are examples of a social cartography typology that combines social and spatial information (*Source* Brazilian Institute of Geography and Statistics (IBGE), 2016. URL: https://agenciadenoticias.ibge.gov.br/en/agencia-news/2184-news-agency/news/18835-one-fourth-of-the-population-lives-on-less-than-r-387-a-month)

cognitive, or mind maps,[9] which organize ideas through text and figurative landscapes.

Paulston and Liebman's approach takes advantage of the overlapping scopes of map types. For them, non-rigidness is potentially insightful. Their social maps can be combined and reinvented: they serve as mere guidelines for creative research. Regarding definitions, the reason why the authors deemed it necessary to designate these conceptual maps as *social maps* is explained as follows:

> (…) social cartography (…) does identify and represent on a two-dimensional plane features perceived to occupy physical space. (…) however, the features are not mountains, rivers and cities, but the networks of humanity built on the variety of understandings and interpretations of numerous socially constructed associations, or cultural clusters' knowledge claims. Because cultural clusters occupy physical space that as often as not is contested, **we believe social cartography often identifies with geopolitical maps because one group's political features are what attract persons to a particular space. The ideological space they choose, their affiliations, directly informs their choice of real space so that when we as social cartographers map our vision of ideologies and social theories we are, in a way, also mapping the isolated pockets of real space** people occupy because of their choices as well as the real spaces they choose not to occupy because of those same choices [emphasis added]. (Liebman & Paulston, 1994, p. 240)

---

[9] Concept maps allow for the exploration of several different ideas, containing labels that express their connection (Novak & Gowin, 1984). Cognitive maps express complex ideas through simple sentences, using multiple links between elements (Ackermann et al., 1992; Eden, 1988). Mind maps, often employed for decision-making, focus on a single element (idea, event, problem, etc.), mapping secondary or related concepts (Buzan & Buzan, 1993).

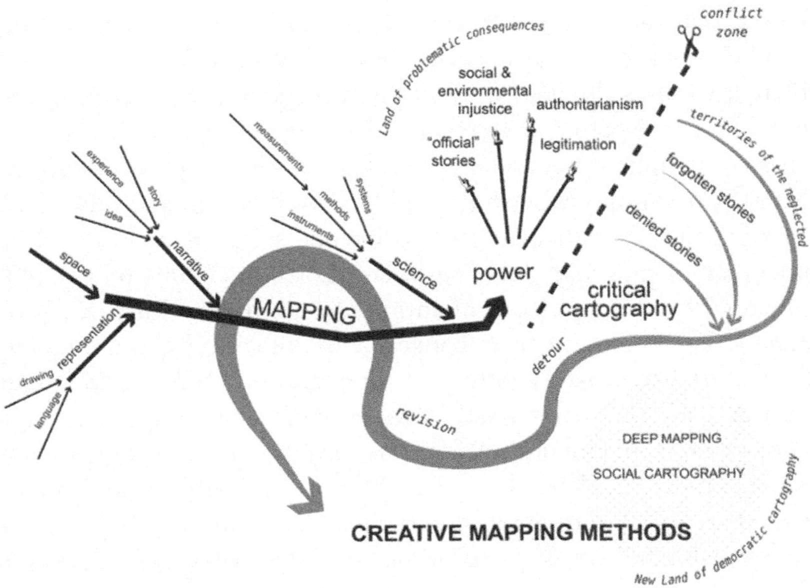

Fig. 11.5 Phenomenographic/conceptual map of critical cartography. This map was inspired by the social maps of Paulston and Liebman. Phenomenographic/conceptual maps use an abstract representation of space to organize ideas and theories. This social cartography method reveals new connections about a topic and the relatedness of its elements (*Source* Talitta Reitz, 2021)

Perhaps the greatest impact of this proposal is the combination of ideas with both concrete and abstract spatial data, a strategy exemplified in the contemporary examples presented in the next section. Paulston and Liebman wanted to generate relational insights—not to "paint a picture". In their model, maps were not used to capture *visual* reality, but to understand *social* reality (Fig. 11.5). They mapped interests, values, and viewpoints that inevitably relate to space, but only through indirect associations. Therefore, this conceptual form of social cartography can assist in understanding how social actors and events occupy territories. It explores *causal spatial relationships*, rather than direct representations of space.

Through Paulston and Liebman, social cartography was established as an interdisciplinary, post-modern, and post-structuralist mapping method that opens up possibilities for inclusive discourses, mini-narratives, and non-binary perspectives (Gorostiaga, 2017, p. 880). Different versions also gained attention in geography and planning disciplines. In participatory design and planning, the use of social maps surpassed disciplinary boundaries and reached the realm of professional practice. Community workshops, public consultations, and other forms of participatory initiatives started to include social maps in the development of private and public projects (Lobatón, 2009, p. 15). As a result, social cartography can oftentimes be confused with *participatory mapping*. For some, this interchangeable use of concepts is unproblematic, because participatory initiatives are necessarily "social" and attempt to break with the *status quo*. However, opposing perspectives argue in favour of a distinction. With the *New Social Cartography of the Amazon Project*, de Almeida et al., (2018, pp. 46–49) explain that, as opposed to social cartography, participatory mapping can perpetuate institutionalized discourses, depending on who initiates and conducts the process:

> They [social maps] are distinguished from participatory maps. Because participatory maps, such as planning instruments, **are defined by planners** in order to incorporate communities in decisions which are made for them or will affect them. Unlike these maps, our social mappings work in favor of social movements and collective identities as well as intrinsic organizational forms (of mobilization and solidarity) adopted by the groups themselves [emphasis added]. (kollektiv orangotango+, 2018, p. 48)

To simplify, both methods are fairly similar and can, indeed, overlap. Both involve the people and groups who receive the resulting impacts of mapping processes and associated projects or events. Both have the objective of concretizing democracy and promoting positive social transformation. However, as discussed above, participatory maps are oftentimes unimpactful and cosmetic, mere checkmarks for the democratic planning cookbook. This deviation occurs because participatory processes frequently include only a handful of participants. They fail to involve key actors and communities, who, dissatisfied with the propagandistic, pre-established outcome of projects, choose (or are compelled) not to opinionate. Why bother, anyway (for further discussion of this point see Ramirez Aranda and Vezzoni, this book)?

Of course, social maps are not exempt from dissatisfaction or irrelevance. Geographers Susana Lobatón (2009) and Ulrich Oslender (2017) warn against social maps that, instead of creating counter-narratives, nurture traditional, inaccessible cartography suited for top-down decision-making. They observe that, in social cartography processes, communities may end up adopting institutional language (such as regarding their territorial relationships in terms of property rights) or even using spatial knowledge to vulnerate the rights of other communities (Oslender, 2017). Furthermore, oftentimes it is not possible to ensure the usability of social maps in retrospective research; this is a reason why Lobatón (2009, p. 16) stresses the importance of accessible codes or captions for out-of-context readings. In other words, as with participatory mapping, social cartography is also criticized for institutionalization tendencies. But their scope is broader. The difference is: social maps are not *just* participatory. As exemplified with Vaughan, Paulston and Liebman, their essential premise is the inquiry about the spatiality of social problems and the defiance of hegemonic cartographies.

For its postmodern character, the method has been accused of nihilism or relativism. Yes, the scope, reach, impact, and obstacles of social maps depend much on their framework. Yes, they are subjective and confrontational by nature. However, as social scientist Jorge Gorostiaga sustains, social cartography—just as the academic critique in which it is inserted—is anti-hegemonic and intends to decline "the emphasis on 'the truth' to highlight the process by which something is considered true" (da Silva, 2001, p. 151, quoted in Gorostiaga, 2017, p. 885). As a result, this "umbrella method" requires certain independence from disciplinary prescriptions, to counteract the rigid ties it rejects in the first place. And so far, it has been heeding fascinating results.

Research and Activism

One example of social cartography that lies within the boundaries of research and activism is the work of American sociologist Nancy Peluso. In 1995, she coined the term *counter-map* to define seditious maps used against the abuse of power (Peluso, 1995, p. 384). The strategy was used by communities and activists in Kalimantan, Indonesia, to reclaim their rights against industrial timber exploitation in local forests. For two decades, companies and government institutions employed "official" maps as means to legitimize exploitative practices. Activists responded with a grassroots process of sketching and marking GPS points to create a counter-map. The instrument helped communities engage in debates, gain support, reclaim the rights to the territory, and secure the preservation of the forest. Peluso talks about *counter-mapping* the mapmaking practice of local people and groups that stand in opposition to authoritative maps, specifically those that support some sort of injustice. This form of mapping can be executed through a re-appropriation of formal techniques and technologies or the legitimization of alternative ones (Peluso, 1995).

Within the rich universe and exciting potential of counter-maps, one transdisciplinary and non-institutionalized project deserves special regard. *This is not an Atlas*, by European/Latin American group kollektiv orangotango+ (2018) brings many examples of social maps, with distinct approaches, aims, and outcomes. The edited book features projects in the intersection of socio-political activism, art, and creative mapping, described as follows:

> The work of social mapping thus includes two aspects: an ethnographic one, which requires academic work, direct observation techniques, detailed descriptions and criteria to select information, and another one carried out by the social agents themselves, defining the use of instruments, their choice, the selection of what is included in the map. (kollektiv orangotango+, 2018, pp. 157–173)

Mapping with children was an exercise used in 2013, by activists Nicolás Frank and Fernanda García in classrooms in rural Uruguay

(Frank & García, 2018, pp. 152–157). Throughout site visits, conversations, and workshops, the instructors asked children of different schools and age groups to create maps of their villages, neighbourhoods, paths, and daily spaces of life. The process included several steps: first, the children produced maps with no help or intervention; then, through debates and the assistance of existing maps and satellite images, they were able to compare and include new references and layers of information deemed relevant; and finally, collaborative sessions guided the creation of group maps. Discussions about their relationships to spaces uncovered important topics:

> An example of this is the relationship between the creek where the children usually play and fish, and the new intensive agriculture that is taking place in its watershed. This agriculture includes the use of high amounts of pesticides and herbicides, which can be harmful to their health. Mapping these spatial configurations isn't enough to establish causal relations about health or environmental problems, but it enriches the collective process of asking new questions. (kollektiv orangotango+, 2018, p. 154)

As an educational exercise, this hands-on approach instructed not only about maps and cartography but about critical cartography. Moreover, it stirred conversations regarding education, identity, the environment, and spaces of childhood—reflections that perhaps remain as legacies for Uruguayan communities as a framework for addressing the necessities of local children.

This collection of social maps starts and ends with provocations. By self-proclaiming "not an Atlas", the editors and collaborators reject traditional cartography and propose something new, oppositional. Even outside of academia, every form of counter-cartography is critical and contributes to critical cartography. And they end with a non-conclusion, by including the perspectives of different cartographers and activists. kollektiv orangotango+ shows how both critical and social cartography are more than a critique and a method, but parts of an ideological movement that extends the agency of cartography to anyone.

Within this democratic momentum, new mapping methods, practices, and products continue to appear. In academic research, participatory practice, and activism, social maps can inform difficult-to-see, unjust social relationships. As described by Lobatón (2009), the role of the activist, practitioner, or researcher is to collect and voice multiple forms of spatial experience. Environmental activists and researchers have one additional responsibility, which is to include more-than-human narratives that can improve relationships with nature. As a result, the social cartographer can bridge communication gaps between society and institutional maps, thus supporting better and more democratic decision-making (Lobatón, 2009, p. 20).

Deep Mapping: A Method for Connection

Imagine the map of a farm. Dimensions are expressed by clearly defined boundaries, area figures, and scale. Contour lines emphasize the topography, and hatches represent different surfaces: pervious and impervious; concrete, water, and soil; cropland, meadow, and pasture. But how to capture the scent of a wheat field, the chirp of a bird, and the patterns of a moving starry sky? How to represent the feeling of home, moist feet touching a marshy ground, or the fear of a wild creature crossing the trail ahead?

The methods of scientific cartography are limited; they fail to express *meaning*. Deep maps come from the urge to represent personal impressions of a place. Sensations, stories, ideas, and memories are difficult qualities to capture in a drawing. To solve this limitation, deep mapping proposes to seek depth. The adjective *deep* contests the conceptual "superficiality" of traditional maps in two directions: (1) deep maps are concerned with qualitative and subjective information, as opposed to quantitative data; and (2) they select unconventional forms of representation, rather than one, strictly visual depiction (drawing). Deep maps come from a humble representational standpoint—one that recognizes the limits of human knowledge and relies on art to translate the essence of a place. Precision is inessential to deep maps, as opposed to multiple layers of meaning. To literary scholar Susan Maher, the amplification of

Fig. 11.6 Tallgrass Prairie National Preserve in Strong City, KS. A view of Kansas' prairie landscapes described in *PrairyErth* (*Source* U.S. National Park Service, 2006)

many stories (human and more-than-human) is their character-defining quality. "(D)eep map makes the 'deeply felt' its forte" (Maher, 2014, p. 133), meaning the principal function of this mapping practice is to convey perception and significance; the past and present feelings of place.

Deep mapping originated amid the rise of critical cartography. In particular, one book received attention for proposing a different form of spatial representation. Published in 1991, *PrairyErth: A Deep Map*, stands out as probably the method's most renowned example (Maher, 2014, p. 92),[10] with two important contributions to critical cartography: the representation of *placeness* and the *literary map*.

From Space to Place: The Process of Connection

The discussion of *placeness* is essential to understanding the contribution of deep maps. *PrairyErth* recognizes an important correlation between space and meaning. In the opening paragraphs, the author and historian William Least Heat-Moon confessed feeling initially estranged to encounter a new landscape (Fig. 11.6) while passing through Chase County at a young age:

> I've probed my memory to find even one detail of that initial passage into the western prairies. What did I see, feel? Nothing now, except our route

[10] In *Deep Map Country*, p. 92, professor and writer Susan Maher discusses *Wolf Willow*, a much earlier (1962) work by Wallace Stegner, possibly the very first known deep map.

> returns. My guess is that I found the grasslands little more than miles to be got over—after all, that's the way Americans crossed Kansas. Still do. (Least Heat-Moon, 1991, p. 27)

Years after his childhood visit, he returned to the Kansas prairie with a novel sense of enchantment. The secret for this new sentiment, this connection, he felt, was a gradual absorption of the place's hidden idiosyncrasies:

> I drove across the prairie again on a visit to California, and the grasslands looked different to me, so alive and varied (…) I began to like them not because they demand your attention like mountains and coasts but because they almost defy absorbed attention. At first, to be here, to be here now, was hard for me to do on the prairie. I liked the clarity of line in a place that seemed to require me to bring something to it and to open to it actively: see far, see little. I learned a prairie secret: take the numbing distance in small doses and gorge on the little details that beckon. Like its moisture, the prairie doesn't give up anything easily, unless it's horizon and sky. Search out its variation, its colors, its subtleties. (Least Heat-Moon, 1991, p. 27)

What could the reader infer of this inner transformation? That the space, the prairie, gained meaning through the author's experience. This notion has been somewhat present in disciplinary debate since the 1970s. Familiarity, emotional connection, and space attachment are central elements in the idea of *placeness*. Examining the topic, humanistic geographer Edward Relph proposed a differentiation between *space* and *place*. To him, the notion of place includes human experiences and memory (Relph, 1976). In other words, a *place is a space with meaning*.[11]

[11] There is a danger in Relph's anthropocentric perspective. If a space happens to lack any sort of human value, it is not, by all means, meaningless or invaluable. Spaces have their own right to existence, regardless of human values placed upon them. Another human geographer, Yi-Fu Tuan, contributes to a different perspective in which both place and space have meanings, but opposing ones. Space is indefinite, vast, difficult to grasp or understand; whereas place implies something more tangible, real, and familiar. Tuan's conceptualization can be considered more environmental because it recognizes all spaces as inherently meaningful. *PrairyErth* and other examples of deep maps seem to rely on this more complex understanding of space and place since they put great significance on ecological relationships and more-than-human activities and

Limited or unpleasant interactions can result in *placelessness*, the absence of bonding or identification with space. This is the sense of detachment described by Least Heat-Moon in his first passage through Chase County. At that initial moment, the prairie meant nothing to him, not yet a place, but a space.

So, why are these concepts relevant to critical cartography? Because systems of oppression reinforce *the silence of maps* (as discussed earlier). Therefore, while some places are extensively surveyed, others remain neglected, "irrelevant". Deep maps re-signify the importance of forgotten or unmapped regions. Using Relph's terminology, deep maps can encourage the transformation of spaces into places, or at least reveal existing narratives and relationships. Individuals attribute spatial meaning through focused attention and embodied practices. Such a process can be conscious or unconscious, deliberate or incidental, prolonged or short—but it is always empowering. Deep maps bring placeness to light. Evaluating the contributions of *PrairyErth*, a Chase County rancher stated: "the book had a positive impact, overall. Because I think it raised our self-esteem. We thought: 'Wow, somebody could see something in us that we didn't see'".[12] His testimony proves that not only did the deep map connect Least Heat-Moon to the prairie, but it also added new values to existing relationships between those Kansas communities and their environment.

William Least Heat-Moon sees the disconnection between humans and land as the main cause of environmental problems. His aspiration for the book was clear: he encouraged Americans to (re)connect with the land. In his mind, if he could show how interesting and deep the natural and human history of Chase County was, then others could start seeing value in their own homes and lands. *PrairyErth* encouraged such connection through the reading and inspired new deep mapping practices. It also attracted passionate readers to visit Chase County and to support initiatives for environmental conservation. By the time of

rhythms. In other words, deep maps can bridge this conceptual gap by blurring the boundaries between "valuable" and "non-valuable".

[12] The documentary *Return to PrairyErth*, by New Truth Films, brings William Least Heat-Moon to Chase County 20 years after the book's publication, in conversation with local ranchers and book enthusiasts.

its publication, in 1991, century-old conversations about the establishment of a national park finally gained motion. The following year, by the efforts of proud ranchers and inspired preservationists, the Tallgrass Prairie National Preserve was created. *PrairyErth* nourished—and still does—local aspirations to protect the ecology, the culture, and the history of the prairie landscape.

Literary Cartography

When Least Heat-Moon regarded *PrairyErth* as a deep map, he left implied a conceptual question: are maps essentially pictorial? The task of conveying tri-dimensional space on flat surfaces was historically central to the visual arts, and Renaissance artists significantly improved perspective drawing. But this challenge was even more complicated in the case of maps, which also required precise and measurable information. The ingenious development of mapping techniques[13] achieved ambiguous results, highlighting precision on the one hand, but distorting realities, on the other—a paradox epitomized by the Mercator projection. Considering the difficulties of visually translating impressions, deep mapping proposes a different tactic: instead of flattening spatial qualities into drawings, it captures them in literature.

In narrative form, maps are powerfully free. *PrairyErth* starts with a collection of quotes about Chase County and the American prairies. These testimonials show how the place is portrayed in people's imaginations. Next, the book presents a black-and-white, very simple visual map of Chase County. So simple in fact, it barely contains relevant information other than the main roads and names of ranches. The drawing helps the reader situate the ranches, but ironically, reveals to be dispensable to understanding the prairie—it tells a shallow story. Perhaps deliberately, it seems that Least Heat-Moon desired to contrast the shallowness of this simplistic map with the depth of his prairie accounts. It is the literature that translates the essence of a place.

[13] Such as conical, planar, azimuthal, or cylindrical projections.

Many authors were inspired by these first deep maps (*Wolf Willow* and *PrairyErth*) and the 1990s saw the expansion of the genre (Maher, 2014, pp. 145–245).[14] Deep maps are environmental literature. But similar to many post-modern methods, there are inquiries concerning their conceptual boundaries. How, then, are deep maps different from other works in environmental history? They focus on spatial narratives. The place is the protagonist, the backbone of the story. But deep maps are also personal. As opposed to other genres, deep maps reject detached, analytical perspectives and bring a situated gaze to cartography. They are prose and poetry, factual and fictitious. Because of this innovative and stylistic freedom, deep maps require authors of mixed talents, abundant creativity, and diverse training. In interdisciplinary writing, researchers can unveil the character of a place through archaeology, ecology, geology, art, anthropology, among many others. A lyrical voice weaves information, storytelling, and meaning. But non-academic authors have also adopted deep maps in their own, talented approaches. Literary cartography brought new meanings to mapping, outside the box of drawings. This parting encouraged new understandings of cartography by emphasizing not the drawing, but the artistic process; not the bare space, but the meaningful qualities and particular stories of a place.

Beyond Literature

Although recent scholarship and independent practices have proposed their approaches to deep mapping, there is a general understanding that, as a creative process, this form of critical cartography is enriched by freedom, plurality, and inventiveness. Deep maps can rely on scientific works, but should by no means be restricted to them, nor by literature. For artist and cultural studies researcher Selina Springett, a deep map can be more than a concept or a method; it can be an aesthetic choice or even a process (Springett, 2015, p. 624). After *PrairyErth*,

[14] Maher makes reference to several deep map authors, such as Sheila Nickerson, Ian Marshall, Matt White, Sharon Butala, and Linda Hasselstrom (Maher, 2014, pp. 145–245).

researchers and activists started to explore creative dimensions of representation beyond writing. The following examples illustrate possibilities for artistic, collective, and urban deep maps.

Brett Bloom and Nuno Sacramento are deep mapping practitioners who recommend a methodology for a *collective process*, with possibilities in writing, storytelling, performances, videos, and art workshops (Bloom & Sacramento, 2017). Their procedure includes nine non-sequential steps and their principles (ibid., pp. 76–78). It goes like this: the first step involves the organization of an immersive experience for spatial recognition, like a retreat or a camp. Gradually, a mutual bond grows between participants, as they start to feel more connected to the place. The next three steps focus on varied explorations. A framework emerges from conversations about personal knowledge and backgrounds. Specific vocabulary, cultural perspectives, and life experiences are some possible topics. Through discussions, walks, workshops, exercises, performances, readings, screenings, lectures, and many other types of activities, the group can explore these frameworks. Deep maps are meant to gather deep, long-ranging, subjective, and specific input; this requires the casting of a wide net to first identify and later select narratives. Casually and formally, in situ and displaced, individually and collectively, the subjects should be woven in "several different, overlapping, layered ways" (Bloom & Sacramento, 2017, p. 76). Only then will the conversation start to open up and reveal hidden gems.

Within their recommendations emerges the creation of a safe space for dialogue. Acknowledging conflicting perspectives in an environment of positivity can be very constructive. For this aim, the authors suggest fostering a spirit of discovery and exchange, the delineation of guidelines, and the careful selection of participants. These decisions are indispensable to the outcome and should be aligned with the purpose of the investigation. However specific the methodology, these considerations can also be relevant to individual deep mapping. Take *PrairyErth*, for instance. To discover the prairie's deep layers, Least Heat-Moon had to establish a meaningful connection to interview local ranchers. Although the results include the filter of his perspective—as opposed to a composition of multiple perspectives—deep maps are hardly strictly individuated works. They encourage a horizontal form of mapmaking.

Another point for consideration, Bloom and Sacramento (2017) believe rural areas to be the ideal settings for deep maps. There is a great potential for achieving group cohesion during immersive experiences because isolation from busy, urban lives can have a powerful unitarian effect. Indeed, even *PrairyErth* followed this model by adopting rural Kansas as a subject. However, a question can be raised about the relevancy of urban deep maps.

Why is urban deep mapping necessary? Most city maps have an orientation function, including transit network, location of businesses and public facilities, urban equipment, and tourist attractions. Despite the contemporary boom of new cartographies, there are, indeed, many challenges to grassroots urban mapping collaborations. One of the obstacles to using deep maps in cities is the multiplicity of narratives. High demographic densities bring too many perspectives on placemaking. Where *PrairyErth* narrates the stories of people living in twelve Chase County ranches, how would one tell the stories and memories of the 10,000 people residing in a single block in Manhattan? Or the millenary history of a city like Rome? The narratives of hidden urban creatures and the many layers of archaeological history? Cities present a daunting complexity. However, contrary to traditional cartography, deep mapping does not presume to portray an absolute picture. It focuses on the common, untold, unnoticed stories: mini-narratives. Urban deep maps are urgent to create bonds and strengthen communities. Metropolises attract low permanence residents who find it hard to connect with a place and to create a sense of community. Therefore, urban deep maps are even more necessary to give cities a holistic, environmental, and inclusive lens (see Humphris et al., this book).

Identifying this methodological difficulty, communication and semiotics professor Daniel Ribeiro suggests three approaches for deep mapping in cities: roaming, archaeology, and montage (Ribeiro, 2019, pp. 45–47). Life in urban centres—especially in predominantly unsafe and unequipped cities—tends to be experienced in enclosed spaces: the home, the car/train/bus, the workplace. But place attachment requires an actual experience of place, with discovery, repetition, memory, connection. Therefore, perambulating (aimlessly) or walking (with destination) are essential ethnographic approaches to the process of recognition and belonging. This idea was incarnated in the *flâneur*, by French poet Charles Baudelaire in the nineteenth century and by landscape historian John Dixon Hunt, with the three types of walking and perceiving the environment: the procession, the ramble, and the stroll (Hunt, 2003). All approaches highlight embodied experiences of pedestrian movement as meaningful city practices.

Ribeiro also argues that urban deep mapping should employ archaeological exploration, uncovering historic artefacts, architecture, landscapes, etc. Environmental narratives can add much value to the process, considering they are frequently dismissed in urban cartography in favour of social and economic topics. Finally, the scholar advises the juxtaposition of different mediums (photography, literature, maps, music, videos, drawings, etc.) through a montage process. Of course, a strictly literary deep map would incorporate such elements within its prose or poetry. But Ribeiro's approach encompasses all sorts of creative expression mediums. In fact, social cartography, for example, can be employed within the entire deep mapping process. The methods of Bloom and Sacramento (whose work involves place-based storytelling, soil analysis workshops, listening sessions in kayaks, among other fascinating activities), reinforce this point about creativity and diversity in deep maps. Conversations are, then, not only deepened but broadened. Both Ribeiro (2019) and Bloom and Sacramento (2017) expand the deep map ontology and its dissemination among activists, researchers, and local communities. They show the immense, yet underexplored potential of creative mapping methods.

A particularly inventive case of non-literary and non-academic deep mapping is that of Dutch theatre company, PeerGrouP (see Davis et al., this book; see also, Van der Vaart, this book). While this creative enterprise does not designate its projects as deep maps, their highly communal and site-specific ways of learning about places correlate to the premises and goals of the method. The theatre group musters professionals of varied art disciplines (actors, sculptors, architects, dancers) to work with immersive experiences and performative projects in rural areas. At the start of every project, professionals mingle with local participants to learn communal stories and embodied practices. "Integrate and infiltrate" is

Fig. 11.7 *Grutte Pier fan Kimswert* procession and performance. In this performative, deep mapping process, the local community was involved in the investigation, conception, reenactment, and celebration of a historical event (*Source* PeerGrouP, 2016. Photo: Reyer Boxem)

their motto. As important as trying to learn about a place, they believe, is to see it through someone else's eyes.

The entire creative process is collaborative: from early conversations, conceptualization, and rehearsals, until the final performance. For instance, *De Affaire Vermaning* (The Vermaning affair) play incorporated an ancient road in their theatre piece about the archaeological discovery of a local farmer (Bruinsma, 2018). In 2016, *Levende Duinen* (Living Dunes) involved "a unique experience in which knowledge of coastal management [was] poetically interwoven with the personal experiences and insights of the islanders of Terschelling" (van der Werf, 2016). The result of this storytelling project is a 45-minute listening walk available for download, which can be appreciated by anyone.[15] Performed by Frisian residents, the 2016 *Grutte Pier fan Kimswert* play and procession (Fig. 11.7) evoked stories of a local hero and a forgotten battle, changing perceptions of place and reigniting a sense of belonging (Bruinsma, 2016). Albeit brief, their contribution is not limited by the ephemeral character of certain performances—most of them were recorded to inspire further works and audiences.

The work of PeerGrouP can be framed as deep mapping because it performs place. That is, it uncovers mini-narratives of spaces with meaning, with memories, and with many (deep) layers of human and

[15] The listening walk is available at: https://www.peergroup.nl/luisterwandeling/.

non-human complexity. In a sense, the collective, immersive practices of both Bloom and Sacramento (2017) and the PeerGrouP exemplify how the essence of a deep map is to expose the inherent but often hidden entanglements between people and the environment. To researchers, artists, and practitioners alike, deep maps offer the opportunity to start an investigation in a different way. In deep mapping, it is the site observation, the lived experience, and the local people who inform the direction of the research. As a consequence, projects and interventions come from a solid base of existing values and concerns, resulting in more meaningful and long-lasting effects.

Conclusions

> Critique has a number of basic principles. First, it examines the (often unexamined) grounds of our decision-making knowledges; second, it situates knowledge in specific historical periods and geographic spaces (rather than being universal fur all time); third it seeks to uncover the relationship between power and knowledge; and fourth it resists, challenges, and sometimes overthrows our categories of thought. The purpose of critique is not to say that our knowledge is not *true*, but that the truth of knowledge is established under conditions that have a lot to do with *power*. Critique is therefore a politics of knowledge. (Cramptom, 2010)

In the 1990s, critical cartography initiated a paradigm shift in mapping theory, its methodologies and application. The critique exposed the pretension, manipulation, and oppression inherent to the legitimized and perpetuated ways of mapping up until that point. To academia, this shift was a great achievement, for it has enabled the surgency of new perspectives continuously contributing to advance knowledge. But most importantly, the new paradigm addressed the disproportional influence of science and knowledge production in power relationships. Critical cartography pointed out that maps rarely stand alone, doing nothing, changing nothing. For better or worse, each new method or approach, each new map holds the potential to provoke an impact.

Even the absence of maps speaks much through its silence. Therefore, surpassing the boundaries of formal education, the critique created a framework for activists and the population in general to demand justice, equality, and responsibility from authorities. Now, it is possible to detect abuse of power more clearly, to elaborate and support criticism with stronger evidence. Moreover, the new framework allows for maps to be deinstitutionalized, re-appropriated. Based on the arguments of geographer Nicholas Blomley, Jeremy Cramptom beautifully summarized three principles of critical geography: "1. It is oppositional: it targets dominant forms of oppression or inequalities; 2. it is activist and practical: it wishes to change the world; 3. it is theoretic: it rejects positivist explanation and enhances critical social theory" (Blomley, 2006, quoted in Cramptom, 2010).[16] Borrowing from these principles, the present chapter has suggested two directions, two creative methods for research within the critical cartography framework.

With social cartography, researchers can, first, decide to put people, groups, and *their* respective problems at the centre of mapping projects. Real-life problems inform research questions and the formulation of theories, so researchers are urged to carefully select whose problems to describe, to investigate, and—perhaps—to solve. A picture is worth a thousand words. Social maps give concreteness and a sense of urgency to issues perceived and criticized by the most vulnerable. There is an intrinsic social justice orientation in this method. It can be used as a compelling tool for protest. This confrontational characteristic is potentialized, secondly, by the active involvement of communities in the process of mapping research. Reciprocally, the emancipation of communities assures that knowledge gaps are denounced and properly amended.

Deep maps target a similar, democratic direction, but operate rather differently. By taking the focus away from the drawing, the visual, the unidimensional object of a traditional map, the method can facilitate

[16] In the following page, Cramptom presents his own four principles of critical cartography. Roughly outlined, he sustains that the critique: (1) challenges unexamined assumptions of knowledge orders; (2) uses historicization and spatialization to convey the problematic around certain mapmaking practices; (3) reveals the intrinsic and unavoidable political nature of maps; and (4) is emancipatory in its orientation.

the recognition of obscured narratives and values. Deep maps identify a stronger and softer side of human-spatial relationships that can unite, seduce, conserve, and transform. Through this practice, it is rather the communities that welcome and involve researchers and practitioners, as opposed to the other way around.

Therefore, these two methods follow the principles of critical geography indicated by Blomley and Cramptom. They aim to oppose exclusive narratives, reform unjust situations, and enrich cartographic knowledge. In short, they reconnect mapmaking to society. This is the true potential of these creative mapping methods.

Acknowledgements This research received funding from the European Union's Horizon 2020 research and innovation programme under the Marie Sklowdoska-Curie Grant Agreement No. 765389.

References

Ackermann, F., Eden, C., & Cropper, S. (1992). *Getting started with cognitive mapping*. https://www.researchgate.net/publication/265411517_Getting_Started_with_Cognitive_Mapping

Blomley, N. (2006). Uncritical critical geography? *Progress in Human Geography, 30*(1), 87–94. https://doi.org/10.1191/2506ph593pr

Bloom, B., & Sacramento, N. (2017). *Deep mapping*. Breakdown Break Down Press.

Bruinsma, D. (Director). (2016, September). *Grutte Pier fan Kimswert by D. Bruinsma* [Play]. Land van Osinga, Kimswert.

Bruinsma, D. (Director). (2018, July). *De Affaire Vermaning* by J. Veldman [Play]. Drents Museum, Hoogersmilde.

Buzan, T., & Buzan, B. (1993). *The mind map book: How to use radiant thinking to optimize your brain's untapped potential* (1st ed.). BBC Books.

Cramptom, J. (2010). *Mapping: A critical introduction to cartography and GIS* (1st ed.). Wiley-Blackwell.

Cramptom, J., & Krygier, J. (2010). An introduction to critical cartography. *ACME: An International E-Journal for Critical Geographies, 4*(1), 11–33. https://www.acme-journal.org/index.php/acme/article/view/723

da Silva, T. (2001). *Espacios de Identidad: una introducción a las teorías del currículum*. Octaedro.

de Almeida, A., Dourado, S., & Bertolini, C. (2018). *A new social cartography: Defending traditional territories by mapping in the Amazon*. In kollektiv orangotango+ (Eds.), *This is not an atlas* (pp. 46–49). transcript Verlag. http://www.notanatlas.org

Eden, C. (1988). Cognitive mapping. *European Journal of Operational Research, 36*(1), 1–13. https://doi.org/10.1016/0377-2217(88)90002-1

Frank, N., & García, F. (2018). *Mapping inside (and outside) the classroom*. In kollektiv orangotango+ (Eds.), *This is not an atlas* (pp. 152–157). transcript Verlag. http://www.notanatlas.org

Gorostiaga, J. M. (2017). Perspectivism and social cartography: Contributions to comparative education. *Educação & Realidade, 42*(3), 877–898. https://doi.org/10.1590/2175-623665366

Harley, J. B. (2001). *The new nature of maps* (p. 2001). The Johns Hopkins University Press.

Harley, J. B., & Woodward, D. (1987). *The history of cartography, Vol. 1: Cartography in prehistoric, ancient, and medieval Europe and the Mediterranean*. The University of Chicago Press. https://www.jstor.org/stable/3106092?origin=crossref

Heat-Moon, W. L. (1991). *PrairyErth (a deep map)*. Houghton Mifflin Company.

Hunt, J. D. (2003). "Lordship of the feet": Toward a poetics of movement in the garden. In M. Conan (Ed.), *Landscape design and the experience of motion* (pp. 187–214). Dumbarton Oaks Research Library and Collection.

Irving, A., & Moss, O. (2018). Imaging homelessness in a city of care. In kollektiv orangotango+ (Eds.), *This is not an atlas* (pp. 270–275). transcript Verlag. http://www.notanatlas.org

kollektiv orangotango+ (Eds.). (2018). *This is not an atlas*. transcript Verlag. http://www.notanatlas.org

Liebman, M., & Paulston, R. G. (1994). Social cartography: A new methodology for comparative studies. *Compare: A Journal of Comparative and International Education, 24*(3), 233–245. https://doi.org/10.1080/0305792940240304

Lobatón, S. (2009). Reflexiones Sobre Sistemas de Información Geográfica Participativos (Sigp) y Cartografía Social. *Cuadernos de Geografía: Revista Colombiana de Geografía, 18*, 9–23. https://doi.org/10.15446/rcdg.n18.12798

Maher, S. (2014). *Deep map country: Literary cartography of the great plains.* University of Nebraska Press.
Novak, D. J., & Gowin, B. (1984). *Learning how to learn.* New York: Cambridge University Press.
Oslender, U. (2017). Ontología Relacional y Cartografía Social: ¿hacia Un Contra-Mapeo Emancipador, o Ilusión Contra-Hegemónica? *Tabula Rasa, 26*, 247–262. https://doi.org/10.25058/20112742.n26.12
Peluso, N. L. (1995). Whose woods are these? Counter-mapping forest territories in Kalimantan, Indonesia. *Antipode, 27*(4), 383–406. https://doi.org/10.1111/j.1467-8330.1995.tb00286.x
Relph, E. (1976). *Place and placelessness.* Pion Limited.
Ribeiro, D.M. (2019). Deep mapping: Uma Introdução Ao Mapeamento Profundo. *TECCOGS: Revista Digital de Tecnologias Cognitivas, 19*, 30–51. https://doi.org/10.23925/1984-3585.2019i19p30-51
Smith, C. D. (1996). Imago Mundi's logo the Babylonian map of the world. *Imago Mundi: The International Journal for the History of Cartography, 48*(1), 209–211.
Springett, S. (2015). Going deeper or flatter: Connecting deep mapping flat ontologies and the democratizing of knowledge. *Humanities, 4*(4), 623–636. https://doi.org/10.3390/h4040623/
Unger, E. (1937). From the cosmos picture to the world map. *Imago Mundi, 2*(1), 1–7. https://doi.org/10.1080/03085693708591828
van der Werf, M. (2016, June 12). *Levende Duinen* by M. van der Werf [Exhibition]. Oerol Festival, Terschelling.
Vaughan, L. (2018). *Mapping society*. UCL Press. https://doi.org/10.2307/j.ctv550dcj
Williams, R. (1985). *Keywords: A vocabulary of culture and society* (Revised ed.). Oxford University Press.
Wood, D. (2003). Cartography is dead (Thank god!). *Cartographic Perspectives, 45*, 4–7. https://doi.org/10.14714/cp45.497

12

'Getting Deep into Things': Deep Mapping in a 'Vacant' Landscape

Imogen Humphris, Lummina G. Horlings, and Iain Biggs

Openings

We begin this chapter with a map (Fig. 12.1). It is not a particularly unusual type of map, in fact it's one that is quite commonly used to

This chapter is written from the first-person perspective of myself, the lead author. I conducted the fieldwork that is described here. The co-authors were instrumental in supporting development of this project and writing of this text with advice and guidance. I use the collective 'we' at various points to refer to the multitude of authors, inhabitants and readers that gather around this text.

The deep mapping work is available to view at http://www.govandeepmap.com.

I. Humphris (✉) · L. G. Horlings
University of Groningen, Groningen, The Netherlands
e-mail: i.t.humphris@rug.nl

I. Biggs
University of Dundee, Dundee, Scotland, UK

Bath Spa University, Bath, UK

A. Franklin (ed.), *Co-Creativity and Engaged Scholarship*,
https://doi.org/10.1007/978-3-030-84248-2_12

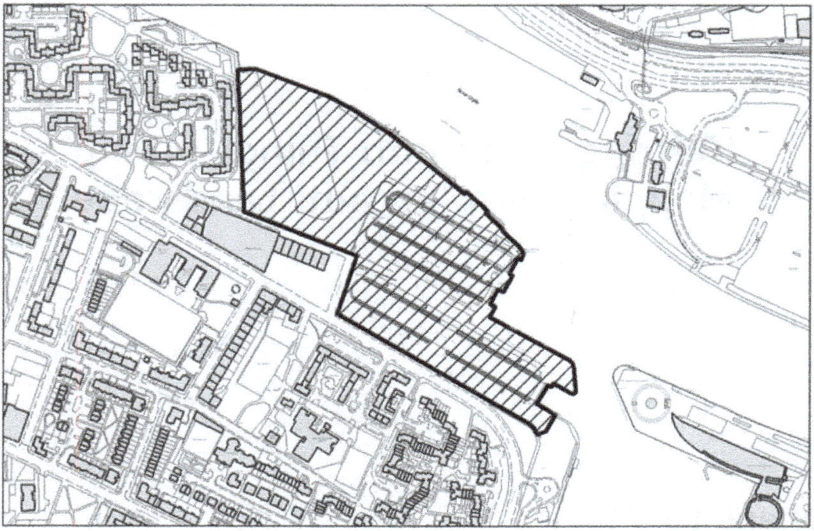

Fig. 12.1 Map of the Govan Graving Docks as used by Glasgow City Council. (Source Glasgow City Council. Published with permission from Glasgow City Council: © Crown Copyright and Database Right 2014. All rights are reserved. OS Licence number 100023379)

depict areas of wasteland, cartographically employing empty outlines or hatched boxes to show a plot available for development, as is the case here. Dotted traces might hint at the remnants of some former life but typically such maps resist telling us any further details of these wasteland places. Their emphasis, rather, is on the plots' status as awaiting development or the reinstatement of formal use.

When we take a walk through such a place however, we find far more that is actively present than passively 'waiting' (see Fig. 12.2). Stepping over that heavyweight boundary-line, we might immediately be confronted with an unabated ecology; grasses that rise to mid-thigh,

Fig. 12.2 (*Source* Author)

unfamiliar spider species disturbed underfoot, and buddleia exploiting the cracks in decommissioned concrete footings. We might also detect the traces of others who have come this way before us. Perhaps dog walkers, children in search of wilderness, or graffiti writers looking for the perfect wall. Striking up conversation with local residents we might well find that the site is alive with memory of former jobs worked in former shipyards, of the friendships formed on factory floors and the smell of the hot roll stand at the end of the street.

These things all elude the kind of instrumental and reductive cartography most commonly utilized in planning and policy-making. The fragmentary, heterogeneous nature defies neat outlines with singular meanings. Further, these things are never still, but a continually shifting tangle of narratives. Like eels, the more tightly we try to contain them, the more readily they slip from us. Despite the certainty with which maps like the one above are presented, they are hollowed out by the reductive instrumentality implicit in their making. It is only when we question what might be missing from such representations, what they avoid telling us, that their air of absolute and objective authority begins to crumble.

Pivotally, it is most commonly those more marginalized narratives that are filtered out from the sanitized map. Given the power that such representations can have, sitting on the desks of planners, speculative developers and policy-makers, omissions can serve to further exclude and invalidate. In this text, I argue that arts-based methods can play a potent role in this context (Kester, 2013) by offering avenues for spatial representation that allow for the multiplicitous, non-aligned and emergent nature of place (Massey, 2005). While the methods I explore are not restrained to any particular type of landscape, I maintain that some of the values that underpin them are particularly helpful and relevant to marginalized landscapes and 'wastelands' as the multiplicity of voices is often overlooked in these spaces.

Despite the existence of multiple, tangential narratives across landscapes such as wastelands, they are often overshadowed by and reframed within dominant meta-narratives. Hegemonic discourse centres around the future use of the land, overlooking uses and meanings in the present. In order to draw out marginalized narratives, there is a need to unsettle rhetorical notions about what the site means and who has the authority to speak about it. It is here that arts approaches can serve as a helpful catalyst to conversation and investigation. Arts practices such as deep mapping seek to harness the capacity of the aesthetic experience to disrupt dominant narratives, creating spaces that allow individuals to speak outside of their commonly held positions (Kester, 2013). This notion of art as a context setting device may be linked to Viktor Shklovsky's (1917, as cited in Lemon and Reis, 1965) ideas around defamiliarization whereby art, as a device, is used to make the 'familiar strange' so that it may be freshly perceived. Through its ability to creatively remake and therefore reframe the present, art practice can situate common, everyday experiences in new contexts, prompting new discussions. Michelle Henning (2020) describes this as 're-presenting the present', in which the reflecting of the present back to the viewer in a way that is 'recognizable', 'negotiable' and 'accessible' creates the conditions for individuals to detect, reflect upon and mediate their being within it.

We consider the field of deep mapping, as an archaeological, geo-social and ethnographically informed art research practice which offers opportunities to generate representations that 'dive into' the heterogeneity and non-aligned multiplicity of place (Modeen & Biggs, 2020; Roberts, 2016; Smith, 2015; see also Reitz, this book). Deep maps embrace, as their starting point, the tensions that exist between incompatible narratives and between one slice of time and the next. They seek to draw out those discordant, micro-narratives that are commonly swallowed up within meta-narratives of a place. This approach also brings into question the role of the artist-researcher who is themself bound up within the present that they seek to *re-present*, calling for a move towards an ensemble of roles and an acknowledgement of their embodied being within the representational process (Bailey, 2018).

In this chapter I reflect back on an ongoing deep mapping process that holds divergent multiplicity as its central motive in depicting a location normally understood as wasteland. My starting point is the tract of land depicted in the opening of this text—the Govan Graving Docks in Glasgow, Scotland. The site formerly lay at the heart of Glasgow's shipbuilding industry which, when it collapsed in the latter half of the twentieth century, brought significant consequences for the local working community (High, 2013; Tovar et al., 2011) from which it still has not fully recovered (Butler et al., 2012). Having been decommissioned in 1987, this site stands as one of the few remaining markers of Govan's industrial past, given that all other shipyards have either been cleared completely or (as is the case with the one remaining yard) unrecognizably modernized and walled off geographically and socially from the Govan community. However, in the 34 years since its closure, the Govan Graving Docks has been almost continually appropriated by local citizens for a broad spectrum of informal purposes. The site is alive with shifting and heterogeneous narratives yet these are commonly precluded by traditional cartographic forms. Through reflections on the empirical and methodological challenges of applying a deep mapping approach to this setting, in this chapter we explore the broader questions around how such a research process can actively seek out and amplify heterogeneous and marginalized narratives in a deindustrialized urban landscape. This holds critical relevance for the recognition and reclaiming of citizen agency within those many European cities now experiencing an upswing in development following the industrial decline of the late twentieth century. While acknowledging implicit tensions around scientific rigour and identifying the alternative forms of exclusion that can serve to hinder the polyvocality of deep mapping, this chapter outlines the helpful contributions that this approach can offer place-based investigations in marginalized landscapes.

The Arts Research Approach of Deep Mapping

The term deep mapping is embraced by a broad variety of artists, creative practitioners and researchers that utilize arts- or performance-based approaches to stimulate conversations *about* and investigations *into* place (Biggs, 2010; Roberts, 2016). Broadly, this approach sets out to contend the totalizing and irreducible nature of traditional cartographic representations, aspiring instead towards the ideal of encompassing 'everything you might ever want to say about a place' (Pearson & Shanks, 2001, p. 65). Clearly practitioners do not regard such an endeavour as a realizable objective as Cliff McLucas (2014), a proponent of deep mapping, described:

> Whilst I can talk about deep maps, whilst I can imagine such things … whilst I can even dream about deep maps, unfortunately, I have to admit that I have never seen one.

This is not to say that McLucas made no attempt to create deep maps (he did in fact leave several precedents), rather, alongside many advocates in the field such as Least Heat-Moon (1999), he recognized that one could never complete the object of a deep map, only engage in *deep mapping*. Instead, we may read the wide-ranging manifestations of deep maps as *efforts to describe* place in its fullness and unending complexity. Implicit in this is an immersive, performative 'dance' between the mapper and the place, as Wood (2015) describes:

> [W]hat's essential is getting out in the field […] and looking hard at stuff. Walking through it and writing it down forces a valuable kind of attention, an irreplaceable kind of attention […] This kind of immersion makes you think about things, dream about them, and this prompts new questions, which send you back out into the field. It doesn't take long to get deep into things when you're paying attention, and mapping focuses attention.

Consequently, examples of deep mapping appear across a broad range of arts' and humanities' disciplines that gain strength in their avoidance of tight definition (Modeen & Biggs, 2020). These include auto-ethnographic texts (Least Heat-Moon, 1999), journeying (Bissell & Overend, 2015; Sinclair, 2017), participatory archaeological digs (Lewis, 2015) and photo collage (Reddleman, 2015) amongst others (see also Reitz, this book). Though diverse in form, they are loosely unified by their effort to lure the mapper into this dance, prizing open discursive interactions with place.

Considering my task of 're-presenting' the Govan Graving Docks, drawing out the many tangential and overlooked narratives that exist there, I identify several pertinent sentiments from discussion on deep mapping. Three themes emerge as offering particularly productive lines of inquiry to the context of the Govan Graving Docks:

Polyvocal

A central motive of deep mapping is to pursue investigations of place beyond that which is immediately evident. Implicit in this notion of 'digging' is the drawing out of multiple, discordant narratives and associated meanings and material manifestations. In a sense, deep mapping removes the confining boundaries of the instrumental, reductive map, creating the space needed for tensions and contradictions in narratives to exist simultaneously. Faced with a polyphony of narratives, I argue that the position of the deep mapper, from a constructivist perspective, is never neutral but rather consciously present and actively seeking to give platform to those voices commonly drowned out in reductive representations. As Cliff McLucas (n.d.) states:

> Deep maps will not seek the authority and objectivity of conventional cartography. They will be politicized, passionate, and partisan. They will involve negotiation and contestation over who and what is represented and how. They will give rise to debate about the documentation and portrayal of people and places.

Such a position offers margin for the individualized and intimately personal to be heard while also providing an arena for collective storytelling and knowledge construction. It carves out room to fully acknowledge lived experiences, not as events confined within the place itself, but rather as the collision between the physical landscape and the continuum of life trajectories that extend far beyond the bounds of the site (Ingold, 2017).

This contesting of 'what is represented and how' (McLucas, n.d.) also holds promise for vacant land in particular. In such locations where strong confluences of meanings are wrapped up in either the period of use prior to closure or the speculative, forthcoming use, deep mapping processes may helpfully redirect focus away from the past or future place and instead create opportunity for the present place to be acknowledged and validated. Giving precedence to this marginalized time frame, the creation of a deep map can incrementally shift speculation about a place away from 'what it should become' and towards 'what it *is*'.

Generative

Deep maps can be interpreted as both catalytic objects and actions taken into the field to stimulate investigation, unearth further questions and open up conversation. In iterative fashion, these 'traces' subsequently prompt reflection and generate the impetus for further, more divergent investigation. The artist-researcher is inextricably embedded in this process as the deep map comes into being through discursive interaction with place. It is through such imperfect and meandering *efforts* to create reflections of place, that deep mapping simultaneously invites debate about both the place and the process of representation. McLucas (n.d.) further points to the collaborative nature of deep maps in this regard: 'Deep maps will bring together the amateur and the professional, the artist and the scientist, the official and the unofficial, the national and the local'. By placing emphasis on the process, the *creating* of a deep map becomes a space of generative exchange, elevating inhabitant knowledge alongside that of the artist-researcher and resetting traditional hierarchies.

Open-Ended

Given the common preoccupation of deep maps with the 'fundamental *unmappability* of the world' (Roberts, 2016, p. 5), the creative outcomes of deep mapping might more effectively be read as 'forever incomplete' processes. Such objects are candid in acknowledging their own inadequacy to embrace the unending and entangled tapestry of narratives surrounding place. However, as Modeen and Biggs (2020) point out, these are the contexts where art comes into its own power: 'It is precisely this inadequacy that enables the arts to evoke our lived experience as always exceeding and falling short of the conceptual definitions central to analytical thinking' (p. 53). In its ability to situate itself within and gesture towards that which we do *not* know or *cannot* express, art can avoid the closing down and flattening out of the continual emergence of place. In this sense, the art research practice of deep mapping suggests a representational approach that tentatively presents a subjective and discursive 'window' in the continual becoming of place.

For our location of interest and, indeed, our particular preoccupations with it, these three sentiments collectively offer fruitful underpinnings for the creation of a conversational, representational research method. As the Govan Graving Docks has majoratively been understood as 'vacant' or 'awaiting development', the many histories of lived experiences and citizens appropriations in the site since its closure have passed largely unrecorded. This lack of formally assigned use can be seen to have created an entirely different kind of space in the urban landscape, one in which many narratives are present, overlapping and confronting each other (Humphris & Rauws, 2020). However, the objective of our representational effort here is not to uncover and pin down meaning in a place where it appears to be missing; in a site that is so actively in a state of becoming, we feel that this would be entirely unconducive. Instead, by putting these sentiments of *polyvocal*, *generative* and *open-ended* into action, we intend to create a representational space that contends the very notion that the place is without meaning. It is an effort to create a representational space around this very different kind of territory that commonly falls out of the urban imagination (Shoard, 2000).

Beyond such attitudes to investigation, the tools and practices used in the actual creation of deep maps are dependent upon the demands of the place and the skills and resources available to the map initiator. This active steering away from any notions of a formalized approach stands central to their offer. As Biggs (2010) proposes, the value of deep mapping partially pertains to its ability to resist 'becoming complicit in its "disciplining"' (p. 21) (see also Modeen & Biggs, 2020; Roberts, 2016). What follows is the development of and reflection upon a deep mapping method that is highly specific to our research interests in the site, and the artistic practice and even personal traits of the lead author who undertook this fieldwork.

Methodology

This first phase of deep mapping the Govan Graving Docks was conducted over a period of two months during which I spent extensive time on the site and in conversation with local individuals. The site is located on the edge of Govan, a formerly prolific boat building community that has seen a high level of vacancy and deprivation since the closure of the majority of the shipyards in the late twentieth century. While the site is privately owned, the owner has failed on several occasions to secure planning permission for a largely private housing development. The site closed in 1987 and in the early 2000s most of the remaining buildings were demolished. The only remaining architectural features are the listed pump house, one tidal dock and three impressive dry docks with stone walls that remain largely intact. Despite this, the site feels relatively 'wild' given the significant amount of flora and fauna that have established there. The gated entrance to the site is almost always left open and there is a general acceptance, including from the police, that individuals may access the land; there are few days that pass where it sees no visitors. Individuals are frequently seen walking, adventuring, graffiti writing, drinking alcohol and nature watching. Over the 2-month period, I spent 19 days on the site, visiting for between 1 and 4 hours. I also hired a studio space in the centre of the community where the map was gradually created and I spent much time walking back and forth between the two locations, observing the surrounding neighbourhood along the way.

I made the recordings for the map using two parallel practices. The first was an iterative interview process of discussion and illustration with individuals closely connected to the site. The second involved regular onsite observational drawing and recording the presence of site users and informal conversations with them.

Interviews

I made contact with the local inhabitants that were interviewed through both snowballing local connections and from chance encounters while onsite. Given that the site is predominantly used by a collection of disparate individuals between whom very little network exists, my sustained presence onsite was a critical factor for making connections with inhabitants. The process involved two interviews; the first taking place on site and the second in my local studio space for a more reflexive conversation that explored emergent themes in greater depth. The primary interview was intended to gather a general overview of the inhabitant's connection to the site. I placed emphasis on how it related to them personally and how their relationship to the place had changed with time. My lines of questioning framed the site as particular and different from other places in the neighbourhood, such as a park and, through this, sought to unearth what the particular qualities were that resonated with them and repeatedly drew them back to the site. As we talked together, we walked around the site itself allowing the conversation about the place to become embodied and thus further illuminated; speaking theoretically about its meanings while physically negotiating its terrain.

From these recordings, I made partial transcriptions, pulling out key quotes and drawing these together into 4–6 emergent themes through a process of inductive coding, categorizing and labelling. The thematic labels were either words taken directly out of the transcript or created by myself in instances where the topic was discussed more abstractly. For each I sought a landscape metaphor to reflect the theme in the particular way that the individual had expressed it. For instance, one inhabitant made several references to change in the community with the inevitable coming and going of development; for this I depicted fishing boats in a harbour, sitting on the mud, waiting for high tide to return and transform the landscape once again. The quotes themselves were then woven around the illustrations to create a landscape of sentiment.

In the second interview, I presented the collection of illustrated themes back to the inhabitant. I gave them time to read through and review their quotes, correct anything that they felt was inaccurate, discuss their thoughts further around the themes and metaphors I had chosen and add any additional themes they felt were missing. Following this, I asked them to consider how the themes might relate to one another by laying the illustrations out spatially to form their own narrative map. Those themes that felt closely related or causational could be placed close together while others could be set down in a more distant location. Pinning these down and drawing strings between them, we discussed the nature of these connections and what the 'spaces in between' might mean. I also asked the inhabitants to consider what the connections might be between themes that they had initially thought were unrelated. Finally, together we laid their map down on top of those maps that had been created by other inhabitants and explored the differences, connections and tensions that lie between them. The intention of this was to empower the inhabitants in the process of compiling the full representation, by electing where to position themselves within it, and to further identify the nuance of their own narratives in the presence of others'.

Observational Drawings

In parallel to these interviews, we routinely spent time on the site making observational drawings and talking with other site users. By making drawings of objects that had been left behind by visitors, I was able to get to know the life of the place more intimately. The practice was often quite sedentary, sitting in one location for 30 minutes to 2 hours at a time in which individuals often felt confident to approach me with passing hellos or asking questions, and, on quiet days, sufficient stillness for resident wildlife to emerge. In line with Causey's (2017) exploration of drawing and ethnography, this practice additionally disciplined me to 'see' the site, to focus my gaze and speculate about objects and markings that I would otherwise overlook. As explained in the findings section, these sketchbook pages of drawings and notations moved from a tool for initial observation to becoming an important component of the overall map themselves as a gathering of traces reflecting my own experiences and encounters on site.

These recordings, notations and illustrations form a growing body of content that constitutes the open-ended deep map (see Table 12.1 that documents the quantity of content so far). Such divergent forms of gathering require different modes of practice. Within these, the role of the researcher becomes dynamic, continually shifting positions throughout

Table 12.1 Content of map collected and created so far (over two month fieldwork November–December 2020)

Interviews	5
Interview drawings	20
Onsite spontaneous discussions	14
Onsite observational drawings	12
Noted site visitors	73
Total days on site	19

Source Author

the research field rather than occupying one static point of observation. At one moment the researcher plays the role of the engaged listener while at other times the passive observer. Sometimes the central focus lies on context setting. At other times the primary role is that of the reflexive academic (Wittmayer & Schäpke, 2014). In this sense, the role of the researcher becomes adaptive, moving between positions in response to the ever shifting circumstances on the ground (see also Franklin, this book). Throughout the project, this 'ensemble-self' approach to the researcher role has afforded the vital flexibility needed to contend with the multiplicitous nature of place.

Uncoverings: Reflections on a Deep Mapping Method in Practice

In my time spent actively deep mapping in the field, I have avoided treating the method set out above as an instructional guidebook, but rather adopted it as a performative tool for investigating and probing within this highly specific context. Thus the method is mediated by the ensemble-researcher; it is alive in response to the landscape in which it is played out. Here I offer reflections that have emerged from this reflexive negotiation between the performative-artist-researcher and the method; drawing out and amplifying those dimensions that proved to be generative and stepping into alternative positions and modes of practice when the method appeared to be reaching its limit.

Divergence

One of the central aspirations of the Govan Graving Docks deep map is to pluralize narrative representations of the site. As explained in the methodology, interviews began with individual discussions that were subsequently woven into an agglomerated, layered form of representation (see Fig. 12.3 depicting many agglomerated narrative representations).

I found that postponing the moment of assembly to the end of the discussion process created a suspended space in which inhabitants were able to delve into, grapple with and eventually articulate their

Fig. 12.3 (*Source* Author)

own nuanced relationships to the site before being confronted with the narrative 'worlds' of others.[1] Together we were able to use this space to follow traces of meaning away from the site itself, out into their broader continuum of life experiences, detecting enduring themes of importance before bringing them back to the context of the Govan Graving Docks. In this way we incrementally 'felt out' their own narrative world.

While this suspended space allowed for greater divergence of narratives, explicitly common themes between the narratives did arise, such as heritage, or the need for sanctuary. However, in employing metaphor and illustration to reflect the sentiments expressed, it was possible to intentionally seek and draw out minutiae differences between the positions held. As such, one theme could appear in several different narrative worlds but with different expressions. Heritage, for instance, was at once depicted as a cliff face, progressively eroded by a hostile sea, while elsewhere it was characterized by a makeshift house continually built upon and adapted through time (see Figs. 12.4 and 12.5). In discussion these variated illustrations appeared to assist individuals in qualifying their own perceptions as being different from others, or even different but partially aligned in some aspects.

Through the process of defining their own narrative 'worlds', several inhabitants made connections between their preoccupation with the site and their own personal traits and preferences, with a sense of being different from the common majority with regard to their needs in the built environment and expectations of urban living. These expressions included the need for quiet and distance within the city, the ability to be alone and unregulated (see Figs. 12.6, 12.7, 12.8 and 12.9).

In my pursuit of multiplicity in the deep mapping process, I detected other knowledges present in the field but requiring other modes of recording and representation. While the interview approach was certainly

---

[1] It is important to note that, despite the fact that this particular deep mapping aims to draw out excluded voices, we cannot assume that this very different kind of investigative process was entirely inclusive. It is also reasonable to assume that the uncommon aspects of this investigation method create exclusions in their own right due to the elitism that is often associated with the arts and the discomfort it causes to many who do not consider themselves to be a part of this world. However, in the creation of the illustrations, I made a concerted effort to avoid overly abstract interpretations and instead generate images that made small, tangible steps over into the metaphorical.

Fig. 12.4 (*Source* Author)

proving successful in deeply investigating personal connections and drawing out diversity in perspectives, it was clear that, culturally and practically, it was not an accessible approach for all, particularly in a community facing multiple disadvantages such as Govan.

The following intersecting cultural factors played a role in the process of deep mapping:

- language barriers: adult illiteracy is a recognized issue in Govan and, with a large migrant population, some experience challenges with English as a second language. We recognized this in some individuals we met on site and found that, while many would pass with a smile and acknowledgement, they often avoided stopping to talk;
- cultural context: after some failed attempts, we also came to understand that for some the interview format (whereby one member in a conversation holds a designated 'power' to ask questions and set the

Fig. 12.5 (*Source* Author)

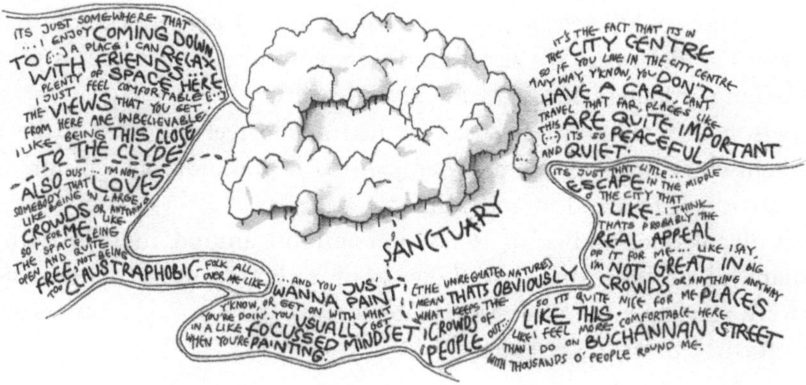

Fig. 12.6 (*Source* Author)

Fig. 12.7 (*Source* Author)

agenda or strategically guide the conversation) was culturally unfamiliar or could even be instrumental in further perpetuating historic power imbalances. We also found different cultural attitudes towards making arrangements, agreements and plans;

- personality: as the site is used by many who feel the need to be by themselves or away from other people, it was not uncommon for individuals to avoid interaction.
- drink: alcohol addiction is not uncommon around the site with a handful of individuals drinking from early morning. While some were very open to discussion, interactions were based upon regular, informal, unstructured chats;

While these may be acute and limited issues, this list encompasses a broad range of individuals whose diverse perspectives could be easily overlooked were the deep mapping relying solely on formally conducted interviews. It seemed my method was beginning to confine rather than

Fig. 12.8 (*Source* Author)

Fig. 12.9 (*Source* Author)

kindle the momentum of the investigation. However, as I spent periods of time on site conducting the observational drawing part of the mapping practice, I found myself increasingly engaging in spontaneous interactions with these inhabitants. Our presence on site and the stillness of the drawing practice lent me a familiarity and approachability, creating space for longer, less pressured conversations. I had originally intended to utilize the drawing practice to make purely visual observations of the site and chiefly as a segue into interviewing inhabitants. However, I came to acknowledge these serendipitous interactions as a crucial touch point between these inhabitants and the mapping process, and thus the sketchbook evolved into a valuable and direct input tool for the map. The recordings I made became richer in content and the pages filled out with details of interactions and stories told alongside the observational drawings (see Figs. 12.10, 12.11 and 12.12). As it grew, I was able to share and reflect on progress together with those who stopped to chat. One regular local even affectionately named this practice as 'going for a doodle'. As such our ethnographic drawing practice became one that not only prompted us to 'see' the site more deeply (Causey, 2017) but also to 'hear' it.

Over time it became clear that these sketchbook pages, as both recordings and mediators of conversation, would exist in the deep map with equal prominence to the content generated through interview. As McLucas (n.d.) eludes to in his statement 'Deep maps [...] will involve negotiation and contestation over who and what is represented and how', deep mapping allows for a 'playing' of value attribution that actively attempts to create representational space for voices that might otherwise be overlooked by those research methods that require more formalized processes of data collection driven by hegemonic calls for scientific rigour and replicability.

Disruption

Given my aspiration to detect and represent diversified spatial narratives around the Govan Graving Docks, I was acutely aware that a necessary objective would be to disrupt commonly held notions about what the

Fig. 12.10 (*Source* Author)

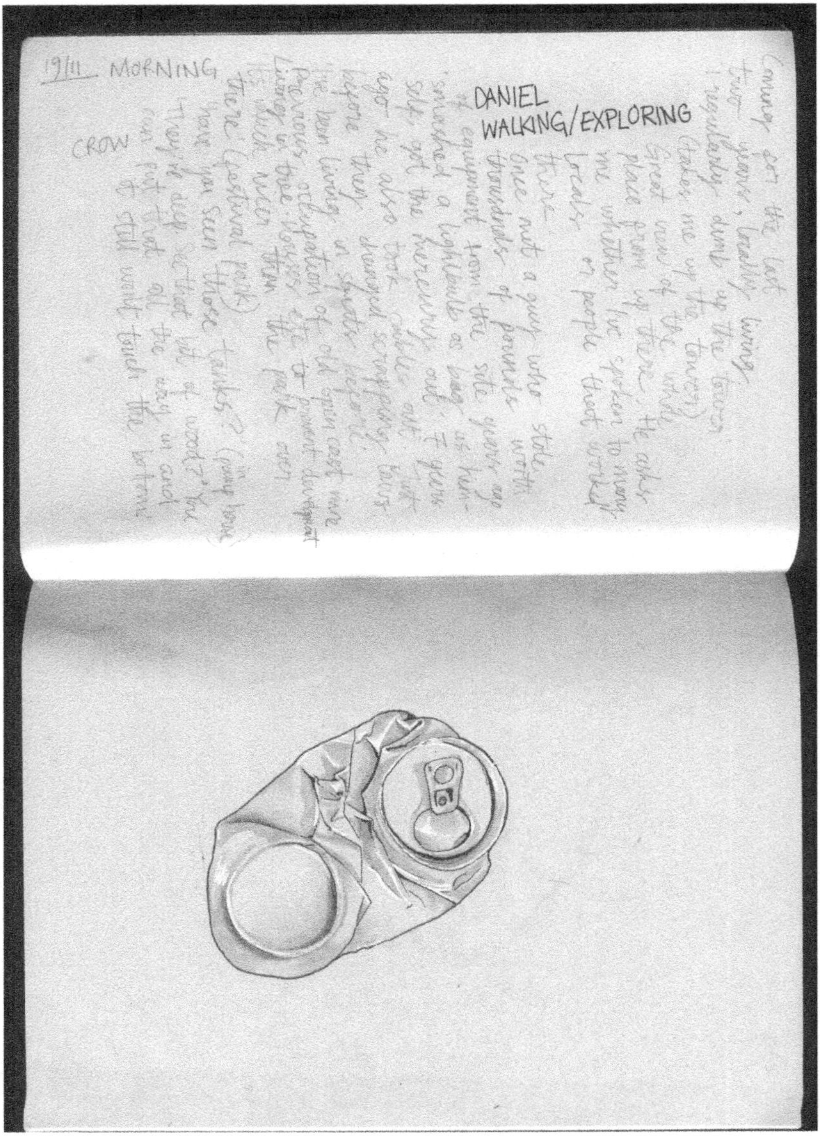

Fig. 12.11 (*Source* Author)

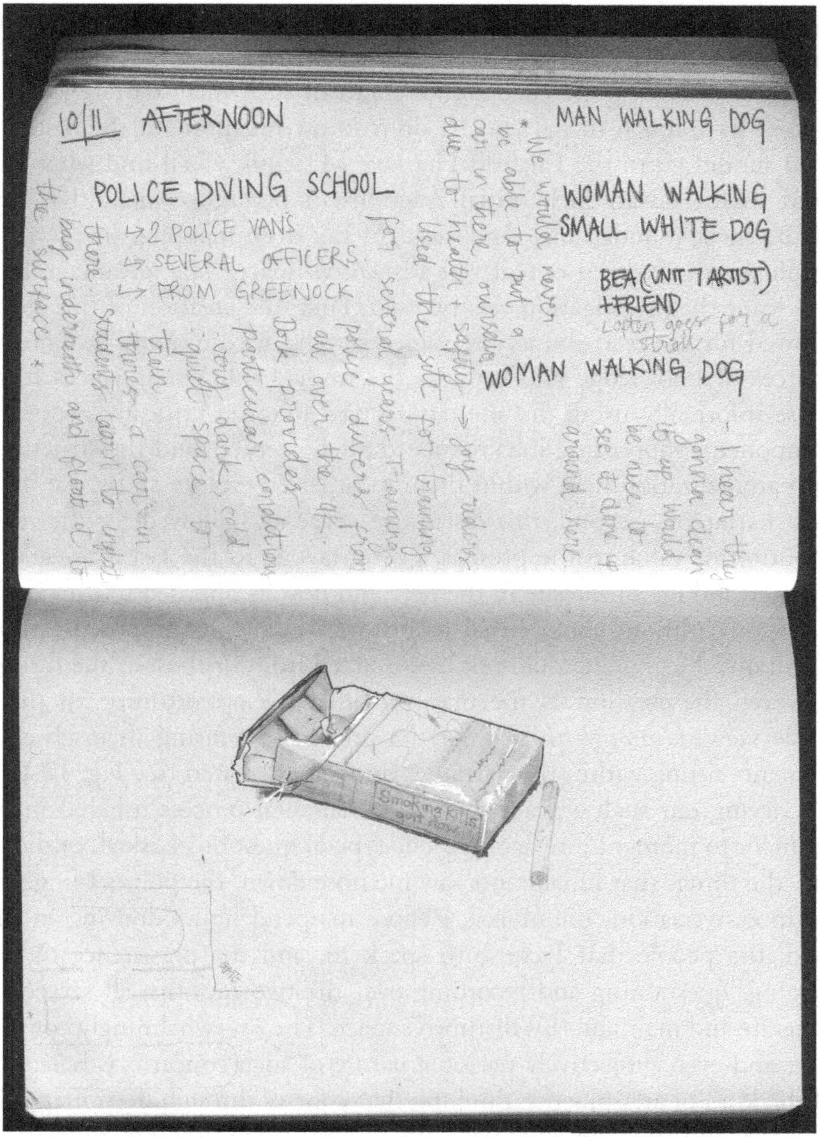

Fig. 12.12 (*Source* Author)

site means and who has the authority to talk about it. Indeed, many times on initial encounter with inhabitants, I was met with responses that reflected the prevalence and strength of such meta-narratives. 'Oh I don't have much to tell you'. 'I don't know much about the history'. 'You should go to the Fairfield Heritage Museum; you'll find what you need there'. Given such initial reactions, it became evident that one of the most foundational daily tasks of my deep mapping in the field would be the carving out of this disruptive space. The advantage that this leant the investigation was twofold. First, the creation of this space allowed for the reattribution of value away from commonly recognized sources of knowledge and towards the myriad inhabitant voices from those informally using the site day to day. It was a critically necessary component, supporting inhabitants to safely step beyond the structures of marginalization held within them in order to explore and voice their own narratives. Second, this disruptive space also allowed for the reattribution of value to the present, giving favour to the last 33 years of vacancy and life of the site in the here and now as opposed to the dominant conversations about either its historical use or its proposed future. Naturally, the present could not be extracted from the past or the future; however, the creation of the map provided the opportunity for them to be viewed *through the lens* of the present, appearing themselves as elements sitting within the narrative landscapes created (see Fig. 12.13).

Carving out such a space was an incremental process ushered in by moment to moment interactions. The type of questions I asked, or didn't ask, the things that inhabitants saw me note down, the points I lingered on in conversation, the objects I chose to spend hours drawing in the field, the people that I chose to speak to, and the persistence of my digging, questioning and recording over the two months, all served to generate and maintain this disruptive space. The overwhelmingly contingent and even subjectively personal nature of such concerns is reflective of the kind of arts-practice thinking that courses through deep mapping ventures and the consequentially broad and irreplicable outcomes they generate.

Combining verbalization (interviews) and visualization (drawings) evidently helped the process of disrupting hegemonic narratives. The

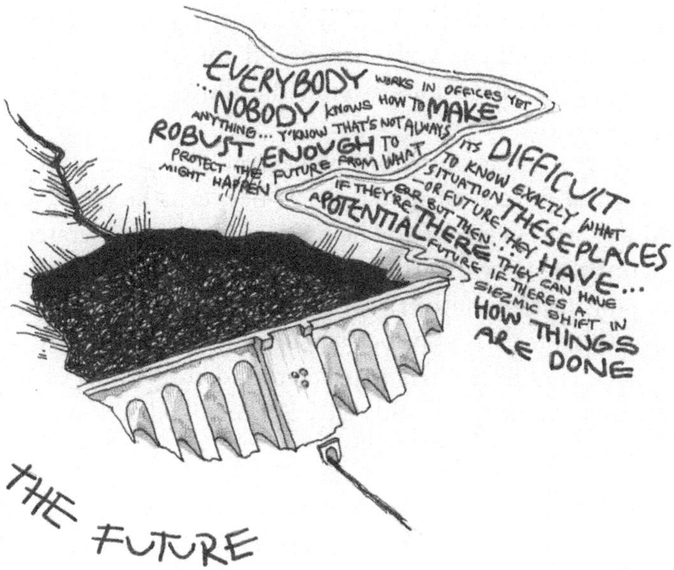

Fig. 12.13 (*Source* Author)

creation of hand-made drawings spun directly from inhabitants' descriptions and storytelling appeared to play a powerful role in expressly attributing value to their personal narratives. When inhabitants were presented with the drawings I had generated based upon their voices, the conversations expanded and almost all began to share their reflections and life experiences, to add them to the map. Indicated by the responses and enjoyment of the drawings as aesthetic objects, it appeared that this time-invested 'making' stage of the mapping process was interpreted by inhabitants as the artist-researcher literally *making time* for them. One inhabitant responded: 'It feels like you're really listening to me and that doesn't happen much in society these days' (J interview 2). While it is not unusual for individuals to respond positively when their viewpoints are given time and attention in interactions such as an interview, the creation of a physical artefact evidenced that this attention was additionally being given outside of the moment of our face-to-face conversation. Thus, we may read this art-production as a kind of 'gifting' in the discussion process. This did raise notable challenges to my aspirations of co-creation as, engaged as they were by the illustrations, inhabitants were mostly hesitant to critique them and challenge them. As such, the majority of the investigative discussions around the illustrations were driven by additive comments.

These written landscape objects became catalytic agents in the reflexive, cyclical process of investigation between the place, the inhabitants and the artist-researcher. The creative outputs generated (be they objects, actions, writings, performances, etc.) acted as 'safe harbour' to the many tangential narratives that revolve around the Govan Graving Docks, creating a rich polyvocality in the process. They therefore serve to incrementally construct space around these wells of inhabitant knowledge like momentary voids that beckon further thought, further questioning and further exchange.

Absence

As the investigation progressed, another necessary dimension of the map emerged. The unfolding COVID-19 pandemic and the travel restrictions in place across Europe put significant limitations on the fieldwork. Although the initial stage of the fieldwork went ahead eventually, the outcomes of this period and the path of investigation it generated still warrant discussion. Unable to physically be present in the field, I made efforts to connect with individuals online; however, this proved challenging for several reasons, including many of those contextual factors that limited the interview process. Gateway community groups and community leaders were also not in a position to collaborate, being overwhelmed by the crisis themselves. I found myself compiling a list of groups and individuals whose inhabitant knowledge would enrich the mapping yet whom I was unable to talk with. This was not only a result of the pandemic but other systemic issues at play, such as a funding crisis for the arts and community organizations and, more broadly, local contextual factors or attitudes held by individuals. This period of limited access made it clear that not only was 'absence' prevalent, but it would also be necessary to represent 'absence' itself as a critical component in the mapping process.

This line of inquiry led me to create a further illustrated map component that would serve to both record and prompt investigation of absent voices and the causes that lie behind them. The illustration entitled *Loch Absence* (see Figs. 12.14 and 12.15) is a simple illustration of a loch with the names of identified absent voices concealed within it by printing black on black. It challenges the viewer to move around the work, in order to position themselves at an angle in which the names may be read. This additional open-ended illustration serves as a task-based device to prompt my own detection and mapping of absence that continues throughout the investigation. As the identities of those voices not present are rarely forthcoming, I adopted tactics for detecting absences such as making conscious records of those declining or unable to talk, opportunities missed, and my own limitations in the field, such as not visiting the site after dark and hesitations about approaching certain individuals. By introducing the open-ended *Loch Absence* illustration into conversations and interviews, and thus the idea of acknowledging voices absent from the discussion, it further became possible to invite inhabitants in as co-researchers in this process.

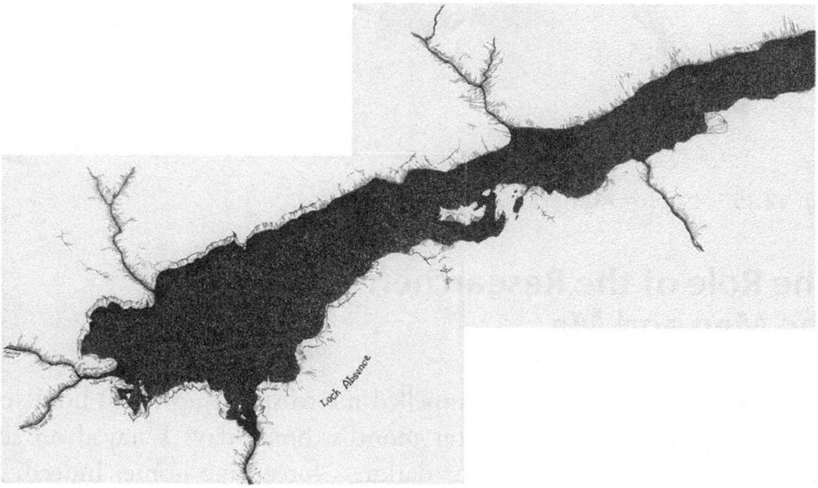

Fig. 12.14 (*Source* Author)

Fig. 12.15 (*Source* Author)

The Role of the Researcher: The Place, the Map and *Me*

The onsite drawing practice compelled me to spend up to 10 hours on site each week during the winter months. Some days I stayed on site until the cold or driving rain or darkness forced me home. Indeed, as Wood (2015) had forewarned, I did dream about it. Beyond observing, listening, recording, I was *living* the site for myself. Committing the whole of myself to the place in this way led me to reflect upon what had brought me to do this particular work in this particular place. Memories surfaced of my own teenage ventures into wild and vacant spaces in my home landscape and the parts of my own identity that had called me there. Over the course of the two months in the field, I found these dimensions of myself increasingly seeping into the research investigation. While the roles I occupied as a researcher were multiplicitous, this effort to understand the place was also mirrored by a reciprocal process of self-understanding. This internal 'movement' alongside the movement of the research investigation drove me deeper into the field. In line with McLucas' assertion that deep maps should be 'politicized and partisan' (n.d.), I began to clarify my own political compulsions and allegiances.

Critically, however, this affected a notable change in the conversations I had with the many inhabitants I encountered. This embodiment, this inviting my own life-world into the mapping process, gave me the ability to meet with and sympathize with the life-worlds of others—to *hear* the experiences of those with whom I spoke. It generated the space within which individuals could express those life experiences that extended far beyond the bounds on the site. Only by relinquishing the position of researcher as neutral, immutable observer could these expressions of place emerge using the conversation space to understand our own experiences together.

My own art practice was central to this gradual embodiment. My employment of illustration was not as a convenient, aesthetically pleasing additive in the creation of a deep map, but rather part of the continuing evolution of my own artistic practice. The nights I spent scouring my thoughts for a suitable metaphor and a suitable image to portray it. The apprehension I felt in the moments before I tentatively offered those objects I had laboured over back to the inhabitants, just hoping I had pulled out the right words from what felt like a cacophony of meanings. The invitation of personal art practice into the making of a deep map is also (to varying degrees) the inviting of oneself into the process. The production of art is inherently rife with vulnerabilities and only in my vulnerability could I step down from the ivory tower and turn my questioning back upon myself.

Conclusions

Deep maps do not present a more accurate representation of place but rather an alternative 'truth' that can resist and confront hegemonic accounts. They do not seek to replace those conventional, instrumental depictions of place such as the map at the opening of this chapter. Instead a deep map may come into an active role when situated alongside existing representations. In such a way they bring into question assumptions that the social context of place is too layered, too contingent, too 'messy' to be included in the debate. The existence of a deep map serves to put a stake in the ground that is difficult to ignore. By *making effort* to represent what would otherwise be left in silence, it breaks through the cloak of omission curbing the momentum of those forces that serve to exclude and invalidate. In this way the deep maps hold the capacity to redefine the boundaries of debate over place, to voice both the 'cry and demand' over the right to the city (Lefebvre, 1991).

These advantages also draw attention to the potential complications and limitations of deep mapping as a highly context-specific approach to investigations of place. Disrupting or suspending commonly held perspectives in order to draw out more marginalized narratives clearly requires considerable time in the field. These are incremental processes contingent upon the building of trusting relationships with inhabitants, and they therefore do not lend themselves so well to research situations in which outcomes are predisposed and time pressured. Further, as is common to ethnographic approaches, it is necessary for the researcher to continually transition through different roles, engage closely with inhabitants and be reflexive about their own positionality in the process. This can be an intensive undertaking, particularly when working in places where socio-economic deprivation is prevalent. With regard to the creative output of the deep mapping process, its efficacy is also dependent upon where it resides. As a living object of representation and dialogue, a deep map would amount to nothing were it extracted entirely and solely enclosed within academic archives or frozen within publications. Resisting this requires partnership at the local level and a transferability of ownership so that it may be alive and in service of those inhabitants it endeavours to give platform to. And while multiplicity and inclusion remain central to the values of deep mapping, the ways in which this less familiar, arts-based approach may create alternative exclusions, demand sensitivity and discussion. This points to the need for further research on deep maps in action within the fields of decision-making, investigating their capacity to be recognized alongside traditional cartographic representations and to advocate for those spaces and peoples they seek to represent.

In the context of place-based research methods, deep maps possess a special ability to foreground absence rather than side-stepping it. As a principal condition of arts approaches are their persistent confrontation with their own inadequacy to describe (Holub, 1990), they decisively leave loose ends untied, inviting the viewer to draw their own conclusions. This permits the researcher to bring all that is missing into focus as a component of the subject matter itself. In this way a deep map may situate conversations about a place within the 'presence of absence', positioning itself with honesty within the fullness of the social landscape. Such a research practice may have resonance in many fields (migration, homelessness or climate adaptation for instance) where singular spatial narratives are held by positions of power at the exclusion of others. They bring us back to and keep alive those vital and difficult questions about how our rural and urban landscapes are produced and for whom. And thus, while deep maps have by no means freed themselves from the dilemmas of representative inclusion, I propose that they offer a far greater contribution than creating an aesthetic, representational object of any sorts. In continually questioning assumptions about who holds the authority to speak about a place, deep mapping empowers marginalized individuals to recognize that they hold within them knowledge that is of value and significance, that without them there is no place.

Acknowledgements This research received funding from the European Union's Horizon 2020 research and innovation programme under the Marie Sklowdoska-Curie Grant Agreement No. 765389.

References

Bailey, J. (2018). Mapping as a performative process: Some challenges presented by art-led mapping that aims to remain unstable and conversational. In *Artistic approaches to cultural mapping*. Routledge

Biggs, I. (2010). *'Deep mapping': A brief introduction.* Mapping Spectral Traces. Exhibition Pamphlet Blackburg. Virginia Tech College of Architecture and Urban Studies, 5–8.

Bissell, L., & Overend, D. (2015). Regular routes: Deep mapping a performative counterpractice for the daily commute. *Humanities, 4*, 476–499.

Butler, R., Curran, R., & O'Gorman, K. D. (2012). Pro-poor tourism in a first world urban setting: Case study of Glasgow Govan. *International Journal of Tourism Research, 15*(5), 443–457. https://doi.org/10.1002/jtr.1888

Causey, A. (2017). *Drawn to see: Drawing as a research method*. University of Toronto Press.

Henning, M. (2020). *Museum media.* Wiley.

High, S. (2013). "The wounds of class": A historiographical reflection on the study of deindustrialization, 1973–2013. *History Compass, 11*, 994–1007.

Holub, M., (1990). *The dimensions of the present and other essays* (D. Young, Ed.). Faber & Faber.

Humphris, I., & Rauws, W. (2020). Edgelands of practice: Post-industrial landscapes and the conditions of informal spatial appropriation. *Landscape Research.* https://doi.org/10.1080/01426397.2020.1850663

Ingold, T. (2017). *Lines.* Routledge.

Kester, G. (2013). *Conversation pieces: Community and communication in modern art*. University of California Press.

Least Heat-Moon, W. (1999). *Prairy Erth (A deep map).* Mariner Books

Lefebvre, H. (1991). *Critique of everyday life* (Vol. I, J. Moor, Trans.). Verso

Lemon, L. T., & Reis, M. J. (Eds.) (1965). *Russian formalist criticism.* Lincoln: University of Nebraska Press.

Lewis, C., (2015). Archaeological excavation and deep mapping in historic rural communities. *Humanities, 4*(3), 393–417. https://doi.org/10.3390/h4030393

Massey, D. (2005). *For space.* Sage.

McLucas, C. (2014). "I was invited to this island", quoted in Anwen Jones and Rowan O'Niel, 'Living maps of Wales: Cartography as inclusive, cultural practice in the works of Owen Rhoscomyl (Arthur Owen Vaughan) and Cliff McLucas'. *International Journal of Welsh Writing in English* (1), 120.

McLucas, C. (n.d.). *There are ten things that I can say about these deep maps.* http://cliffordmclucas.info/deep-mapping.html

Modeen, M., & Biggs, I. (2020). *Creative engagements with ecologies of place: Geopoetics, Deep mapping and slow residencies.* Routledge.

Pearson, M., & Shanks, M. (2001). *Theatre/archeology.* Routledge.

Reddleman, C. (2015). The deep mapping of Pennine street: A cartographic fiction. *Humanities, 4*(4), 760–774. https://doi.org/10.3390/h4040760

Roberts, L. (2016). Deep mapping and spatial anthropology. *Humanities, 5*(1), 5. https://doi.org/10.3390/h5010005

Shklovsky, V. (1917). *Art as technique* (L. T. Lemon & M. J. Reis, Trans.). In L. T. Lemon & M. J. Reis (Eds.) (1965), *Russian formalist criticism.* Lincoln: University of Nebraska Press.

Shoard, M. (2000). Edgelands of promise. *Landscapes, 1*(2), 74–93.

Sinclair, I. (2017). *The last London.* Oneworld

Smith, J. (2015). Anticipating deep mapping: Tracing the spatial practice of Tim Robinson. *Humanities, 5*, 283–303.

Tovar, F. J., Arnal, M., de Castro, C., Lahera-Sánchez, A., & Revilla, J. A. (2011). A tale of two cities: Working class identity, industrial relations and community in declining textile and shoe industries in Spain. *International Journal of Heritage Studies, 17*(4), 331–343.

Wittmayer, J. M., & Schäpke, N. (2014). Action, research and participation: Roles of researchers in sustainability transitions. *Sustainability Science, 9*, 483–496. https://doi.org/10.1007/s11625-014-0258-4

Wood, D. (2015). Mapping deeply. *Humanities, 4*(3), 304–318. https://doi.org/10.3390/h4030304

13

Engaging 'Future Generations' in Meaning Making through Visual Methods: An Alternative Approach to Defining City-Regions

Lorena Axinte

Introduction

Global environmental campaigns such as *Fridays for Future* and *Extinction Rebellion* have been incited and led by young people. Carrying banners that said 'if the climate was a bank, it would be saved by now', 'systems' change, not climate change' or 'if you'd be doing your job, we'd be in school right now', young people 'took to the streets in an estimated 185 countries to demand action' in 2019 (Laville & Watts, 2019).

These movements demonstrated that while young people might have become vote-apathetic (Agger, 2012; Collin, 2015) and disengaged from traditional politics, they are not apolitical. Instead, the youth has chosen alternative spheres of political action such as protests, advocacy, rallies and online campaigns (Agger, 2012; Chou et al., 2017; Collin,

L. Axinte (✉)
Sustainable Places Research Institute, Cardiff University, Cardiff, Wales, UK
e-mail: axintel@cardiff.ac.uk

A. Franklin (ed.), *Co-Creativity and Engaged Scholarship*,
https://doi.org/10.1007/978-3-030-84248-2_13

2015; Farthing, 2010). Through these actions, young people have been calling upon their representatives to act faster and adopt more ambitious strategies for climate justice.

Because of the COVID-19 pandemic, the 2019 movements seem to have lost momentum and their legacy is uncertain. Furthermore, planning and governance systems have chronically failed to reflect and engage young and future generations. This is particularly true in the case of British city-regions, an intermediary administrative level that sits between the national and the local. Established all over the UK, city-regions represent collaborative projects between various local councils, often aiming to drive economic development. While claiming to devolve power from the central government, city-regions have been criticized for their elitist and opaque governance structures, which rarely involve the civil society or a wider public (Axinte et al., 2019; Beel et al., 2018; Flinders et al., 2015).

In Wales, two city-regions were established in March 2016. The research project detailed below looked at the possibilities of balancing the narrow economic rationale with broader social, cultural and environmental aims in developing city-regions. Focusing on Cardiff Capital Region (CCR), the study positioned the city-region within the *Wellbeing of Future Generations Act*—a legislative framework that requires Welsh public bodies to place sustainable development at the core of every future action.

Nolt (2017) debated whether current generations hold any responsibility for future, unpredictable events, and thus, for future generations. Finally, he concluded that our rapid 'acquisition of knowledge and power', which could lead to a mass extinction, is a strong enough argument for intergenerational ethics and accountability (Nolt, 2017, p. 11). To operationalize the concept of *future generations* and help simplify such debates, this research defined them as 'the first humans and non-humans that will be affected by policies and decisions currently made, having to live the longest with their consequences'. Considering that Cardiff Capital Region is a 20-year long scheme, the first future generations affected by it are actually the young people living in South East Wales today. Therefore, doing research *with* future generations and not only *for* or *about* them became an important research aim.

Starting from these ideas, the research tried to engage young people in a discussion about the city-region area and their aspirations for its development. The fieldwork began soon after the CCR was created, leading to a major research challenge: most of the research participants had never heard of the city-region, or had little interest in it. Nonetheless, the city-regional plans were going to affect young people in spheres such as transport and mobility, jobs and education, and digital connectivity. The questions leading this investigation became: How could a research project stimulate a conversation with the *future generations* about the areas where they live, and how could it encourage meaningful reflections on previously unfamiliar concepts, such as city-regions?

It is worth mentioning that this research was developed within the SUSPLACE Marie Curie ITN, which explored pathways for sustainable place-shaping. Within SUSPLACE, the notion of *sense of place* played an important role. Often used in social sciences and increasingly within sustainability studies, it refers to the 'collection of meanings and emotions that people assign to a particular setting', as well as 'the way people experience, use, and understand place' (Grenni et al., 2019). In general, research has shown a positive correlation between sense of place and pro-environmental behaviour, because people might act more responsibly towards their immediate environment when they feel a certain attachment and can ascribe meaning to a place (for a detailed discussion, see Grenni et al., 2019; and/or Kudryavtsev et al., 2012).

Kevin Lynch was the first scholar who referred to *sense of a region*, suggesting that spatial planning and design must strive to harness human perception of the physical form of cities and regions in order to improve their qualities, and thus, people's and places' well-being (Lynch, 1976). The following pages explain the approach used to uncover young people's *sense of the city-region* in Cardiff Capital Region—a first step to understanding their perceptions of the city-region's physicality, as well as their lived realities. Initially, the research also aimed to create a forum for young people and city-regional leaders to interact directly, allowing youth to offer their input to the city-region's plans. Despite failing to accomplish this final step, this research offers useful lessons for anyone interested in using technology-enabled visual methods, working with youth, or accessing alternative definitions of city-regions (as interpreted by young people who live in South East Wales).

Visual Methods and Participatory Research

The methods chosen to conduct this research were web-mapping and Photovoice. Both are visual research methods, as they require participants to generate visual data (Lorenz & Kolb, 2009) in the form of maps and photographs. They were selected due to their strengths: (i) the potential to shift the power differential between researcher and participant, as the latter can be in control of their contribution, (ii) the capacity to stimulate reflections and conversations in a less intrusive manner than direct questioning and (iii) the use of technology that might be more appealing for the young participants. Popular among researchers working with youth (Driskell, 2002; Wang, 2006; Ward et al., 2015), these two less conventional, creative methods seemed most suitable in allowing participants to apprehend the concept of a city-region, elicit their own interpretations and provide insights into their lived experiences without being too prescriptive. In addition, the participants' input was facilitated by technology (Geographical Information System in the case of web-mapping and phones or digital cameras in Photovoice), an aspect that the participants seemed to appreciate.

Both (web-)mapping and Photovoice have been used in participatory research *with* young people (Driskell, 2002; Literat, 2013; Wang, 2006), which flourished in the beginning of the 90s, when the *United Nations Convention on the Rights of the Child* (UNCRC) was first ratified. The Convention marked a paradigm change, as children and youth started being regarded as potential collaborators, with individual agency and valuable experiences (Adams et al., 2017; Tisdall, 2016). The researchers' task became to harness this knowledge, and efforts concentrated on creating more participative approaches that would allow young people to engage and affect decision-making. Many participatory research projects used visual techniques because they support gathering 'information concerning whether lives are meaningful and fulfilling', that might otherwise remain invisible through conventional and non-participative approaches (Chawla, 2002).

Although visual techniques do not guarantee an increase in participation, scholars often cite their potential to place participants and researchers on an equal footing, as active co-creators of knowledge (Trell & van Hoven, 2010). Some also choose these methods for their emancipatory potential and capacity to affect change (Driskell, 2002; Wang, 2006). Although out of scope for a wider debate here, it is worth mentioning that engagement and participation were not conceived as intrinsically 'good' in this research project. Therefore, the wider investigation went beyond engaging young people, and aimed to also understand if city-regional leaders consider youth engagement possible and valuable at city-regional level, and why/why not. This series of 16 semi-structured interviews was inspired by Farthing (2012), who condemned the 'heroic claims made to justify participation', and highlighted the need to ask more critical questions, such as 'why do we engage young people?' In this way, normative judgements about 'what are good things for young people and what a good society looks like in the first place' (Farthing, 2012, pp. 91–92) can surface. Issues related to participation and the participatory qualities of the two methods applied in this research will be critically analysed in the reflection section.

Nonetheless, this study demonstrates how two visual methods can be combined to obtain rich depictions of young people's everyday lives, and to overcome initial engagement barriers, adding to the methodologies in youth studies literature. Thematically, this is one of the few research projects that have sought to engage young people—a section of the population generally deemed 'hard to engage' (Flinders et al., 2015)—in conversations about their city-regions. This research could easily be replicated in other contexts to understand people's lived experiences in relation to certain administrative boundaries, test these demarcations' legitimacy, as well as their effects.

The next section critically discusses the ways in which scholars have applied mapping and Photovoice, expanding on the characteristics that also made them suitable for this project. The chapter then presents an overview of the research project, as well as a detailed account of the application of the two methods. The final section represents a reflective account of the entire research experience, the process and the results.

Using Maps and Photos in Research—A Brief Literature Review

Although boundaries—such as the administrative ones of a city-region—are a constantly changing social construct, they have real effects for people and places, both within and outside those borders. A city-regional level established primarily using commuting patterns might seem rather arbitrary for some individuals—not only young people—as they might not identify with that particular demarcation. Therefore, to engage young people in a discussion about a previously unknown or little understood concept, this research combined two creative visual methods—web-mapping and Photovoice—and used them in a series of workshops.

Maps—whether hand drawn or digital—embody a wealth of information, well beyond their functional spatial indications, and often represent the 'worldview and particular interests of dominant powers' (Literat, 2013, p. 198; see also Humphris, this book, and Reitz, this book). Participant-created maps symbolize a specific understanding of a certain location, marking personally relevant facts for the place's development. In addition to harnessing individuals' knowledge, such approaches invite participants 'to take an active stake in the visual representation of their spatial environment' (Literat, 2013, p. 199).

To highlight the antithesis to formal maps and the deeply political character of the participatory ones, Lee Peluso (1995) coined the term *counter-mapping* in a study that showed the intricacies of maps as consolidators of state control over Indonesian forests. Simply put, counter-mapping allows any actor, especially disempowered ones, to use cartographic tools and maps for an alternative (spatial) representation, often contesting the official one. Headrick Taylor and Hall (2013) used counter-mapping to study young people's personal mobility and to improve participants' spatial literacy. They described the result of counter-mapping as a 'thirdspace' where personal interpretations and experiences supplement or contradict existing, official knowledge. Although characterized by conflicts, this interaction holds potential for positive change, both within communities and for the participants' self-development. In fact, the pedagogical potential of mapping has been highlighted in other youth studies; this is because the production of maps supports critical thinking and reflection and can be an introduction to GIS and digital tools when maps are technology-enabled (Literat, 2013; see also Ramirez Aranda and Vezzoni, this book).

For many young people who were part of the research described in this chapter, participating became an occasion to use the map creation tool available in Google Maps for the first time. Since the research relied on online GIS, the method is referred to as *web-mapping*. It is inspired by Lynch's first experiments with cognitive/mental mapping (Lynch, 1977) and the aforementioned *counter-mapping*. The use of technology aimed to facilitate the participants' contributions, as they could access a digital map and produce their own layers (after a brief training session), irrespective of their geographical knowledge or spatial thinking capacities.

Studies using a variation of participatory mapping stress that, while important in itself, the resulting visual output becomes a means to elicit conversations and should only be interpreted after a (group or individual) discussion. In addition to dialogues, certain projects combined mapping with other research methods to enrich findings, including participant-led walks (Driskell, 2002) or bike rides (Headrick Taylor & Hall, 2013), as well as interviews and photos (Dennis et al., 2009).

To enhance findings and to bridge between the young participants and the wider public (including city-regional leaders), this project combined web-mapping with Photovoice. Photovoice is a method developed by Wang et al. (1996) to reveal lived experiences and empower people, particularly marginalized ones, to voice their needs and take part in shaping their environments. It draws from Paolo Freire's *critical pedagogy* and from participatory action research (PAR). The former is a philosophy that supports students to engage in critical thinking, reflect on and discuss their own life conditions, as opposed to traditional teaching methods that 'bestow knowledge upon students' (Derr & Simons, 2019, p. 361). PAR aims to develop practical knowledge through collective reflection and action that can lead to positive individual and community change (Reason & Bradbury, 2008). As the following parts will show, this project did not necessarily lead to positive change, and its PAR characteristics remained rather limited.

Generally, Photovoice requires participants to capture and record specific issues through the means of photographs. The photos taken are then discussed during focus groups, allowing the development of narratives and themes. The results can be further communicated to the wider community and to decision-makers, so new perspectives can affect change. From its design, therefore, Photovoice had three main goals: (i) to enable people, particularly those coming from marginalized groups, to record and reflect on their community's strengths and concerns; (ii) to promote critical dialogue and knowledge about important issues through group discussion of photographs; and (iii) to reach policymakers (Wang et al., 1996, p. 1391). Its creators defined a complex nine-step strategy to help researchers apply Photovoice, as well as a series of questions to facilitate group discussions and identify themes within the photos taken (see Wang, 2006). Nonetheless, the method has been modified and adapted according to context and needs (for a pertinent critique, see Derr & Simons, 2019), and the research described in this chapter is also a variation.

Conceptualizing young people as 'competent citizens and active participants in the institutions and decisions that affect their lives' (Wang, 2006, p. 152), Photovoice projects have been used for a variety of purposes: to explore children's perceptions and representations of nature (Adams et al., 2017), to elicit youth's perspectives on issues within their neighbourhoods, to find ways to address them (Wang, 2006), to improve the built environment in cities based on children's lived experiences (Driskell, 2002) and to understand the impact of commuting on teenagers' well-being (Ward et al., 2015).

This method's strengths lie in its capacity to shift perspectives, as it allows the researcher to experience the world through the participant's view, reducing preconceptions about what might count as important (Chawla, 2002). Nonetheless, scholars have also raised awareness that 'real, structural barriers' can sometimes persist, reducing the participants' ability to act as co-collaborators in research (Packard, 2008). Indeed, methods alone cannot reduce power differentials, and researchers should reflect critically on how knowledge was generated (see Franklin, this book).

Even so, photos can facilitate the expression of views and are also a prompt for group discussions. Furthermore, Photovoice encourages people to depict not only weaknesses and needs, but also assets and potential improvements (Wang & Burris, 1997). However, Photovoice also implies serious ethical considerations regarding privacy, representation and safety (Wang & Redwood-Jones, 2001), requiring initial training for participants. Another difficulty raised is the analysis and summary of photos, considering that images contain complex information. To avoid bias in this phase, Photovoice projects should end through a facilitated discussion that allows participants to offer a wider context for their images.

Although some scholars assert that Photovoice could be worthwhile in itself for participants, due to the opportunity to express views, practice critical thinking and be part of group discussions (Lorenz & Kolb, 2009), Derr and Simons (2019) criticized the lack of interaction with policy-makers. Half of the 32 studies they reviewed failed to bring information to decision-makers and bridge between groups, not achieving one of the original tenets of Photovoice: that of enacting positive change within communities by challenging power structures and the *status quo*. For this reason, the aforementioned scholars emphasize the need for more honesty and critical appraisal of using Photovoice to discuss the desired vs. the achieved outcomes, and to understand how well-intended projects can turn into tokenistic approaches.

Concurring with the need for more honesty within academic accounts, as well as the potential of learning from less successful projects, the following part offers a detailed account of how Photovoice and web-mapping were combined and applied to elicit the views and aspirations of young people living in the Cardiff Capital Region. Despite several unsuccessful attempts to reach city-regional leaders and policy-makers, the project offers valuable lessons for researchers interested in creative research methods.

Exploring Cardiff Capital Region with Its Future Generations

Paradoxically, Wales might be both the place that 'led the global transition to a carbon economy' (Eames et al., 2014, p. 3) through the industrial revolution, and one of the first nations in the world that integrated sustainability principles in its constitutive act (Williams, 2006). The Welsh sustainability agenda progressed over the years and culminated with the *Wellbeing of Future Generations Act*—a legislative framework meant to safeguard the well-being of both current and succeeding inhabitants. Since its adoption in April 2016, the concept of 'future generations' became central, without necessarily becoming better defined.

A month before the Act's ratification, Wales also established its first city-regions—a new administrative scale deemed suitable to make Welsh cities more economically competitive. The decision was aligned to a national and international trend of defining cities as 'engines of economic growth, and catalysts for creativity and innovation' (Welsh Government, 2012, p. 5).

Cardiff Capital Region (CCR) became a collaboration between ten local authorities in South East Wales, the Welsh and UK Governments. Its ultimate purpose is to achieve economic growth through physical and digital infrastructure improvements, as well as upskilling and enhanced employment opportunities. Interviews conducted for this research have shown that although many of the city-regional programmes are targeting youth, the leaders considered businesses as the main stakeholders. As a result, any other groups (volunteer groups, trades unions, charities, etc.) and individuals remained disenfranchised, without the opportunity of influencing the city-region's development.

CCR is currently home to 50% of all young people aged 16–24 living in Wales. The city-region's proportion of youth (12.4%) is, in fact, higher than the average of Wales and the UK (Welsh Government, 2016). This project aimed to establish a dialogue with young people, as the first generation to be affected by the decisions made on their behalf by the CCR, and the ones who would have to bear the consequences for the longest. Because of its novelty, CCR was largely unknown and required a strategic methodological choice to encourage meaningful reflections.

Participant Selection

The 16–24 band age was chosen to simplify the research process, since research with underage persons has specific administrative and ethical requirements that can prove lengthy and complicated. Finding young people and convincing them to participate has been a major hurdle, requiring numerous emails, phone calls and visits to youth groups, youth fora, clubs, organizations and some schools. While some of the adults contacted have been extremely helpful, others acted as gatekeepers. Nonetheless, in two cases—Caerphilly Youth Forum and Bridgend College—participants were recommended by their officer and teachers. In all other cases, recruitment relied on personal connections and snowballing. In total, 29 people were recruited; however, only 24 of them got involved in both methods.

Web-Mapping Workshops

The fieldwork lasted for 10 months and the two research methods chosen were applied in a workshop format. The first session used web-mapping and elicited young people's perceptions of their (personal) city-regional span, and of the assets and liabilities within it. In simple terms, participants created online maps of their own city-regions, marking also the significant places within those boundaries. These workshops aimed to gather young participants' depiction of their personal geographies without being an obtrusive investigation. Each session allowed participants to explore their region's online map, reflect on their experiences within their city-region(s) and superimpose their personal layers and pointers.

All sessions used laptops or computers connected to the Internet and participants created an online map using the map creation feature in Google Maps. Their participation required no previous knowledge and only basic computer skills, which all attendants had. Each person received a step-by-step written guide and continuous assistance, and every session started with defining and discussing the concept of city-regions. Participants were instructed to think of their city-region as the area that expands beyond their hometown, where they might travel occasionally (for leisure, shopping, education, medical services, etc.) and to which they felt connected in some way. They were also told that boundaries did not have to be very precise, that each map is a personal artefact, and no answer was wrong.

After inputting general information (name, age, place of residence), each participant had to complete three tasks: (i) mark the boundaries of their city-region based on the aforementioned discussion; (ii) mark places of personal significance that they appreciated, indicating in a comment why; (iii) mark places they disliked and would like to see change, explaining why and how (Fig. 13.1).

Workshops varied in length, depending on the number of attendees (between one and nine) and their familiarity with the topic and tool used. After finishing the mapping task, each individual presented their contribution and the groups briefly discussed the differences and similarities between results. As expected, web-mapping showed that individual

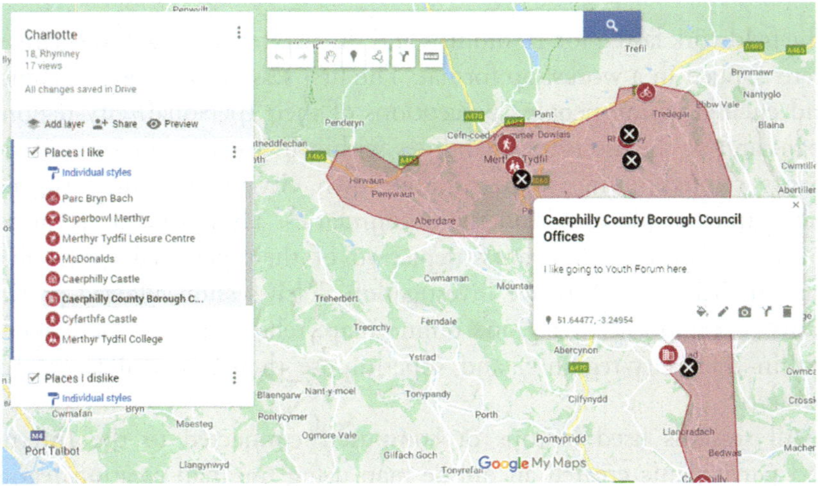

Fig. 13.1 Web-mapping using the Google Maps interface

city-regional boundaries take a myriad of forms. As Fig. 13.2 shows, they do not overlap with the official ones, challenging their relevance particularly in the context of extremely limited public engagement.

All the web-mapping workshops ended with short feedback sessions. These revealed that, in general, participants appreciated the web-mapping workshops for three reasons: (i) they had not heard about CCR and thought it was important to gain more awareness about future plans; (ii) they enjoyed expressing themselves geographically and considered this was an innovative way of eliciting their views; (iii) they were happy to have learned to make digital maps and the younger ones were actually hoping to use this skill in school projects. Therefore, the web-mapping technique enhanced young people's capacity to express spatial information, and to reflect on the suggested subject and the emerging themes. At the same time, it supported a shared language and understanding among all workshop participants, including the researcher. For instance, one of

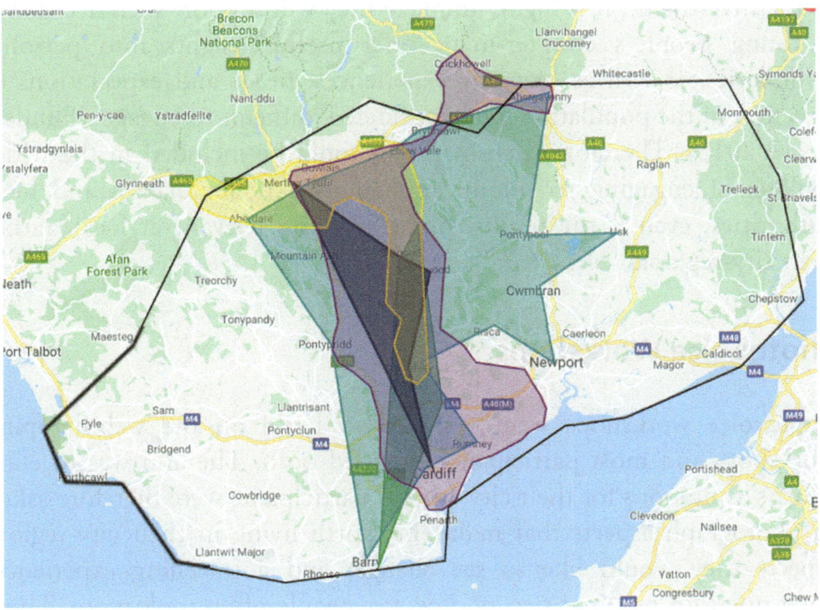

Fig. 13.2 Example of five city regional representations within CCR (marked with different colours) and the official boundaries (in black)

the participants marked a place from their community where they had come in conflict with local police officers, due to 'loitering'. This led to a thorough discussion about the lack of leisure spaces for young people, which proved to be a common theme across workshops.

On the other hand, apart from the one person who was familiar with CCR (as a student in Geography and Planning), nearly all participants had some difficulties at first in understanding what the city-region referred to. This was an alien concept for them, demonstrating the lack of information coming from CCR during its first years of existence. The narrower spatial representation (compared to the official boundaries) showed that the participants' sense of place is much more localized, and that when prompted to describe significant places, these would most often be situated in the proximity of their homes. In addition, the different representations demonstrate that city-regional identities are rather fragmented, and this could turn into significant impediments for future projects and policies. If CCR's inhabitants do not see the relevance of the investment, it is highly unlikely they will support such decisions.

Young people′s counter-maps offer useful insights into personal boundaries, as well as everyday experiences, needs and expectations of a section of the population long considered 'difficult to engage' (Flinders et al., 2015). Therefore, web-mapping could be included in the wider toolbox of engaging techniques available for policy-makers and local authorities, even if the results might be more relevant at local, rather than city-regional level.

Photovoice Workshops

Photovoice workshops were designed to build on the web-mapping workshop, and most participants attended both. The aim was to elicit youth's aspirations for their city-region; participants were therefore asked to photograph aspects that make life worth living in their city-region, aspects they would like to see change, and a travelling experience. Each question required two photos, yet people often took more. These instructions and ethics-related issues were discussed at the end of the web-mapping session, and each participant also received a briefing email.

Participants could choose to use their own gadgets or be provided with a digital camera they returned after use. After completing the task, each person wrote captions for their photos and returned everything by email. This allowed the researcher to become familiar with the visual outputs produced and to prepare ahead of the workshop, when the group met to present and discuss results. Nonetheless, these workshops did not have a predefined structure, therefore departing from Wang's method of identifying themes (see Wang, 2006). Still, as certain topics re-emerged either in the form of captions or during discussions, it has been possible to form key themes together with the participants.

These workshops also varied in length depending on the number of attendees. Although initially designed to follow precisely the Photovoice nine-step strategy (see Wang, 2006), the method was condensed to reduce the number of meetings, and therefore, the risk of tiring participants. As mentioned, briefing was done at the end of the web-mapping workshops, and then repeated through email. In addition, as none of the participants required training on camera use, this step was entirely skipped. The reality is that although recruitment was the most difficult part, commitment was also fluctuating and was sensitive to weather, exam periods, or personal events. Nonetheless, the compromise found yielded rich findings, without sacrificing the quality of the research process.

The guidelines received encouraged participants to reflect on both desirable and undesirable features of their city-region. Transport—a recurrent theme stimulated by the last aspect required to document—rarely elicited any positive feelings (Fig. 13.3) and shifted the balance from appreciative inquiry towards criticism. The participants who lived outside of Cardiff felt disadvantaged because of the unreliable public transport services in the South Wales Valleys and the impossibility of driving (determined by age or costs). Those from Cardiff declared they mostly walk or cycle, despite the lack of safe conditions for active travel. Given the car-centric nature of CCR, this is an important message for policy-makers who are trying to reduce transport emissions and encourage people to replace private vehicles with more sustainable alternatives. Without improved infrastructure and services, it will be impossible to achieve less carbon-intensive lifestyles.

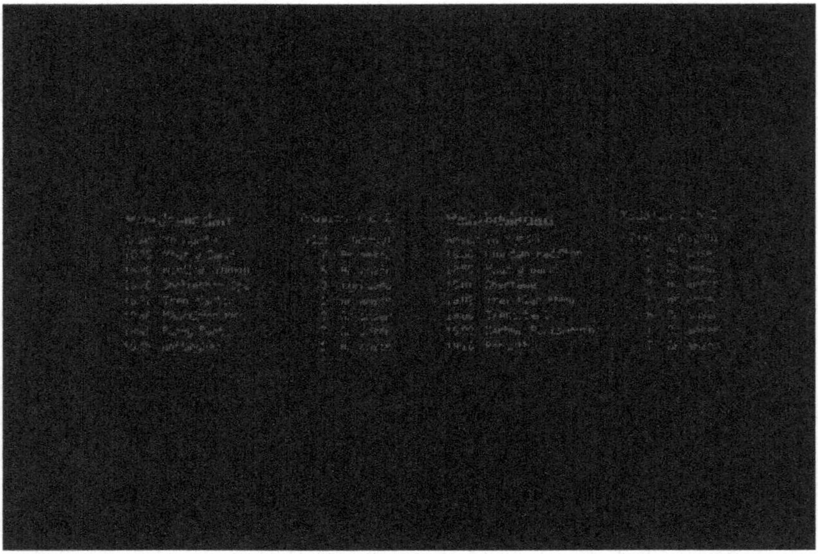

Fig. 13.3 Broken train timetable board (C. Thomas)

Overall, photos and subsequent discussions unveiled different layers of inequality (in terms of access to leisure, employment and transport facilities), and most young people living in Cardiff felt strongly about the growing levels of homelessness in the capital. While trying to respect rough sleepers' privacy, various participants decided to document the homelessness crisis in Cardiff, as one of the most critical issues that required local authorities' immediate attention. In addition to the large number of homeless persons in the city centre, they mentioned a shelter located on the same street with a suite of corporate offices (where some of the participants actually worked). This juxtaposition demonstrated young people's astute observations of their surroundings, and a certain sense of social justice. While aware of the authorities' limited budgets, participants thought investment was often going to the wrong places. This argument also surfaced when mentioning new developments, and some participants decried the opening of a new shopping facility (Trago Mills in Merthyr Tydfil), instead of support for the smaller shops on the High Street.

Among the positive aspects, the most common were related, unequivocally, to CCR's environment, its diverse natural landscapes and the opportunities to spend time outdoors. Regardless of their place of residence, participants felt a strong connection to nature, in both urban and peri-urban areas. The natural environment was also used as a hook to talk about more delicate subjects, such as the stigma attached to certain areas of the CCR. Referring to the Welsh Valleys, various participants from two different groups cited the lack of place appreciation among individuals and communities who had come to internalize the deprecating narratives portrayed in the media.

Photovoice workshops offered a multi-layered understanding of young people's sense of place through the combination of photos, captions and group discussions. On the one hand, this would have been impossible to capture fully through more conventional research methods, or non-visual ones. As photos were analysed during workshops, they became prompts and supports for richer discussions. On the other hand, photos alone would have not been enough to grasp the participants' ideas, as 'coincidental things can be overrated' (Trell & van Hoven, 2010, p. 101) and misinterpreted. Furthermore, following Photovoice's emancipatory goal, this research project aimed to create a bridge between the young participants and city-regional leaders, through an open exhibition and/or a roundtable. Unfortunately, this event did not materialize, and the results did not reach policy-makers (detailed in the following section).

Despite this shortcoming, the combination of web-mapping and Photovoice formed a well-rounded strategy through which young people could share their lived experience in relation to an unfamiliar topic—the city-region. The wider study also included an initial quantitative analysis that was useful to understand general trends concerning youth; however, the majority of the data was obtained through these two creative methods. They provided a rich amount of information, which generally escapes surveys and statistical accounts. Workshops enabled participants to designate the city-regional span they identified with (via maps), show how positive and negative aspects looked in their communities (via photos) and clarify how these affected their lives (via discussions).

The following paragraphs reflect on the experience of using a combination of two creative visual methods, on the process and the results yielded.

Reflecting on the Experience

This exploration validated Wang's (2006, p. 152) view, showing that young people can be 'competent citizens and active participants in the institutions and decisions that affect their lives', if given the chance, and that visual methods that allow a more creative form of expression can be useful engagement tools. Some of the most remarkable findings were young people's critical thinking skills and their strong connections to their communities. Participants invested time to reflect on the tasks and questions asked, and were able to contribute with both objective and subjective arguments, as experts of their local environment's conditions that support or inhibit their well-being (Chawla, 2002; Chou et al., 2017; Driskell, 2002; Wang, 2006).

However, this project also demonstrated that methodology alone cannot overcome other, more structural barriers in levelling the field between the researcher and the research participants (Packard, 2008), or in bridging between participants and policy-makers. In general, the workshops were successful in terms of attracting youth and in generating rich findings. However, like other Photovoice projects, this one also failed to reach policy-makers and to accomplish the method's emancipatory goal (Derr & Simons, 2019). Various blog posts, a photo essay (Axinte, 2018), a video (Axinte, 2019) and an invitation to an exhibition (Axinte et al., 2018) held in November 2018 in a popular arts centre in Cardiff were shared with various contacts from CCR. Despite all these efforts, there is no proof that city-regional leaders have taken notice of the project and its findings, although they were made aware of the ongoing research during interviews. Therefore, it became necessary to manage participants' expectations as many young people were enthusiastic that their voices could be heard and could make a difference. Participants were promised that their messages would be passed on, yet they were informed that the capacity to influence the uptake was minimal.

In addition to this shortcoming, another caveat worth discussing is the 'participatory' nature of this research, which has been limited. In a fully participatory project, the participants would be involved throughout all stages, from choosing the topics, to selecting methods, analysing data, formulating conclusions and choosing the dissemination format (Driskell, 2002; Wang & Burris, 1997). Although this was the initial intention, it became unfeasible given the long time needed to recruit participants and to run the workshops, as well as the limited (human and financial) resources. The different dissemination activities mentioned above, and in particular the exhibition, might have had a bigger outreach had participants been directly involved instead of mere invitees. While every single session was flexibly moulded depending on the participants, the researcher chose the main topics, the data collection and the analysis methods, as well as the dissemination formats. Participants were, of course, in full control of their contributions and the conclusions formulated during each occasion, participating in knowledge creation via research (Driskell, 2002, p. 98).

A further thorny issue for this project, and for visual research in general, is the danger of misanalysing and misinterpreting data. Figure 13.4 offers an example of an image that can be misleading when examined in isolation. The research participant used it to express her admiration for the Valleys' natural capital, as well as her regret that these places (and their inhabitants) are victims of entrenched social stigma. These ideas cannot be 'read' by simply looking at the photograph, a conclusion also reached by Trell and van Hoven (2010; see also Baimukmedhamedova, this book). Therefore, to avoid any misunderstandings, participants were always asked to comment on their outputs in writing, and to discuss them during focus groups.

Some other reflections arising from this research relate to the abilities, tasks and roles required to run creative methods workshops. Skills such as empathy, deep listening and facilitation are often taken for granted, yet the effort of applying them simultaneously during fieldwork should not be underestimated (see also Moriggi, this book). In addition, it can be particularly challenging to play two roles at the same time, as

Fig. 13.4 Rhondda Valley—a place of great beauty, but also one with stigma and bad press (C. Howson)

a researcher and a facilitator. Although this topic is rarely discussed in the literature, facilitation skills are crucial for a successful participatory workshop. Employing a second person can reduce the burden of hosting sessions and collecting data at the same time; in this project it was not possible. To simplify the process, the researcher did a lot of preparation and planning, and tried to complete some tasks in advance. For instance, participants received all the information (including instructions and ethics-related issues) by email before the workshop. This allowed them to become acquainted with what was going to happen, and helped reduce the time spent on organizational matters at the workshop itself. Furthermore, every session was voice recorded, with the participants' approval. Listening to the recordings allowed the researcher to take notes, revisit preliminary findings and conduct the analysis. It also permitted the capture of any information that might have been missed during the workshop.

Despite certain challenges arising, the two creative research methods used have yielded a wealth of information regarding young people's lived experiences in CCR. Employed in a workshop format, the methods complemented each other, eliciting spatial, visual and narrative information. In return, these workshops had a strong educational component, first by raising awareness regarding CCR's existence and future plans, and second, by allowing young people to gain new skills, such as basic web-mapping, and the capacity to express themselves geographically. Young people were encouraged to adopt a critical, analytical understanding of their environments. Their visual and textual outputs demonstrate their engagement and interest in contributing to a more complex depiction of CCR, beyond its formal socio-economic designation.

Furthermore, the two methods nurtured a more balanced researcher-participant relationship, allowing the latter to choose how they preferred to participate and what they wanted to prioritize. In return for their participation, young people could gain new skills and expand their understanding of the city-region, and of other people's insights. This less extractive researcher-participant relationship avoided the situation in which participants feel they have not gained enough in exchange for their time and contribution. As each workshop ended with a brief feedback session, some participants expressed content with the newly acquired web-mapping skills, and others appreciated the invitation to reflect on their surrounding environment. Although not included in this research due to limited time resources, a final feedback session with all participants could have been useful to compare experiences across groups.

Conclusions

This chapter has discussed the application of two creative participatory research methods—web-mapping and Photovoice—in a research project that explored young people's lived experience within a newly created administrative layer—Cardiff Capital Region. Despite the participants' limited knowledge of CCR, the approach facilitated their understanding and their engagement with the topic.

The two methods successfully complemented each other. Through web-mapping, participants were invited to mark the larger territory to which they felt attached, as well as the places that seemed significant for them. The findings suggest that the official city-regional boundaries have little relevance for young people, whose personal demarcations cover smaller territories, showing a much more localized *sense of the city-region.*

For the Photovoice session, the participants were asked to photograph some of the places they had pinpointed on their maps. They captured some of the aspects that made life worth living in the city-region, as well as those they would like to see change. Finally, they discussed differences and commonalities during focus groups, getting exposed to their peers' perspectives.

Youth's voices have been captured through GIS, visual and textual means, forming a rich body of data. Therefore, the project was successful in highlighting the young participants' ideas and concerns with their communities, and in promoting critical and analytical discussions during focus groups. In this sense, it maintained two of the original aims established by Photovoice's creators (Wang et al., 1996). However, as the project did not succeed in bridging between young people and city-regional leaders, the findings did not affect or 'infuse with young perspectives' (Wang, 2006, p. 159) the development of the city-region. As Derr and Simons (2019) showed, this is a common issue across various Photovoice projects: the method's adaption led to losing one of its original tenets—that of helping to emancipate participants' voices and needs. Moreover, although both (web-)mapping and Photovoice have been used elsewhere as participatory techniques, allowing children and young people to be fully engaged in all research stages (Driskell, 2002; Wang, 2006), this study has been rather limited. Therefore, it serves as a cautionary tale for anyone looking to design participatory research projects, confirming that 'democracy and engagement cost' time, effort and money (Flinders et al., 2015).

Nonetheless, this exploration offers relevant lessons for socio-environmental researchers because of (i) its capacity to attract research participants generally deemed harder to engage, (ii) the innovative use of two creative research methods, enabled by technology, (iii) the results that support previous claims that young people are experts of their local environments, with pertinent input for the place's future development, (iv) some honest reflections on the achievements and failures, as well as advantages and disadvantages of employing these methods. These lessons will hopefully inform future researchers in their endeavours, and contribute to critical discussions on visual creative methods, as well as engagement issues within youth studies.

Acknowledgements This research received funding from the European Union's Horizon 2020 research and innovation programme under the Marie Sklowdoska-Curie Grant Agreement No. 674962.

References

Adams, S., Savahl, S., & Fattore, T. (2017). Children's representations of nature using photovoice and community mapping: Perspectives from South Africa. *International Journal of Qualitative Studies on Health and Well-Being, 12*(1), 1333900.

Agger, A. (2012). Towards tailor-made participation: How to involve different types of citizens in participatory governance. *Town Planning Review, 83*(1), 29–45.

Axinte, L. F. (2018). Cardiff capital region and its future generations. In M. Barry et al. (Eds.), *Metro and me* (pp. 47–50). Retrieved March 7, 2021 from https://www.iwa.wales/wp-content/media/2018/10/MetroAndMeDigiFinal.pdf

Axinte, L. F. (2019). *Engaging youth in city-regional development*. Retrieved March 7, 2021 from https://youtu.be/gDYu_-78dM0

Axinte, L. F., Quinn, M., Rebelo, C., Taherzadeh, A., & Giambartolomei, G. (2018). A summary of sense(s) of place. Retrieved March 7, 2021 from https://blogs.cardiff.ac.uk/sustainableplaces/2018/12/07/a-summary-of-senses-of-place/

Axinte, L. F., Mehmood, A., Marsden, T., & Roep, D. (2019). Regenerative city-regions: A new conceptual framework. *Regional Studies, Regional Science, 6*(1), 117–129.

Beel, D., Jones, M., & Jones, I. R. (2018). Elite city-deals for economic growth? Problematizing the complexities of devolution, city-region building, and the (re)positioning of civil society. *Space and Polity, 22*(3), 307–327.

Chawla, L. (2002). "Insight, creativity and thoughts on the environment": Integrating children and youth into human settlement development. *Environment and Urbanization, 14*(2), 11–22.

Chou, M., Gagnon, J.-P. R., Hartung, C., & Pruitt, L. J. (2017). *Young people, citizenship and political participation: Combating civic deficit?* Rowman & Littlefield.

Collin, P. (2015). *Young citizens and political participation in a digital society: Addressing the democratic disconnect*. Palgrave Macmillan.

Dennis, S. F., Gaulocher, S., Carpiano, R. M., & Brown, D. (2009). Participatory photo mapping (PPM): Exploring an integrated method for health and place research with young people. *Health & Place, 15*(2), 466–473.

Derr, V., & Simons, J. (2019). A review of photovoice applications in environment, sustainability, and conservation contexts: Is the method maintaining its emancipatory intents? *Environmental Education Research, 26*(3), 359–380.

Driskell, D. (2002). *Creating better cities with children and youth: A manual for participation.* Routledge.

Eames, M., De Laurentis, C., Hunt, M., Lannon, S., & Dixon, T. (2014). *Cardiff 2050: City regional scenarios for urban sustainability*. Cardiff University.

Farthing, R. (2010). The politics of youthful antipolitics: Representing the 'issue' of youth participation in politics. *Journal of Youth Studies, 13*(2), 181–195.

Farthing, R. (2012). Why youth participation? Some justifications and critiques of youth participation using New Labour's youth policies as a case study. *Youth & Policy* (109).

Flinders, M., Ghose, K., Jennings, W., Molloy, E., Prosser, B., Renwick, A., Smith, G., & Spada, P. (2015). *Democracy matters. Lessons from the 2015 citizens' assemblies on English devolution* (78 pp.). University of Sheffield.

Grenni, S., Soini, K., & Horlings, L. G. (2019). The inner dimension of sustainability transformation: How sense of place and values can support sustainable place-shaping. *Sustainability Science, 15*(2), 411–422.

Headrick Taylor, K., & Hall, R. (2013). Counter-mapping the neighborhood on bicycles: Mobilizing youth to reimagine the city. *Technology, Knowledge and Learning, 18*(1–2), 65–93.

Kudryavtsev, A., Stedman, R. C., & Krasny, M. E. (2012). Sense of place in environmental education. *Environmental Education Research, 18*(2), 229–250.

Laville, S., & Watts, J. (2019). Across the globe, millions join biggest climate protest ever. *The Guardian.*

Lee Peluso, N. (1995). Whose woods are these? Counter-mapping forest territories in Kalimantan, Indonesia. *Antipode, 27*(4), 383–406.

Literat, I. (2013). Participatory mapping with urban youth: The visual elicitation of socio-spatial research data. *Learning, Media and Technology, 38*(2), 198–216.
Lorenz, L. S., & Kolb, B. (2009). Involving the public through participatory visual research methods. *Health Expectations, 12*(3), 262–274.
Lynch, K. (1976). *Managing the sense of a region.* MIT Press.
Lynch, K. (1977). *Growing up in cities: Studies of the spatial environment of adolescence in Cracow, Melbourne, Mexico City, Slta, Toluca, and Warszawa.* MIT Press.
Nolt, J. (2017). Future generations in environmental ethics. In S. M. Gardiner & A. Thompson (Eds.), *The Oxford handbook of environmental ethics.* Oxford University Press.
Packard, J. (2008). 'I'm gonna show you what it's really like out here': The power and limitation of participatory visual methods. *Visual Studies, 23*(1), 63–77.
Reason, P., & Bradbury, H. (2008). Part one: Groundings. Introduction to groundings. In *SAGE handbook of action research* (pp. 11–15). Sage.
Tisdall, E. K. M. (2016). Participation, rights and 'participatory' methods. In A. Farrell, S. Kagan, & E. K. M. Tisdall (Eds.), *The SAGE handbook of early childhood research* (pp. 73–87). Sage.
Trell, E.-M., & van Hoven, B. (2010). Making sense of place: Exploring creative and (inter)active research methods with young people. *Fennia: International Journal of Geography, 188*(1), 91–104.
Wang, C. C. (2006). Youth participation in photovoice as a strategy for community change. *Journal of Community Practice, 14*(1–2), 147–161.
Wang, C. C., & Burris, M. A. (1997). Photovoice: Concept, methodology, and use for participatory needs assessment. *Health Education & Behavior, 24*(3), 369–387.
Wang, C. C., Burris, M. A., & Ping, X. Y. (1996). Chinese village women as visual anthropologists: A participatory approach to reaching policymakers. *Social Science & Medicine, 42*(10), 1391–1400.
Wang, C. C., & Redwood-Jones, Y. A. (2001). Photovoice ethics: Perspectives from Flint Photovoice. *Health Education and Behavior, 28*(5), 560–572.
Ward, A. L., Freeman, C., & McGee, R. (2015). The influence of transport on well-being among teenagers: A photovoice project in New Zealand. *Journal of Transport and Health, 2*(3), 414–422.

Welsh Government. (2012). *City regions final report.*
Welsh Government. (2016). *Population estimates by local authority and year, mid-year 2015*. Retrieved March 7, 2021 from https://statswales.gov.wales/Catalogue/Population-and-Migration/Population/Estimates/Local-Authority/populationestimates-by-localauthority-year
Williams, P. (2006). The governance of sustainable development in Wales. *Local Environment, 11*(3), 253–267.

14

Technology as a Tool for Environmental Engagement. The Case of Digital Participatory Mapping (DPM)

Nohemi Ramirez Aranda and Rubén Vezzoni

Introduction

Digital innovation opens up new possibilities of exploring scientific methods beyond what is traditionally accepted in research. This process can be smooth and linear, but often it creates tensions, misunderstandings, and unmet expectations. The means to access and create data are rapidly changing, and so is the knowledge creation process,

---

N. Ramirez Aranda (✉)
Flanders Research Institute for Agriculture, Fisheries, and Food (ILVO), Merelbeke, Belgium
e-mail: nohemi.aranda@ilvo.vlaanderen.be

Department of Geography, Ghent University, Ghent, Belgium

R. Vezzoni
LUKE (Natural Resources Institute Finland), Helsinki, Finland

University of Helsinki, Helsinki, Finland

A. Franklin (ed.), *Co-Creativity and Engaged Scholarship*,
https://doi.org/10.1007/978-3-030-84248-2_14

pushing the boundaries of academic work into new grounds. Crowdsourcing online platforms and Geographical Information Systems (GIS) are good examples of this. In this chapter, we reflect on some original ideas concerning the genesis and direct consequences of adopting digital tools in research, with a particular focus on new forms of participation in defining geographies of space, and planning activities from the bottom-up.

Section 14.1 touches on the progression of participatory mapping, its origins, and its incursion in the digital era. Moreover, it argues how its digital and creative nature can be an asset in making participatory processes more democratic and inclusive. Section 14.2 describes two real applications of this technology used in case studies in Belgium and the Netherlands. Section 14.3 presents the challenges and opportunities faced in the case studies, giving special attention to *forced creativity*, data quality matters (such as precision and accuracy), and digital participatory mapping's complementarity with traditional research methods. Lastly, Sect. 14.4 provides a comprehensive conclusion and discussion of the critical points emerging in the chapter.

Participatory Mapping Origins

Amid the 1930's modernist movements for rethinking cities and housing in the United States, the foundations of participatory mapping were laid through community surveys and hearings (Guldi, 2017). An early example shows rudimentary but effective participatory maps as court documents by the Cree people to protect their land from developers (Poole, 1995). Ever since, participatory mapping has been slowly evolving and migrating to more digital means through software such as GIS (Chambers, 2006). GIS (programs for collecting, storing, and integrating spatial data analysis) have since found extensive research applications. Several criticisms have, however, emerged in the last decades, calling for a more socially aware type of GIS. Dunn (2007) suggests that this would mean paying greater attention to local spatial knowledge than the technological aspect. The context of the application and the people involved in the process would then be the core elements of this new type of GIS that give legitimacy to research's social aspects. This new type of GIS comes when the silent majority is in dire need of tools that bridge their necessities and local knowledge with practitioners and policymakers. If used effectively, these bridging tools can feed government actions, so their benefits can be equitably distributed while reflecting people's needs more accurately. Consequently, these alternative approaches have eventually established a well-defined scientific niche as participatory mapping (PM).

Putting Participation in GIS

In the mid-1990s, two major academic events paved the way for combining GIS technologies with social issues. The first was the 1993 "GIS and Society" workshop by Poiker (National Center for Geographic Information and Analysis—NCGIA). Second, the 1996 workshop organized by Onsmd, Schroeder, and Lopez at the University of Maine. Both events focused on improving GIS access, especially for those who have been historically under-represented (Obermeyer, 1998). These early attempts to harness the potential of GIS for democratic participation were first labelled under the term Public Participation GIS (PPGIS). PPGIS gradually metamorphosed into a multidisciplinary concept involving different stakeholders and goals (Sieber, 2006). This has led to several variations of the term, such as Participatory GIS (PGIS) or Volunteered Geographic Information (VGI), to be interchangeably used with PPGIS. These terms often share the same essence and their mild differences are not always clear-cut (McCall, 2004).

Brown and Kyttä (2014) propose a characterization of the different types of participatory mapping based on their essential characteristics, as shown in Table 14.1. Although some distinctions can indeed resonate with the history of the tools or their context of application (e.g., PPGIS and PGIS started to be developed in different parts of the globe, respectively in industrialized countries vs. peripheral countries), we argue that current power dynamics and technological advancements have blurred some of these distinctive traits. A strict definition of them is, therefore, of little value for this chapter. Alternatively, we adopt the definition that the

Table 14.1 Characteristics of PPGIS, PGIS, and VGI (extracted from Brown and Kyttä (2014))

	PPGIS	PGIS	VGI
Process emphasis	Enhance public involvement to inform land use planning and management	Community empowerment Foster social identity Build social capital	Expand spatial information using citizens as sensors
Sponsors	Government planning agencies	NGOs	NGOs, ad hoc groups, individuals
Global context	Developed countries	Developing countries	Variable
Place context	Urban and regional	Rural	Variable
Importance of mapped data quality	Primary	Secondary	Primary
Sampling approach	Active: probability	Active: purposive	Passive: voluntary
Data collection	Individual (e.g., household sampling)	Collective (e.g., community workshops)	Individual
Data ownership	Sponsors of the process	People and communities that created data	Shared (e.g., data commons license)
Dominant mapping technology	Digital	Non-digital	Digital

same authors proposed in a more recent paper (Brown & Kyttä, 2018), suggesting identifying all the nuanced alternative concepts mentioned above under the umbrella term of participatory mapping (PM). More specifically, we will hereafter use the term digital participatory mapping (DPM) to identify PM's non-physical practices.

Since its first conceptualizations in the 1990s, the adoption of DPM has been gradual due to technological and cost limitations. The early 2000s tech revolution brought cheaper, faster, and more resourceful computers, a variety of software, data availability, integration of remote sensing technology, and the launch of new satellites (GIS Geography,

2020). This combination of events has turned DPM into a cost-effective technique potentially reachable from everywhere and by anyone.

The groundwork laid by decades of technological advancement has indeed increased DPM's capabilities and accessibility to the masses, making its use in diverging contexts by academics possible (Mccall & Dunn, 2012). Nonetheless, DPM is still not widely adopted by governmental organizations. The rationale behind this is that its results contain potentially sensitive issues that can create dilemmas for policymakers (Carton, 2005; Sulistyawan et al., 2018). Therefore, it is perhaps no surprise that the main ambassadors and advocates of DPM are academics who have tested, modified, and re-tested DPM tools in real-life scenarios where power dynamics are skewed against the underprivileged masses in society.

One of the highest ambitions of researchers using DPM, especially those with a sociological background, is to help break down power hierarchies (Guldi, 2017). Using DPM, less privileged groups in society can be empowered through access to tools previously only accessible to government officials, practitioners, professionals, and policymakers (Sieber, 2006). It is a context-specific application of GIS technology meant to harness indigenous knowledge through community engagement and social learning. Concerning its relationship with traditional research methods in spatial studies, according to Dunn (2007, p. 616), it "celebrates the multiplicity of geographical realities rather than the disembodied, objective and technical 'solutions' which have tended to characterize many conventional GIS applications".

In many participatory mapping studies, especially in those contexts of low levels of digital literacy, communities' knowledge is gathered through physical, low-tech data collection methods, e.g., drawing ephemeral maps in the sand (Poole, 1995), building physical models (Rambaldi, 2010), or stickers on paper maps (Brown & Pullar, 2012). To their advantage, these methods are straightforward and do not require the respondents to have any technical mapping or IT skills. Accordingly, from the pool of academic literature regarding PM, there is a predominance of articles referring to physical data collection methods. This imbalance emphasizes the shortage of articles focused on collecting data solely using creative and online DPM methods, despite their surge in the past ten years.

Whereas some literature already exists concerning the advantages and drawbacks of these forms of traditional (i.e., physical) PM, DPM and its contextual application remain a rather new subject (Brown & Reed, 2009). Digital technology may support innovative ways of empowering marginalized social groups, giving voice to unheard local communities, or strengthening the methodological baggage of emerging transdisciplinary sciences (see also Axinte, this book). In this chapter, therefore, we intend to explore the possibilities offered by DPM as visual methods. Moreover, we try to bring to light its capabilities for remote participation, integration of additional input by participants, and the spare the burden of double-entry of information. For the reasons outlined above, we deliberately do not cover other kinds of research methods involving physical activities.

Description of Digital PM Tools

Thanks to technological advancements (e.g., widely available Internet access, user-friendly GIS interfaces, and mainstream use of mobile devices), DPM has become a cost-effective approach for spatial data collection and a promising method for map co-creation (see also Axinte, this book). However, it is not a panacea for all participation and democratization problems. In Sect. 14.3, we further discuss the challenges and opportunities of DPM.

The tech revolution of the early twenty-first century and a rising interest in the academic environment have paved the way for PM to establish itself as a scientific discipline. These technological advancements provided different platforms upon which the foundations of DPM were built. In the last two decades, scientific literature shows a transition from spatial data collected through low-tech physical methods to digital online alternatives. In the online category, three main types of GIS tools can be distinguished: software, native apps, and web apps.

Software packages are on the heavier end, although there are alternative options, including extensions on standard GIS software to adapt it for the specific type of participation (Allen & Christensen, 2001; Dieber, 2003). Notwithstanding these are quite flexible tools, they rely heavily on participants' technical skills and capable hardware access. The incursion of community map-based services (e.g., open street maps), open-source GIS software (e.g., QGIS), and the mainstream of mobile devices have helped DPM tools to become cheaper, lighter, and friendlier. This has also enabled a shift from bulkier packages to lighter web versions on mobile devices. Native apps are built based on the specific device's OS on which they will be installed, e.g., Android, IOS, and can be built on the research's specific needs, offering limitless possibilities of customization. On the downside, they might require a skilled developer to build and update them properly, which can be pricey.

On the other hand, web-based apps can be customized but do not need to be installed, require fewer resources since they run through the Internet browser, and are cheaper as their architecture and update require less effort. DPM web apps' success has led companies such as ESRI and Maptionnaire to capitalize on it by offering GIS survey services to governments and the general public. They offer DPM tool "templates" that can be customized based on customers' needs as well as access and handling of the collected data in a simple manner.

DPM and Transdisciplinary Research: Towards more Democratic Participation Practices

In the 1960s, Jane Jacobs was the Cassandra of spatial planning, fighting rationalist planners driven by a capitalist boom and stubborn politicians who deliberately disregarded science and people. Back then, opportunities and channels for citizens to be included in planning processes and decision-making were just not there yet. Moreover, planning trends were mostly intra-disciplinary, leading to segregated solutions for problems that required an holistic approach. This was reflected in the countless "urban renewal programs" (Jacobs, 1994) that affected entire neighbourhoods and destroyed the social fabric in and around them.

However, technological advances in the last two decades have enabled people to increase their role in decision-making and co-creation of cities together with planners and practitioners. The GIS technology used in DPM offers valuable capabilities that can be put at the service of trans-disciplinary research to tackle the major challenges we face as a species. Think about climate change, natural hazards, cultural heritage, security, health, or poverty. All these challenges require spatial information on a variety of scales if they are to be addressed. Moreover, they all require a comprehensive understanding of the place and interaction of (countless) things and people, in space and time, that a few decades ago would have been impossible to acquire.

More than ever, connecting with strangers across gender, race, class, and space is fairly easy. This is especially true when 60% of the world population uses the Internet and possesses a mobile device (Statista, 2020). DPM tools, such as the examples given in the following case studies section, can provide insightful information reflecting the many different perspectives each of its contributors had across populations' width and length over a vast range of disciplines, formats, and languages. When this content is put into planners' and policymakers' hands, DMP becomes a channel that boosts representative democracy. It creates or reinforces legitimacy in governance by providing a ground for stake-holders to interact in the twin processes of executing various mapping actions and then analysing the visualized results.

Moreover, DPM creates opportunities to visualize the interests, needs, and potentials of disparate groups in participatory spatial mapping. In this way, those governing are provided with a mechanism to recog-nize and appreciate the legitimacy of the governed's local interests. DPMs can also foster community participation, allowing capacity-building, management, and planning initiatives by the communities. These include skills development and increased confidence in dealing with external economic and political powers, professionals, and technical experts, and heightened attitudes of community "ownership" of the data and maps produced, giving them a better opportunity to control the data and entitlement to any projects making use of the mapping.

In the following sections, we describe two case studies of DPM web apps: "My Green Place" and the "Greenmapper".

Case Studies

Case Study 1: My Green Place Ghent

My Green Place/Ghent was the pilot test of a DPM tool that is the centrepiece of the doctoral research, with the working title of "Public Participation GIS as a tool for improving community management and preservation of green open spaces in Belgium" by Nohemi Ramirez Aranda. The joint research between the Institute for Agricultural, Fisheries, and Food Research (ILVO) and the geography department of Ghent University focuses on improving the management and preservation of green open spaces by including communities' social values in the planning process. Moreover, the ongoing project aims for a tool that allows a more inclusive participation process, giving special attention to those groups that current processes do not reach, such as teenagers, migrants, and older adults.

It was necessary to create a web tool to map communities' social values attached to green open spaces. We wanted to make it as remote, fast, and cheap as possible while keeping the possibility of people of different backgrounds using it. Budget and time/space limitations finally drove us to a digital option. Moreover, a digital tool could (in theory) help combat the "self-selection bias" issue by reaching those citizens that represent the majority but are rarely present in the participation process (Bhattacherjee, 2012, p. 81; Fung, 2006; Renn et al., 1993). Let us take an example of a consultation process where people are invited four days in a row for planning workshops. Expectedly, not everyone would be able to miss four days of work to attend the consultation. Self-selection bias then happens when participants choose whether or not to participate in the consultation, and the ones who do are not representative of the population targeted by the study. Nevertheless, if we take away the fact that people would not have to skip work and could "participate" from their homes and at a time convenient for them through DPM, we could mitigate this specific case of bias.

We operationalized social values using the concept of the cultural practice from the cultural ecosystem services framework in Church et al. (2014) as indicators. To get an accurate reading of what these cultural practices were and where they happened, it was very important to provide a suitable geographic entity for participants to indicate them digitally in a map. A *geographic entity* is the drawing tool participants use to provide input. In physical PM, participants tend to provide their input drawing points and polygons in printed maps using pens, markers, or stickers of different colours and sizes depending on the mapped attribute. Despite their limitations, points, and in a lesser proportion, polygons dominate publications (Brown & Fagerholm, 2015). The selection of a suitable geographic entity to map specific attributes such as cultural practices was a critical concern in the research since depending on how participants interacted with it, the output could vary greatly. Unfortunately, within literature only a few academics had approached the geographic entity dilemma, doing so in a way that was considered insufficient for our study (Brown & Pullar, 2012). Therefore, the pilot tool tested which geographic entity (point, polygon, and marker) delivered better performance when mapping specific cultural practices.

To maximize our tool's exposure with our target groups, we designed a promotional campaign called "We Love Gent". This consisted of a whole branding image with a very identifiable logo, slogan, and flyers distributed across social media, WhatsApp, and stakeholders' communication channels such as Ghent University, neighbourhood centres, and religious facilities (see Fig. 14.1).

The data collection strategy included testing sites such as schools, care homes, and migration training centres through mobile devices such as computers, smartphones, and tablets. Additionally, we invited people in public areas to participate in the exercise by using their phones. Due to the variety of devices on which we needed the tool to operate and the exercise's one-time nature, installing something on their devices seemed very unlikely. Reasons varied from privacy issues, to the hassle of installing something that will take storage space on their devices and consume data in the background.

In addition to the tool's development cost and the need for it to be cross-platform, these issues played a critical role in choosing a web app over a native app. Moreover, a web app offered additional advantages against a native app. These included: easy maintenance due to a common code base across multiple mobile platforms, compatibility with any older

Fig. 14.1 Flyer used for diffusion in the We Love Gent campaign (*Source* Author)

mobile device (a particular advantage considering older adults might have outdated mobile devices), and release at the developer's discretion since no app store approval is required.

In the particular case of My Green Place, its development was based on java language with PHP (no framework) using Leaflet JS for the maps, Open Street Map (OSM) as a background layer, and Postgres/PostGIS as database. It runs on a virtual server in Apache, hosted by Ghent University. The layout was made using bootstrap CSS, and all data collected are stored in a secure server from where it can be examined via PgAdmin or QGIS (Fig. 14.2).

The suggestions displayed at the end of the survey were used to maximize the sharing of the tool's link and therefore participant numbers. By providing this map with suggestions of places, participants could share it on their social media and compare their friends' results and suggested places (Figs. 14.3, 14.4, 14.5).

Since the tool could be used on any mobile device with an Internet connection through their Internet browser, the diffusion was relatively simple and cheaper than mail or workshop-based approaches. The link was shared through the Ghent University network, social media, religious, and neighbourhood centres. Moreover, the way the tool itself and collected data are stored provides easy access to fix or adapt the tool in case of need and at a very low to no cost compared to those of a native

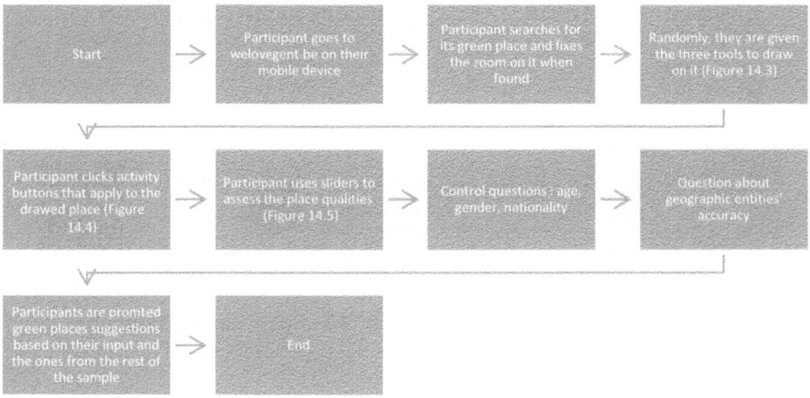

Fig. 14.2 My Green Place demo chart (*Source* Author)

Fig. 14.3 Use of different geographic entities to map cultural ecosystem services in the PPGIS online tool My Green Place / Ghent (*Source* Author)

app. The main goal of the tool was to test which geographic entity better mapped specific cultural practices. By doing so, we could be sure participants' input provided an accurate representation of their preferences. After five months of data collection and the input of 449 participants, the data collected allowed us to analyse the three geographic entities' accuracy through quadrat analyses, regressions against a constructed collective truth,[1] and a survey on the three geographic entities' perceived accuracy among respondents. The results showed points performing the weakest, and markers performing the strongest. This was noted, especially when mapping for dynamic cultural practices, implying a displacement across space such as running, biking, and walking. The performance of the polygon was similar to that of the marker, although slightly weaker. The marker not only provided a more precise image of respondents' input, it was also simpler to use and had less risk of spatial errors than offered by the polygon. Therefore, it was concluded that the marker was a suitable

[1] To test the suitability of each geographic entity to map cultural practices, we compared the three datasets, namely the aggregation of points, polygons, and markers, to a 'true' representation of respondents' favourite green open spaces per type of cultural practice. Therefore, an estimate of that 'true' representation, denoted as the "Collective Truth" (CT) was built by aggregating the average of points, polygons, and markers per quadrat cell.

Fig. 14.4 Button section with activities for participants to indicate (*Source* Author)

alternative to the point and the polygon when collecting spatial data in future cultural ecosystem services research (Aranda et al., 2021).

Another goal of the tool was to test how easy it was to reach teenagers, migrants, and older adults. In this regard, the data showed that the biggest group to use the tool was teenagers, followed by young adults. Regarding migrants, the criteria used to test this was "nationality"; this was proven ineffective for our purposes since, e.g., participants who were second-generation Turkish or had just recently settled in the country will still identify themselves as "Belgians". Several tests were made to get the

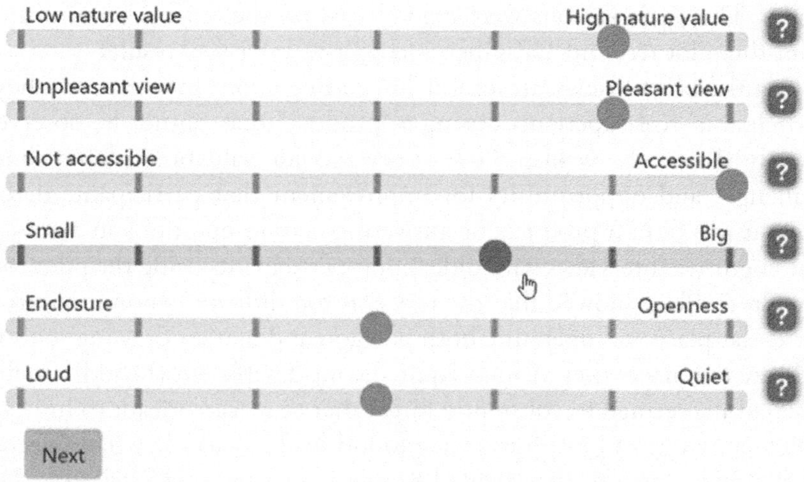

Fig. 14.5 Slider section on My Green Place tool to assess the place qualities (*Source* Author)

right question to provide what was needed without being too provocative or incurring major GDPR requirements. After the pilot test and the later trials on this specific point, we added two complementary questions about parents' nationality to track migrant backgrounds in the participant's sample. Tracing the migration background proved to be more challenging than expected since asking this is a very sensitive topic in the face of the many ex-pats that have arrived in Europe during the refugee crisis. Even when participants were second or third-generation descendants of migrants, the question posed something uncomfortable that not many were willing to answer.

The 60+ year age group was barely present in the sample, which led to changes for the tool's final version (My Green Place/Woluwe). Since the tool management was done via Pgamin, changes based on the pilot test's feedback were included relatively easily in the tool's next version. Thanks to this versatility, a twin version of My Green Place was used in a research project from the University Hospital of Ghent (UZ Ghent) about how older adults use green open spaces. This twin version was made via a copy of the main My Green Place that kept the tool's essence, such as the layout, buttons, and questions, while adapting it for older adults. These adaptations were made based on the feedback we secured from the pilot test and UZ Ghent's cognitive interview results.

Cognitive interviews are useful for getting more information about participants' real experiences using a "product" that cannot be observed directly (Beatty & Willis, 2007). They provide valuable insights about challenges and opportunities for improvement that participants detect while using certain products or answering certain questions in a survey. The cognitive interviews with older adult participants using the pilot tool My Green Place showed that the tool was too difficult. Zooming in to a particular place and operating both polygon and marker drawing tools to indicate the places they visited required considerable effort. Additionally, older adults seemed to constantly forget that they were meant to indicate which activities and attributes were found in the place they had selected on the map. Instead, they tended to select activities and attributes they wished or liked to do but not particularly in the place they selected. To handle these issues, the twin version of My Green Place highlights in green all areas within Ghent in the category of "green open spaces", e.g., parks, cemeteries or farmlands (see Fig. 14.6). Then, before continuing

Fig. 14.6 Twin version of the My Green Place tool, offering a preselected view of Ghent's green places for the older adult group (*Source* Author)

with the interview, the system asks for confirmation from participants concerning the selected area.

To address the issue of them providing general preferences (instead of attributes specific to the place they had marked), selected additional pop-up messages were strategically inserted to constantly remind respondents that their input should be based on what they did or found in the place they had selected (see Figs. 14.6, 14.7 and 14.8). These small but focalized adaptations to the original tool proved to be effective in the second round of cognitive interviews.

In tests where bigger groups were reached at once, such as classrooms or mass events, we carried out a series of feedback sessions. In

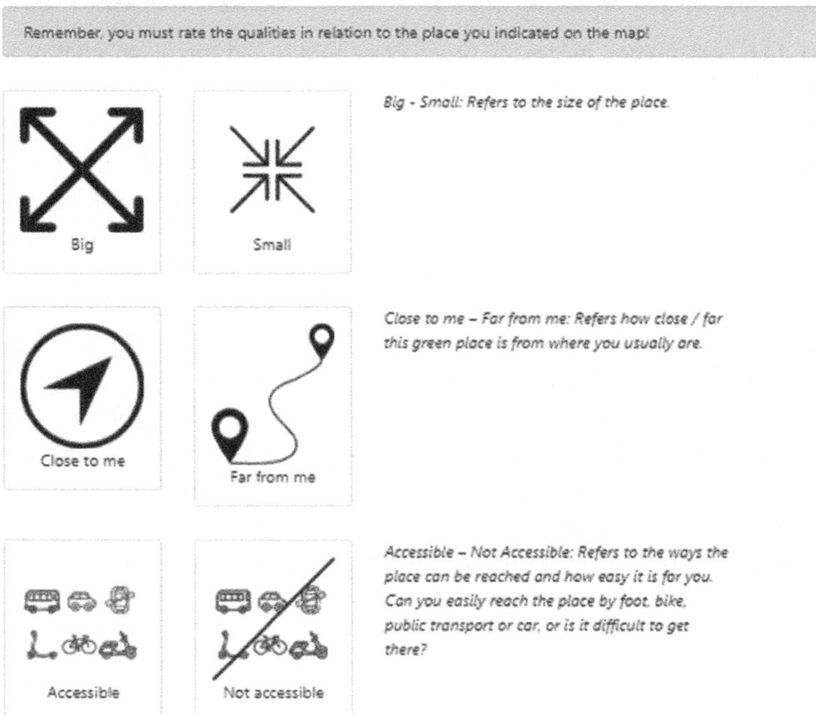

Fig. 14.7 Adaptation in the My Green Place tool to aid the older adult group to remember what attributes and activities to select (*Source* Author)

Fig. 14.8 Adaptation in the My Green Place tool to aid the older adult group to remember which places to select (*Source* Author)

these sessions, participants could, using "post-its", provide direct feedback about issues such as how easy or difficult it was to use the tool, if they found it relevant, or if it was the first time they were consulted in such a way. In addition, we could get valuable information about things we missed regarding activities, elements people found important, and survey questions that could help us better understand the bond and attachments people have with particular green places.

Through these sessions, we made citizens content creators by using the tool and developers, since it was through their feedback and interest that the next iteration of the tool was shaped. By actively participating in shaping the tool, they also become more committed to sharing it and using the outputs as grounds for shifting the direction of any plans that do not fit the collective interest.

Case Study 2: Multi-Level Governance and Partaking in Research on the Greenmapper

According to environmental psychology, the unique bonds that human beings develop with natural places have a key regulatory role in self-identity, such as self-esteem maintenance (Korpela, 2002). This is what Williams and Vaske (2003) call place identity, namely "the symbolic importance of a place as a repository for emotions and relationships" (*ibid.*, p. 831). Place identity occurs both at the individual and the aggregate level (i.e., community level) and across various geographical scales, well beyond the area of residence (Cuba & Hummon, 1993; Hidalgo & Hernández, 2001). Thanks to easier international traveling and increasing globalized imaginaries, Bijker and Sijtsma (2017) have demonstrated that citizens—at least in Western countries—have a clear "portfolio of natural places" spanning from the local scale to favourite places at the planetary level. Consequently, citizens from different parts of the world may be attached to the same natural place, constituting what Sijtsma et al. (2019) refer to as the "community of fans".

Therefore, the "community of fans" is a heterogeneous social conglomerate in which different particular interests coexist. Simplifying for the sake of clarity, a primary distinction exists between those people living in or next to the natural place and the rest of the fans. The demands of these two groups can diverge, sometimes to the extent of being antithetical (e.g., usage vs. conservation). Therefore, multi-level governance is needed to functionally coordinate social bodies for delivering mutually beneficial outcomes (Brondizio et al., 2009). In the presence of different and stratified interests,[2] multi-level governance through DPM platforms is used to reveal the otherwise invisible needs, motivations, and participation preferences of the actors involved. This is what Cox (2014) has referred to as the orchestrations of "management activities carried out by different sets of actors at different spatial scales" (*ibid.*, p. 312).

---

[2] The diversity of particular interests represents the dissimilarities among citizens living in different places (e.g., in preferences or requests) as well as those between public authorities and other actors involved in the management of the natural resource.

In addition to being characterized by a variety of particular demands, the members of a "community of fans" may also have different possibilities and preferences for contributing to the natural area's maintenance. For example, citizens living far away from their favourite places may have difficulty in providing physical support (e.g., collecting garbage); nonetheless, they may be willing to provide remote help (e.g., funding). The potential of financial support for natural places from online communities (e.g., crowdfunding) has already been investigated in literature (Bijker et al., 2014), although usually limited to specific areas or ecological services. The relevance of the subjective value attributed to the place and the geographical scale has often been overlooked. The Greenmapper.org provides a DPM platform for exploring the different ways in which citizens want to participate in support of their favourite natural areas, considering both the uniqueness of their individual contributions and the importance of the geographical scale.

The Greenmapper software (www.greenmapper.org) is a crowdsourcing platform developed at the University of Groningen, the Netherlands. It is a web-based participatory GIS platform based on Google Maps, previously known as HotSpotMonitor (HSM) (Sijtsma et al., 2012). Users can register with their home location and then mark their favourite natural places at four geographical levels: local, regional, national, and global. The data collected shed light on the participants' value to the natural places they find more attractive. After marking a natural place, respondents are asked to specify the type of connection they have with the place and why it is important for them (Fig. 14.9). This process is repeated at the local, regional, national, and global levels. Bijker and Sijtsma (2017) also used these four geographical scales in their study on the "portfolio of natural places", as shown in Fig. 14.10. The nature of these scales is obviously arbitrary. They come from the researchers' need to systematize the study, and they are interpreted based on individual perceptions of the participants' space. Nonetheless, it is useful to frame all the studies according to the same structure as it eases data collection and comparative analysis.

Following the previously introduced definition of a digital "community of fans" (Sijtsma et al., 2019), the Greenmapper is designed to

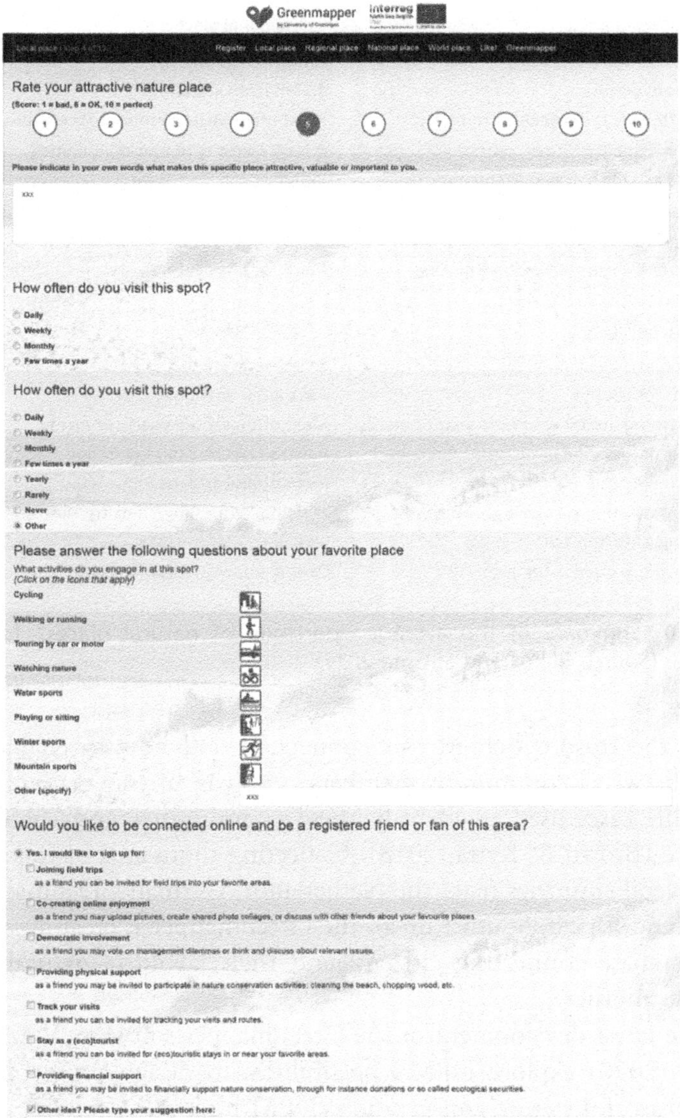

Fig. 14.9 Basic steps for registering for the Greenmapper survey. The figure shows the questions for the local place. The iterative process is repeated for every geographical level, i.e., local, regional, national, and global (*Source* Author)

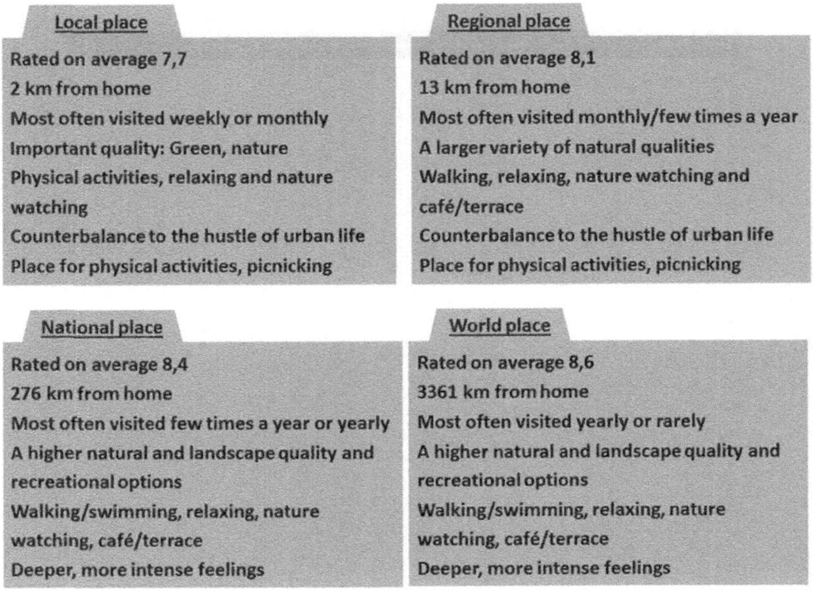

Fig. 14.10 Summary of the average "portfolio of natural places" for urban residents. (*Source* Bijker and Sijtsma 2017)

explore the reasons behind this connection and how to enable interaction between community members. Crowdsourcing through DPM is, in this case, used to identify these communities and enable their existence (Brown & Kyttä, 2018). Collecting thousands of respondents from several countries (e.g., the Netherlands, Germany, Denmark, Italy, Switzerland, Brazil, South Korea), the Greenmapper is creating a worldwide database connecting citizens, and their favourite natural places, with one another.

In the latest development of the Greenmapper software, Vezzoni and Sijtsma (2020) explore DPM's possibilities to design online financial mechanisms for nature conservation. DPM tools are a suitable method for transdisciplinary research. In this case, for instance, they have been used to combine research methods from different disciplines, such as contingent valuation from environmental economics, with the study of multi-level governance from the institutional analysis. Additionally, the

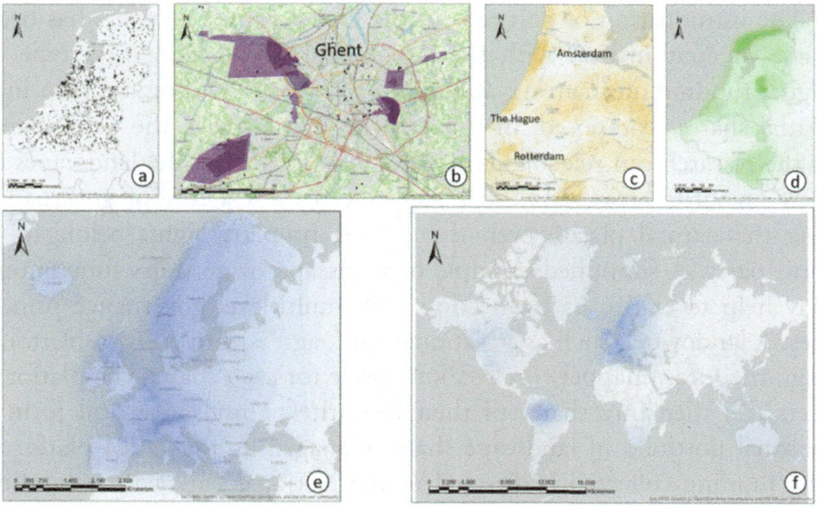

Fig. 14.11 Spatial results of the Greenmapper survey (*Source* Vezzoni and Sijtsma CCBY): (a) location of respondents, (b) local polygons in Ghent, (c) regional polygons in the Netherlands, (d) national polygons in the Netherlands, (e) and (f) global polygons

research's participatory element allows the respondents to define first-hand the locations in which the experiment[3] takes place. This has both the advantage of improving the accuracy of the results and enabling the collection and analysis of novel types of data pools.

Vezzoni and Sijtsma (2020) also introduce a new feature of the Greenmapper, namely the possibility of drawing polygons around favourite natural areas instead of marking them with a point. In line with My Green Place and other studies using participatory mapping (see Table 14.1), polygons deliver additional information concerning overlapping areas of common interest. For example, while points only allow for the clustering of interesting regions, with polygons the authors can create heat maps in which the intensity of the colour indicates the overlying of respondents' interests (Fig. 14.11).

[3] The experiment here refers to the contingent valuation section of the study. In this case, contingent valuation methods were used to infer on the willingness to pay expressed in euros per km^2.

The introduction of polygons in the Greenmapper software has been made in combination with the design and testing of a new type of digital funding mechanism: *landscape ownership*. Dividing the map into spatial shares (composed, in this case, of 1km^2 each), the participants in the research can select and virtually *buy* (i.e., fund) the landscapes in their favourite areas. *Landscape owners* can exert a role of governance over the natural places, even though the property rights belong to a third party. A simplified example of landscape ownership's functioning may help to clarify its connection with multi-level governance principles. A landowner, such as a National Park, agrees with a DPM platform (e.g., the Greenmapper) on a €/km^2 price for each share. The platform users can then buy shares of their favourite natural areas and form a personal portfolio of landscape shares managed by the DPM platform. The funding collected is directed to the National Park with a binding description of how and where to spend the budget. In return, whenever a major change will affect the area's landscape, the National Park agrees to communicate it to the *landscape owners* and let them partake in the decisional processes. In line with the principle of multi-level governance, *landscape ownership* allows different actors' voices to be heard and enables the coordination of multiple and stratified interests, making the average citizen an active player rather than a passive spectator (Arnstein, 1969).

The innovative potential of digital technology in research lies in the malleability of online tools such as DPM to adapt and complement traditional research methods. In the case studies presented in this chapter, digital tools have been used to stratify the richness of results by dramatically enlarging the sampled population and the diversity and complexity of each participant's responses. DPM methods allow Greenmapper users to partake in the design of research experiments instead of relying on given research settings.

Building on the Case Studies: Challenges and Opportunities of DPMs

> *"Cities have the capability of providing something for everybody, only because, and only when, they are created by everybody".*
>
> *Jane Jacobs*

This section will address some of the challenges and opportunities that the DPM tools offer. These have been selected based on the authors' experiences in their case studies and the limited links between them that literature offers. We will therefore discuss issues related to the dangers of pushing DPM as a solution to ease and defer conflicts emerging in society and communities. Moreover, we will address data quality issues that arise when collecting and analysing digital data from a diverse audience. Finally, we will explore DPM's role in planning consultations compared to "traditional" (physical) participatory methods.

Forced Creativity in DPM

In Chapter 2, "The Managerial University" (see Leitheiser et al., this book), we warned against the danger of *forced creativity. Forced creativity* refers to the co-optation of creative research methods by the dynamics characterizing the institutional setting in which they are mobilized. In the context of a highly managerial university, identified by increasing competition for funding and precarious job positions, DPM and other creative and visual methods may even end up exacerbating the problems they try to address. To start this section on the challenges of DPM, we expand on the idea of *forced creativity* by illustrating the case of *artwashing*.

A*rtwashing* is the strategy of managing discontent by giving an "artsy" touch to solutions that impact the appearance but not the problem's substance (Novak, 2019). It has an anaesthetic effect on the transformative potential of using creativity in research. *Artwashing* practices relate more to marketing strategies than scientific methods, potentially converting community involvement into a promotional campaign[4]. The narrative of participation in governance advanced by researchers using DPM is exposed to the danger of engaging with communities without actually having any possibility of empowering them (see also Reitz, this book, and Axinte, this book). DPM is an effective tool for eliciting the individual values and preferences of the participants. However, while social and cultural *recognition* is an important aspect of democratic practices, *redistribution* of power is often at risk of being neglected (Fraser, 1997). Even if the development of means, such as DPM, that can facilitate the share of information to, from, and with citizens, is consistent with democratic principles, it does not automatically enable the emergence of emancipatory actions. On the contrary, it could be a counterproductive exercise, creating the illusion of democratic involvement on paper and deception in practice.

According to Friedmann (1992), participation is more effective if it occurs early enough in the planning process. This kind of strategy can serve as a safety net to prevent the process from being co-opted by a particular political agenda. A practical example will help to clarify the concept. Once citizens become involved, the logical next step would be to expect that change will come to their local green place. This was a very sensitive topic among those citizens using My Green Place. More often than not, participants would directly address the researcher, asking when their park would be renovated or intervened. Although the researcher has actively worked to put local issues in the spotlight using a variety of resources, expectation management has to be addressed at the very start of any participation session. The researcher is responsible for always being clear with citizens participating in the study about the intentions of the research so that false expectations are avoided.

In the case of *landscape ownership* presented in the Greenmapper case study, the boundary between empowering mechanisms and deceiving practices is a blurry one. Without a clear (contractual) agreement between the landowner, the platform, and the share buyer, chances exist that the DPM's result remains limited to yet another way of collecting

[4] For example, the citizens of Boyle Heights (Los Angeles, US) created a grassroot organization precisely to prevent the cultural gentrification that goes under the label of *artwashing*. http://alianzacontraartwashing.org/en/bhaaad/.

funding from citizens. It adds innovation to the method's appearance for the simple fact that it uses digital means, but it does not transgress the traditional structure of the study. The researcher would still be able to objectify the results, the participant would remain the passive object of the study, and the outcome would simply be a one-way financial contribution to nature conservation.

Data Quality: Precision and Accuracy

Accuracy and *precision* are two key factors that have to be addressed to ensure the quality of collected spatial data and avoid possible bias. They rely on a series of factors from which choosing a suitable geographic entity is a critical one. A geographic entity's suitability to map certain attributes can compromise the collected data and any results coming from it if neglected. A clear example of this can happen when mapping dynamic activities such as running. Imagine people are asked to map places where they run using two different geographic entities: a point and a polygon. Although mapping the same item, results will be different since points are incapable of mapping beyond one place, whereas the polygon allows tracing better routes and complete surfaces. If participants were only given the point option, that would inevitably lead to biased data, faulty analysis, and mistaken conclusions.

Precision refers to how participants place a geographical entity (e.g., point, polygon) in a map, either printed or digital. *Precision* when mapping can be determined by factors such as the size of the object used for mapping (e.g., marker or a pen's width, stickers size), the map's scale, participants' visual capacity and dexterity (Brown & Pullar, 2012). The last two factors can weigh heavily in people who suffer an impairment or are advanced in age, causing a potential bias in this group (Gottwald et al., 2016). *Precision* is more significant when using physical data collection methods than digital ones. The number of open variables in the process, e.g., quality of the printed map, marker width or sticker size, and later processing to digital format, leaves much room for spatial error. Brown and Pullar (2012) discussed this issue and how particular types of marker colours, printed maps, and stickers can produce defective input, which has to be discarded since it is not up to quality for analysis.

When using DPM, current map services like Google Maps or open street maps that provide flexible map scales may (in theory) increase *precision* by allowing participants to adjust better the area where they will point out an attribute on their screens (Brown & Pullar, 2012). It is nonetheless important that the researcher guides the participants in the process of interacting with the maps. In the case of the Greenmapper, for example, data have often been collected through large online surveys. Considering the different geographical scales of the software (i.e., local, regional, national, global), it has been noticed that participants tend to select larger areas the bigger the scale is. Favourite places at the global level are several times larger than those selected at the local level. The possibility of improving precision through, for example, the zooming option is not well understood unless explicit guidelines are provided. This highlights a selection bias due to the scale that can easily distort results if it is not properly treated.

Accuracy can be determined by factors such as how well instructions are provided, the nature of what is being mapped, and participants' map literacy. Mapping specific attributes, especially those perceived as abstract, e.g., "pleasant view" or "relaxation" using a single point, can lead to an inaccurate representation of input as a direct consequence of using an unsuitable geographical feature for its capture. This brings to light the importance of choosing a suitable geographic feature for the particular attribute to be mapped to avoid this kind of bias. Recalling the case studies presented above, the My Green Place pilot in Ghent addresses this particular situation by comparing every participant's input for the same attribute using three different geographical features (point, polygon, marker). Similarly, the Greenmapper has recently introduced the possibility of drawing polygons instead of points.

When data collection for the "We Love Gent" pilot test was done, large groups of people gathering together in one place for any occasion was something feasible, safe, and granted. However, after the COVID-19 pandemic hit the world and restrictive measures were put in place from March 2020 onwards, gatherings of any type were unfeasible, at least physically. This meant that one of the advantages of doing PM physically (the capability of bringing people together, working, and discussing one topic) was temporarily lost. In this way, the fact that DPM allows remote data collection without gathering people in one place at a specific time is an asset to the current scenario. Furthermore, the quarantine's mental and physical effect on people worldwide has led many to increase their need for and appreciation of green open spaces. This has made many realize the value that DPM tools such as My Green Place can bring to people trying to find a pleasant piece of nature nearby them while engaging in the process of management and conservation.

Digital Tools: Complementary to rather than Substitutes for Traditional Methods

Intuitively, digital platforms give researchers new possibilities, including, but not limited to, increasing the number of respondents reached by a survey, the size of datasets, and improving the dissemination of results. However, a few aspects should be considered that may have counterproductive consequences. First, as we have shown above with the examples of *precision* and *accuracy*, increased quantity does not imply increased quality. On the contrary, it may have distortive effects on data reliability (Al-Salom & Miller, 2017). The researcher provides the first filter in the collection of data during, for instance, fieldwork activities. Substituting it with digital tools for large-scale surveys may lead to distortive effects. In fact, respondents may have fewer incentives to provide meaningful answers and may be exposed to a higher range of distractions (Hardré et al., 2012). In the case of research studies directly rewarding respondents (e.g., students participating in research studies as part of university courses), DeRight and Jorgensen (2015) show that poor quality data accounts, on average, for 10% of all digital surveys directly rewarding respondents. Similarly, participation in scientific data collection can be reduced to a mere utilitarian calculus, as it is common practice for Internet panel services to pay participants to respond. The risk of neglecting the quality of in-person data collection compared to web-based methods has to be considered.

Second, researchers using DPM as a creative method should also be aware that they are not an unequivocally time-effective tool. As mentioned above, the time spent on cross-examining and filtering large datasets of dubious quality can balance out the time saved on in-person interviews or other fieldwork activities. However, the perhaps greater limitation of digital technology is in the imperfect substitutability of data from digital tools and real-life exchanges. Particularly in the case of research methods for which communities' involvement is a core element, like DPM, we stress that the researcher's physical presence can determine the reliability and soundness of the study. In these cases, the study's technological or productivity-oriented character should not prevail on its anthropological attributes.

Third, digital tools have high start-up and running costs. For instance, an individual researcher is unlikely to have the skills and time needed to design, create, and maintain online platforms. These processes often involve research groups and outsourcing of services to consultancy companies or academic spin-offs. It is not a simple task to predict the value of the research results a priori, as well as being in contradiction with a rigorous scientific approach. However, assuming that the results and the digital tool itself will be shared with the scientific community (and beyond) may also make the effort of investing in a new digital tool worthwhile. The high costs and time effort required argue for sharing access to the digital tool to the broadest audience possible. This is in line with recent calls for *open-science* to use digital technology to democratize access to science instead of creating an elitist tool, i.e., accessible only for those in a financially privileged position.

Fourth, the funding structure of DPM is often challenging. Founders of projects using DPM frequently forget the importance of disseminating and maintaining the digital tool. Suppose the intentions are creating a successful tool. In that case, a DPM platform needs not only financial support for the development phase (which represents the most substantial part of the funding structure anyway) but also budget specifically allocated for reaching the target audience and for keeping the tool working overtime. The "We Love Gent" campaign effort to reach our target audience was made via the delivery of printed media such as flyers in concurred places across the city and invitations by email through the university network. Additionally, a couple of secondary schools shared the link and allowed us to talk to students about the project and the importance of people being engaged in managing their green places. Posting weekly invitations on Facebook groups, as well as sharing the link to the tool via WhatsApp groups from certain neighbourhoods and religious centres, also proved to be a good strategy to engage with more participants at relatively low cost.

However, in the second stage of the tool, the new campaign called "We Love Woluwe" has had to invest more in remote advertising due to the impossibility of reaching groups of people, e.g., in parks, stations, schools, or care homes due to the concurrent COVID-19 restrictions. The strategy includes paid advertising of the tool's link in social media such as Facebook and Instagram in the eight municipalities where the research is being conducted. Moreover, to address the older adult group, we mailed postcard invitations. Although the cost itself of sending a postal card via local Belgian post is relatively cheap (€0.09 per 25gr), the rental service for accessing the database of residents and filtering our target groups can vary from €500 to €5,000 or more, depending on the complexity of the filter and the group size.

Many older adults rarely have access to a computer or the knowledge to use it. Nevertheless, now more than ever, aging populations are turning to technology in mobile phones and tablets (AARP Research, 2020). Therefore, by reaching them with a postcard invitation to participate, they can do so via their mobile devices. Social media platforms offer a wide array of options and budgets to promote any product. Furthermore, depending on the research needs, it can hardly be a fraction of the price and time required to hire field researchers to do the collection manually.

Conclusions

The almost century-long tradition of participatory mapping is entering a new era. In the last decades, information technology has radically changed the field, mostly through the adoption of GIS techniques. Drawing on the review presented by Brown and Kyttä (2018), in this chapter we focused our analysis on digital practices of participatory mapping, or DPM. Digital technology is enabling citizens to increase their stake in the decision-making process. Although its adoption is not a cure-all, at the very least it gives scholars and practitioners an unprecedented opportunity to expand participation in the co-creation of spaces and planning policies. The solution to most of the challenges we face in the twenty-first century, from climate change to control of pandemics, requires spatial information on different scales. This is not to mention the democratic gap brought about by rampant increases of inequality and economic opportunities, both at the intra- and inter-national levels. DPM can be used to extend the scope of public policy. The concerns of the many can now partake in the decision-making process beyond the vested interests of the few.

However, as already mentioned, it is critical to acknowledge that DPM tools are not a panacea. For example, they can be an easy target for tokenism, as in *artwashing* practices. Practitioners and researchers may be incentivized to trumpet new participation methods, such as DPM, instead of making the participation process effective and influential. As Fraser (1997) suggests, we stress the relevance of considering *redistribution* of power together and beyond social *recognition*. The researcher is often not in a position of changing the composition of institutionalized power in the short term, and is surely not alone. This is not in itself a weakness of DPM. Nonetheless, it should be openly acknowledged not to generate false expectations. For DPM to be truly effective, it is important to avoid the paradox of "empowering without power".

Assuming technological progress to be neutral per se, both advances and downsides may accompany the adoption of digital tools. For instance, we have shown that data quality collected through DPM is not unequivocally better in terms of *precision* and *accuracy* than in person collection. While precision refers to the technical aspects that influence the mapping activity, accuracy identifies the subjective factors that influence the data's interpretation from both the participant and the researcher. Brown and Pullar (2012) show the relevance of assessing participants' visual capacity and dexterity, which we have explored in this chapter's two case studies. At the same time, the researcher or practitioner should also be aware of their own positionality. The way they give instructions, the subjective nature of what is being mapped, or the expectations that arise from participating in this process are likely to influence the study's accuracy.

Concerning the composition of DPM datasets, digital technologies represent a major improvement in information accessibility. However, while it may be a powerful tool to improve representativeness and engage with socially marginalized groups, a certain bias still exists concerning access to online and digital platforms. This is the case, for example, of older adults. The My Green Place case study shows how early recognition of the target audience and planning can help to mitigate this selection bias.

Digital tools are indeed a major improvement in mapping, but the examples above mark the importance of not throwing the baby out with the bath water. Instead of being perfect substitutes, DPM should be considered as a complementary tool to in-person activities. Fieldwork can mitigate the distortive effects of large-scale online surveys, such as recognizing problems of selection bias or inaccurate framing of the questions, e.g., not fitting the context-specific characteristics of the place (Al-Salom & Miller, 2017). For example, in the data collection phase, the researcher often provides a first filter that can significantly improve the study's soundness. In any case, participatory mapping is transdisciplinary as it rests at the intersection of different fields, from geography to humanities. The technologically driven performance of these new digital tools should not neglect their anthropological characteristics.

DPM and What Lies Ahead

Technological developments keep making GIS cheaper and more accessible. The potential of DPM tools will keep growing, and with it the possibilities to make participatory planning practices more democratic. DPM provides a bridge between citizens and public stakeholders that fosters transparency, accountability, and legitimacy (McCall, 2003). Therefore, there is little doubt that these tools will help to improve citizen participation in planning processes.

While writing this chapter, the COVID-19 pandemic has dramatically changed how we interact with others, at least for the time being. Standard participation practices, where face-to-face interactions were the norm, became suddenly unfeasible. This presented a challenge but also a great opportunity to test DPM capabilities. For example, data collection with My Green Place (in the "We Love Woluwe" campaign[5]) also continued during the pandemic thanks to its digital character. The same goes for the Greenmapper that has always been operating globally, collecting data from remote locations, even before most research was forced to move online.

DPM tools can provide means for legitimizing local demands by collecting quantitative and qualitative data (Poole, 1995) However, this is not enough to achieve multi-level governance and truly democratic participation (McCall, 2003). A paradigm shift is needed for solving the mistrust that many decision-makers still have in participatory mapping processes, such as DPM. We recognize that DPM tools alone cannot make participatory planning better or more influential for the future. This would require a joint effort from the bottom-up with the support of citizen organizations. At the end of the day, citizens are the main source of information feeding the decision-making process.

Social media and the Internet offer an easily accessible means of sharing information and propel causes across different scales. Notwithstanding a certain institutional stiffness, the world is changing. The radically new ways of living dictated by the digital era, climate change, and increasing demands for democratic representation will eventually sweep away old habits of "doing planning". Some researchers and planners are already aware of this trend, and they are adapting their approaches accordingly, together with the institutions they represent.

Acknowledgements This research received funding from the European Union's Horizon 2020 research and innovation programme under the Marie Sklowdoska-Curie Grant Agreement No. 765389.

[5] "We Love Woluwe" is the third iteration of the My Green Place tool. It was the result of the lessons learned in the "We Love Gent" campaign and the beta version. "We Love Woluwe" collected data from July 2020–March 2021 in the eight municipalities surrounding the Woluwe river, located between Brussels and the Flemish region.

References

AARP Research. (2020). 2020 Tech trends of the 50+. Available at: https://www.aarp.org/content/dam/aarp/research/surveys_statistics/technology/2019/2020-tech-trends-survey.doi.10.26419-2Fres.00329.001.pdf.

Allen III, W. L., & Christensen, J. (2001, July). Managing a land trust with geographic information systems. In Presentation at the Annual ESRI Users' Conference (pp. 9–15), San Diego, CA.

Al-Salom, P., & Miller, C. J. (2017). The problem with online data collection: Predicting invalid responding in undergraduate samples. *Modern Psychological Studies, 22*(2).

Aranda, N. R., De Waegemaeker, J., Venhorst, V., Leendertse, W., Kerselaers, E., & Van de Weghe, N. (2021). Point, polygon, or marker? In search of the best geographic entity for mapping cultural ecosystem services using the online public participation geographic information systems tool, "My Green Place." *Cartography and Geographic Information Science*. https://doi.org/0.1080/15230406.2021.1949392

Arnstein, S. R. (1969). A ladder of citizen participation. *Journal of the American Planning Association, 85*(1), 24–34. https://doi.org/10.1080/01944363.2018.1559388

Beatty, P., & Willis, G. (2007). Research synthesis : The practice of cognitive interviewing. *Public Opinion Quarterly, 71*(2), 287–311. doi: https://doi.org/10.1093/poq/nfm006.

Bhattacherjee, A. (2012). *Social science research: Principles, methods, and practices* (2nd ed.). Tampa.

Bijker, R., Mehnen, N., Sijtsma, J. F., & Daams, M. (2014). Managing urban wellbeing in rural areas: The potential role of online communities to improve the financing and governance of highly valued nature areas. *Land, 3*(2), 437–459. https://doi.org/10.3390/land3020437

Bijker, R. A., & Sijtsma, F. J. (2017). A portfolio of natural places: Using a participatory GIS tool to compare the appreciation and use of green spaces inside and outside urban areas by urban residents. *Landscape and Urban Planning, 158*, 155–165. https://doi.org/10.1016/j.landurbplan.2016.10.004

Brondizio, E. S., Ostrom, E., & Young, O. R. (2009). Connectivity and the governance of multilevel social-ecological systems: The role of social capital.

Annual Review of Environment and Resources, 34(1), 253–278. https://doi.org/10.1146/annurev.environ.020708.100707

Brown, G., & Fagerholm, N. (2015). Empirical PPGIS/PGIS mapping of ecosystem services: A review and evaluation, . *Ecosystem Services, 13*, 119–133. https://doi.org/10.1016/j.ecoser.2014.10.007

Brown, G., & Kyttä, M. (2014). Key issues and research priorities for public participation GIS (PPGIS): A synthesis based on empirical research . *Applied Geography, 46*, 122–136. https://doi.org/10.1016/j.apgeog.2013.11.004

Brown, G., & Kyttä, M. (2018). Key issues and priorities in participatory mapping: Toward integration or increased specialization? . *Applied Geography, 95*, 1–8. https://doi.org/10.1016/j.apgeog.2018.04.002

Brown, G., & Reed, P. (2009). Public participation GIS: A new method for national park planning. *Landscape and Urban Planning, 55*(2), 166–182. https://doi.org/10.1016/j.landurbplan.2011.03.003

Brown, G. G., & Pullar, D. V. (2012). An evaluation of the use of points versus polygons in public participation geographic information systems using quasi-experimental design and Monte Carlo simulation. *International Journal of Geographical Information Science, 26*(2), 231–246. https://doi.org/10.1080/13658816.2011.585139

Carton, L. (2005). How to cope with map controversies in deliberative policy making. *Mapping for Change (PGIS)*, Nairobi, September, (March 2002), pp. 1–24. Available at: http://www.iapad.org/publications/ppgis/linda_carton.pdf.

Chambers, R. (2006)/ Participatory mapping and geographic information systems: Whose map? Who is empowered and who disempowered? Who gains and who loses? *The Electronic Journal of Information Systems in Developing Countries*, 1–11. doi: https://doi.org/10.1002/j.1681-4835.2006.tb00163.x.

Church, A., Fish, R., & Haines-Young, R. (2014). UK national ecosystem assessment follow-on. Work package report 5: Cultural ecosystem services and indicators.

Cox, M. (2014). Applying a social-ecological system framework to the study of the Taos valley irrigation system . *Human Ecology, 42*(2), 311–324. https://doi.org/10.1007/s10745-014-9651-y

Cuba, L., & Hummon, D. M. (1993). A place to call home: Identification with dwelling, community, and region. *Sociological Quarterly, 34*(1), 111–131. https://doi.org/10.1111/j.1533-8525.1993.tb00133.x

DeRight, J., & Jorgensen, R. S. (2015). I Just want my research credit: Frequency of suboptimal effort in a non-clinical healthy undergraduate sample. *The Clinical Neuropsychologist. Routledge, 29*(1), 101–117. https://doi.org/10.1080/13854046.2014.989267

Dieber, M. (2003). Paint the town: Lessons learned. Presentation at the second annual conference on PPGIS, 20–22 July, Portland.

Dunn, C. E. (2007). Participatory GIS: A people's GIS?. *Progress in Human Geography*, 616–637. https://doi.org/10.1177/0309132507081493.

Fraser, N. (1997). From redistribution to recognition? Dilemmas of justice in a "post-Socialist" age. In *Justice interruptus. Critical reflections on the "Postsocialist" condition* (pp. 68–149). Routledge. doi: https://doi.org/10.1002/9780470756119.ch54.

Friedmann, J. (1992). *Empowerment: The politics of alternative development.* Blackwell Publishing Ltd. Available at: https://www.wiley.com/en-be/Empowerment%3A+The+Politics+of+Alternative+Development-p-9781557863003#download-product-flyer.

Fung, A. (2006). Varieties of participation in complex governance. *Public Administration Review, 66*(SUPPL. 1), 66–75. https://doi.org/10.1111/j.1540-6210.2006.00667.x

GIS Geography. (2020). The remarkable history of GIS where did GIS begin ? Stages of GIS development. Available at: https://gisgeography.com/history-of-gis/.

Gottwald, S., Laatikainen, T. E., & Kyttä, M. (2016). Exploring the usability of PPGIS among older adults: Challenges and opportunities . *International Journal of Geographical Information Science, 30*(12), 2321–2338. https://doi.org/10.1080/13658816.2016.1170837

Guldi, J. (2017). A history of the participatory map. *Public Culture, 29*(1), 79–112. https://doi.org/10.1215/08992363-3644409

Hardré, P. L., Crowson, H. M., & Xie, K. (2012). Examining contexts-of-use for web-based and paper-based questionnaires. *Educational and Psychological Measurement, 72*(6), 1015–1038. https://doi.org/10.1177/0013164412451977

Hidalgo, M. C., & Hernández, B. (2001). Place attachment: Conceptual and empirical questions. *Journal of Environmental Psychology, 21*(3), 273–281. https://doi.org/10.1006/jevp.2001.0221

Jacobs, J. (1994). *The death and life of great American cities.* Penguin Books.

Korpela, K. (2002). Children's environment. In Bechtel, R. & Churchman, A. (eds.), *Handbook of environmental psychology*. Wiley. doi: https://doi.org/10.1016/0307-904X(80)90124-9.

McCall, M. K. (2003). Seeking good governance in participatory-GIS: A review of processes and governance dimensions in applying GIS to participatory spatial planning. *Habitat International, 27*(4), 549–573. https://doi.org/10.1016/S0197-3975(03)00005-5

McCall, M. K. (2004). Can participatory-GIS strengthen local-level spatial planning ? Suggestions for better practice, Skudai, Johor, Malaysia.

Mccall, M. K., & Dunn, C. E. (2012). Geo-information tools for participatory spatial planning: Fulfilling the criteria for "good" governance? *Geoforum, 43*(1), 81–94. https://doi.org/10.1016/j.geoforum.2011.07.007

Novak, D. (2019). The arts of gentrification: Creativity, cultural policy, and public space in Kamagasaki. *City and Society, 31*(1), 94–118. https://doi.org/10.1111/ciso.12195

Obermeyer, N. J. (1998). Evolution of public participation GIS. *Cartography and Geographic Information Systems, 25*(2), 65–66. https://doi.org/10.1559/152304098782594599

Poole, P. (1995). Land-based communities, geomatics and biodiversity conservation. *Cultural Survival Quarterly, 18*(4), 1–4. Available at: http://www.iapad.org/publications/ppgis/landbased_communities_geomatics_and_biodiversity.pdf.

Rambaldi, G. (2010). *Participatory three-dimensional modelling: Guiding principles and applications* (2010th ed.). ACP-EU Technical Centre for Agricultural and Rural Cooperation (CTA). Available at: https://bit.ly/38NdkCe.

Renn, O., Webler, T., Rakel, H., Dienel, P., & Johnson, B. (1993). Public participation in decision making: A three-step procedure. *Policy Sciences, 26*(3), 189–214. https://doi.org/10.1007/BF00999716

Sieber, R. (2006). Public participation geographic information systems: A literature review and framework. *Annals of the Association of American Geographers, 96*(3), 491–507. https://doi.org/10.1111/j.1467-8306.2006.00702.x

Sijtsma, F. J., Daams, M. N., Farjon, H., & Buijs, A. E. (2012). Deep feelings around a shallow coast. A spatial analysis of tourism jobs and the attractivity of nature in the Dutch Wadden area. *Ocean & Coastal Management, 68*, 138–148. https://doi.org/10.1016/j.ocecoaman.2012.05.018

Sijtsma, F. J., Mehnen, N., Angelstam, P., & Muñoz-Rojas, J. (2019). Multi-scale mapping of cultural ecosystem services in a socio-ecological landscape: A case study of the international Wadden Sea Region. *Landscape Ecology, 34*(7), 1751–1768. https://doi.org/10.1007/s10980-019-00841-8

Statista. (2020). How many smartphones are in the world ? Number of mobile phone & smartphone users, Bankmycell. Available at: https://www.bankmycell.com/blog/how-many-phones-are-in-the-world.

Sulistyawan, B. S., Verweij, P. A., Boot, R. G. A., Purwanti, B., Rumbiak, W., Wattimena, M. C., Rahawarin, P., & Adzan, G. (2018). Integrating participatory GIS into spatial planning regulation: The case of Merauke District, Papua, Indonesia. *International Journal of the Commons, 12*(1), 26–59. https://doi.org/10.18352/ijc.759

Vezzoni, R., & Sijtsma, F. J. (2020). Multilevel governance in the digital age: Exploring online connectedness and willingness to pay for favourite local, regional, national and global nature places. Manuscript submitted for publication.

Williams, D. R., & Vaske, J. J. (2003). The measurement of place attachment: Validity and generalizability of a psychometric approach. *Forest Science, 49*(6), 830–840. https://doi.org/10.1093/forestscience/49.6.830

15

Living Labs: A Creative and Collaborative Planning Approach

Maria Alina Rădulescu, Wim Leendertse, and Jos Arts

Introduction

On a rainy afternoon in November 2020, in the midst of the coronavirus pandemic, a group of inhabitants of the Hegewarren area, located in the province of Friesland, in the northern part of the Netherlands, gathered in an online workshop; they were accompanied by residents from neighbouring villages and by representatives of water sports, nature conservation, and agriculture organizations active in the area. Despite

M. A. Rădulescu (✉) · W. Leendertse · J. Arts
Department of Spatial Planning and Environment, Faculty of Spatial Sciences, University of Groningen, Groningen, The Netherlands
e-mail: m.a.radulescu@rug.nl

M. A. Rădulescu · W. Leendertse
Rijkswaterstaat, Dutch Ministry of Infrastructure and Water Management, The Hague, The Netherlands

J. Arts
Northwest University, Potchefstroom, South Africa

A. Franklin (ed.), *Co-Creativity and Engaged Scholarship*,
https://doi.org/10.1007/978-3-030-84248-2_15

the lack of a physical environment, the team of facilitators warmed up the atmosphere with an ice-breaker that led to a screen full with virtual colourful sticky notes about each participant's feelings and expectations. Afterwards, the participants shared their knowledge and interests about the area, therefore making each other aware of their different perspectives. Over the following months, with help from the team of facilitators, experts in engineering and landscape architecture, the participants created, compared, analyzed, and refined alternative solutions for the future of the area. These alternative solutions, together with supporting arguments, were to be later presented to the decision-makers who would then decide which, if any, was both attractive and financially feasible.

This short vignette about the first co-creation workshop of the Hegewarren Living Lab provides a tangible example of the practice of co-creation in the water infrastructure and spatial planning domain (in this Chapter referred to as 'planning'). At the initiative of the Province of Friesland, the Hegewarren Living Lab was developed to provide a 'safe environment', in which different stakeholders could explore alternative future scenarios for the Hegewarren polder (Fig. 15.1), a low-lying, flat tract of peat meadow land that faces severe challenges such as soil subsidence, CO_2 emission, difficult water management of the quays and nature destruction. However, this is just one of the many examples of Living Labs (LLs) that emerge today as a promising planning approach for addressing 'tangled problems' through "experimentation on suitable scales and with multiple stakeholders" (Borgström-Hansson, as cited in McCormick & Hartmann, 2017, p. 4).

The growing adoption of LLs in the planning domain, and in the public sector as a whole, comes as a reaction to the many claims about the benefits of LLs that stem from the private sector, where they have been extensively used in open innovation (Chesbrough, 2003) and user-oriented innovation (von Hippel, 2005). According to Molinari (2011, p. 133) a LL can be considered as a multi-stakeholder platform "comprising different stakeholders, who perceive the same problem, realize their own respective interdependencies, and come together to agree on the best action strategies for solving it". LLs hold a promise for bringing to light innovative solutions for the numerous existing 'wicked'

Fig. 15.1 Panoramic view of Hegewarren polder (*Source ©Siebe Swart* https://www.siebeswart.nl/)

challenges (de Roo et al., 2012; Liedtke et al., 2012; Zivkovic, 2018) that communities face. However, their potential as a planning method has only recently started gaining attention. Consequently, and given the fundamental differences between product—and planning-oriented LLs, "the conceptual and methodological understanding of living labs remains focused on technology-based innovation processes rather than socio-spatial research questions" (Franz, 2015, p. 55).

Other researchers (Bergvall-Kåreborn & Ståhlbröst, 2009; Følstad, 2008) have argued that current theories and methodologies, methods and tools, as well as analyses and reflection on LL practices are limited. This is further substantiated by Leendertse et al., (2016, p. 403) who stated that "literature on actual implementation and experiences in a project context is very scarce". Reflecting these arguments, Rosado et al., (2015, p. 181) argued that there is a:

> "need for more specific descriptions of the practice of running a living lab, i.e. how to organize a living lab's activities, how to involve different

> stakeholders, ways of collaboration, co-ordination, etc., combined with a more conceptual concern with the possibility of reconciling the interest of these different stakeholders".

Therefore, in this Chapter we aim to position LLs as a creative and collaborative planning method. To elaborate LLs as a planning method, we first provide a theoretical overview of LLs, looking at their interpretation, their characteristics, and typology. This chapter is based on a review of relevant literature in the field of LLs and co-creation. The following section discusses LLs as a planning method, and is based on a literature review, empirical research in relation to the planning of water infrastructure and spatial development projects that adopted a LL or a co-creative approach, as well as our own experiences in observing and joining LLs.

While striving for clear steps and a flowing text in this section about LLs as a planning method, we also wanted to provide empirical substantiation. Therefore, we have chosen to use text boxes to illustrate the practice-oriented aspects of using LLs as a planning method. These text boxes provide illustrations from the empirical research we carried out in three projects in the Netherlands—the Overdiepse polder, the Essenburg Park from Rotterdam, and the Hegewarren LL. The latter is an ongoing project, in which the first author conducted participant observation over a period of one year. This method helped to gain a deep insight perspective into the LL's organization and procedures, to understand the evolution of the process, the roles of the different stakeholders and the way their interactions shaped the process; it also laid the foundation for subsequent interviews. As a basis for this case, a rich source of information consisted of the 'thick description' of the meetings and project documentation. The first two case studies were based on an analysis of planning documents and 15 in-depth, semi-structured interviews with key players. In the final sections, we reflect on the use of LLs as a planning method and conclude with recommendations for the application of LLs as a method in the water infrastructure and spatial planning domain.

Theoretical Aspects of Living Labs

Given their increasing popularity in various fields, both in the private and the public spheres, the concept of LLs has now morphed into a buzzword. To the proliferation of LLs in the European context have contributed the many streams of funding that encourage, or even demand the application of such an approach (Voytenko et al., 2016). Furthermore, the growing appeal of LLs stems from their use and study environments, various forms of experimental governance, user-centric research methodologies, multi-stakeholder platforms and collaborative experimental approaches. In theory, a LL resembles an almost 'magical concept' (Pollitt & Hupe, 2011), that fills a gap by contrasting the traditional, siloed and expert-driven approaches that are no longer deemed suitable as a response to the complex and 'wicked challenges' of our society.

The Origins of Living Labs

One of the first uses of the concept of a 'living laboratory' was by Bajgier et al., (1991, p. 701) to describe the potential of urban neighbourhoods as learning environments for students who are interested in solving real-world issues. Later it was further developed by William J. Mitchell from MIT Media Lab who was interested in investigating the application of smart home systems in day-to-day human activities (Eriksson et al., 2005). Subsequently, the concept has spread rapidly all over the world and gained popularity in various domains as a new innovation approach—see Box 15.1.

Box 15.1 Early example of a LL as a planning method.

In 1993, Rijkswaterstaat introduced the Infrastructure Laboratory (InfraLab), described as an approach to experiment with interactive and open planning procedures aimed at a speedy and creative development of new and innovative solutions for infrastructure projects. In the InfraLab, traffic planners worked directly with user communities and other stakeholders to define transport problems and their solutions (Evans et al., 1999; van den Brink, 2009; Woltjer, 2000).

"In Europe, the concept attracted interest and led to a number of scattered experimentations" (Dutilleul et al., 2010, p. 63). A milestone was reached in 2006 when the concept was officially introduced during the Finnish presidency of the European Union (EU) through the Helsinki Manifesto. In the same year the European Network of Living Labs (ENoLL) was founded, a "European platform for collaborative and co-creative innovation, where the users are involved in and contribute to the innovation process" (European Commission, 2006, p. 4). Seen as a starting point for "a new European R&D and Innovation System, entailing a major paradigm shift for the whole innovation process" (Molinari, 2011, p. 131), this represented the approach taken to tackle Europe's declining economic competitiveness and increasing societal challenges (Dutilleul et al., 2010). The widespread emergence of LLs in a large variety of domains is reflected in the evolution of ENoLL, which initially consisted of 19 LLs from 15 EU member states and today has over 150 active LL members worldwide. However, the popularity (and fuzziness) of LLs is also emphasized by the numerous definitions and applications, which will be explored in the following sub-sections.

What are Living Labs?

Living Labs can be included in the larger category of real-world laboratories (Schäpke et al., 2018) together with other types of experimental approaches such as urban living labs, design labs, city labs, smart city initiatives, innovation hubs, community-based initiatives, social innovation labs and other niche experiments. Given the relative novelty of the concept, the numerous applications it has in practice, and the various perspectives that are taken to research it, there is no widely accepted definition of a LL (Leminen, 2015a). Based on our literature search, Table 15.1 presents the most relevant definitions of water infrastructure and

Table 15.1 Overview of LL definitions relevant to the water infrastructure and spatial planning domain

References	Living Lab definitions	Key characteristics
Frissen and van Lieshout (2004)	"Consciously constructed social environments in which the uncontrollable dynamics of everyday life are accepted as part of the innovation environment which enables designers and users to co-produce new products and services"	Environment Innovation Users Co-production
Ballon et al., (2005, p. 3)	"An experimentation environment in which technology is given shape in real life contexts and in which (end) users are considered 'co-producers'"	Environment Technology Real-life context Co-production
Lama and Origin (2006, p. 6)	"User-centric research methodology for sensing, prototyping, validating and refining complex solutions in multiple and evolving real life-contexts"	Research methodology User-centric Real-life context
Bergvall-Kåreborn et al., (2009, p. 4)	"User-centric innovation milieu built on every-day practice and research, with an approach that facilitates user influence in open and distributed innovation processes engaging all relevant partners in real-life contexts, aiming to create sustainable values"	Milieu User influence Open innovation Real-life context
Pallot et al., (2010, pp. 2–3)	"An open innovation ecosystem frequently operating in the context of competitiveness clusters and public development agencies within social innovation environments engaging local authorities in territories such as cities, agglomerations, regions"	Ecosystem Open Innovation Social innovation Geographical territories

(continued)

Table 15.1 (continued)

References	Living Lab definitions	Key characteristics
Westerlund and Leminen, (2011, p. 20)	"Physical regions or virtual realities, or interaction spaces, in which stakeholders form public–private-people partnerships (4Ps) of companies, public agencies, universities, users, and other stakeholders, all collaborating for creation, prototyping, validating, and testing of new technologies, services, products, and systems in real-life contexts"	Physical regions/virtual realities/interaction spaces Public–private-people partnerships Collaboration Real-life context
Dell'Era and Landoni (2014)	"A design research methodology aimed at co-creating innovation through the involvement of aware users in a real-life setting"	Research methodology Co-creation Innovation Real-life setting
European Network of Living Labs (2015)	"User-centred, open innovation ecosystems based on a systematic user co-creation approach integrating research and innovation processes in real life communities and settings"	Ecosystem User-centred Open innovation Co-creation Innovation Real-life setting

spatial planning practice.

Table 15.1 substantiates the findings of Leminen (2015a) who identified that there are three layered streams of LL studies: LLs as a context, LLs as a method, and LLs as a conceptualization. Accordingly, this chapter strives to contribute to the second stream of studies and, based on the key characteristics of LLs, we define a LL as an iterative, experimental and user-centric planning method in which multiple stakeholders co-create innovative solutions for planning issues (see also the keywords in Table 15.1).

Unsurprisingly, the numerous definitions and wide-ranging utility of LLs indicate not only their versatility, but also a main drawback of the LL concept. The many definitions and interpretations are not necessarily a bad thing in itself as it highlights the multiplicity of approaches for dealing with various kinds of challenges. However, a problem is that LLs have come to mean many different things to many different people, from various domains. Consequently, in the search for consistency and unifying features, the attention of academics and practitioners moved towards identifying the key characteristics of LLs.

The first characteristics stem from the term itself. 'Lab' comes from laboratory, and even though in the case of planning it does not refer to a traditional type of laboratory—with chemical substances, test tubes, funnels, and varied types of flasks—but refers to a *real-life setting* in which *experimentation* (such as innovative integrated spatial designs) is encouraged, and where room for failure is provided. Another key feature of LLs is the involvement of *multiple stakeholders,* from public institutions, private stakeholders, academics and research institutions, NGOs, individuals and groups of citizens. From this inclusive engagement derives the core feature of LLs, *co-creation*; this refers to "any act of collective creativity that is experienced jointly by two or more people [...] where the intent is to create something that is not known in advance" (Sanders & Simons, 2009, p. 27), and which "highlights the potential impact of collaborative interaction on the ability to foster new and innovative solutions to intractable problems" (Puerari et al., 2018, p. 804). Consequently, two other main features of LLs, which result from the multi-stakeholder collaboration, are *innovation* and *learning*.

The aforementioned characteristics are becoming increasingly important in identifying LLs because there are many examples of collaborative initiatives that do not use the terminology, but fulfil all the criteria of a LL (IIIEE at Lund University, 2015). In the case of such initiatives, the extent to which each of the main characteristics is found may vary. However, to be considered a LL, each of the features should at least be present to some extent—see Box 15.2. This flexibility in terms of the degree of specific characteristics allows for a wide variety of LL and a need for the development of typologies based on different aspects.

Box 15.2 The case of Western Harbour in Malmö.

The Malmö western harbour project, which set itself the very ambitious goal of being a 100% energy renewable neighbourhood, does not wear the LL name, but fulfils all its criteria: it took place in a real-life context, it had numerous design competitions that demanded experimentation, exploration and entrepreneurship, and used collaborative working methods (IIIEE at Lund University, 2015).

For example, Leminen et al. (2012) proposed four types of LLs based on the type of stakeholder who drives the activities and plays the most active role in the innovation process: *the utilizer-driven, the enabler-driven, the provider-driven* and *the user-driven LL*. Similarly, when considering LL as an environment, Ståhlbröst and Holst (2012, p. 6) identified five main types:

> "*research LLs* focusing on performing research on different aspects of the innovation process; *corporate LLs* that focus on having a physical place where they invite stakeholders to co-create innovations; *organizational LLs* where the members of an organization co-creatively develop innovations; *intermediary LLs* in which different partners are invited to collaboratively innovate in a neutral arena; *a time limited LL*, as a support for the innovation process in a project".

The latter refers to the situation in which the LL closes when the project ends.

An interesting categorization is proposed by Neef et al. (2017), who identified two main types: *Product Oriented Labs,* which stem from the open innovation paradigm and where the main goal is innovation, and *Urban Transition Labs,* which stem from the transition management paradigm and where the main goal is to facilitate a transition in, for example, sustainability, the lab being considered a niche in inducing transitions (Geels, 2005).

A relevant typology of LLs related to the planning field is proposed by Marvin et al. (2018) who defined three ideal types of Urban Living Labs: *strategic, civil* and *grassroots.* Extrapolating this typology to the more generic category of LL in the water infrastructure and spatial planning domain, we consider the following three types of LLs: 1. *strategic*, which are led either by the national government or by large private actors and operate on a large scale, sometimes with multiple projects under one umbrella; 2. *Civic*, which are led by regional or local authorities, higher education and research institutes or local companies, and focus on economic and sustainable development on the regional and local scale; 3. *Grassroots*, which are led by members of civil society, communities, NGOs, or groups of residents, and focus on specific issues through micro-projects or single issues projects (Marvin et al., 2018; McCormick & Hartmann, 2017).

Box 15.3 LL typologies reflected in practice.

An example of a *strategic LL* is SmartwayZ.NL, an umbrella programme for eight sub-projects, in which the Dutch Ministry of Infrastructure and Water Management, Rijkswaterstaat, the provinces of North-Brabant and Limburg, numerous municipalities, companies and knowledge institutes work together towards improving accessibility and promoting innovation in the field of mobility.

An example of a *civic LL* is the Hegewarren LL, initiated by the Province of Friesland, the waterboard and the Smallingerland municipality, and uses the Hegewarren polder as a lab for exploring different future alternative scenarios in response to the spatial and environmental challenges that peat areas are facing.

An example of a *grassroots LL* is the Essenburg Park project from Rotterdam; this started with a group of inhabitants who wanted to improve the sustainability prospects of the neighbourhood and prevent the build-up of an old railway area by transforming it into a publicly accessible green space with a natural water retention area. Initially, this started as a civic initiative, but it later gained the support of the Rotterdam city council and municipality, of the waterboard and of local health and educational institutions, all of whom worked together on defining and implementing the development plan, and who still continue to work together on the maintenance of the park.

From the increasing number of studies related to LLs, there seems an increasing trend in considering LLs as a 'magic recipe' for experimentation and development of innovative and creative solutions for the numerous environmental and societal challenges that communities are facing. However, beyond their attractiveness, LLs pose many practical and operationalization challenges because "a wide variety of activities are carried out under the umbrella of living labs, and they feature many different methodologies and research perspectives" (Leminen, 2015b, p. 29). Therefore, more attention needs to be given to the practical aspects of LLs as a planning method, because they play an important role in practicing co-creation and experimentation with multiple stakeholders. For this reason, the next section describes the method itself, including the 'ingredients' and the 'how' aspects of a LL as a planning method. We look at the 'living lab way of working' (Steen & Van Bueren, 2017) by emphasizing the phases of a LL, the conditions that make a LL successful, and the main roles played by a LL's stakeholders.

Living Labs as a Planning Method:

There are numerous studies that propose different stages of a LL. For example, in relation to ICT design, Pierson and Lievens (2005) proposed that an LL has four phases: contextualization, concretization, implementation, and feedback. Another LL staging is proposed by Malmberg et al. (2017) who identified three main phases: exploration, experimentation and evaluation. A more detailed explanation of a 'living lab way of working' was provided by Steen and Van Bueren (2017), who identified eight phases of a LL: initiation, plan development, co-creative design, implementation, evaluation, refinement, dissemination, replication.

Based on the 'living lab way of working' proposed by Steen and Van Bueren (2017), and the insights emerging from our empirical research of the three cases, we propose the following five phases when using LLs as a planning method: initiation, preparation, co-creative design, evaluation and link with decision-making, and feedback. In the sub-sections below we will refer to the empirical base by giving illustrations from the cases studied.

Phase 1: Initiation

Start from an idea or a problem

The adoption of LL as a planning method is usually triggered by complex, tangled problems that cannot be solved with the traditional, siloed approaches (see Box 15.4), but that demand cross-level and cross-sectoral collaborative approaches that show "explicit appreciation of complexity and uncertainty, likelihood of surprise and need for flexibility and adaptive capacity" (Kemp et al., 2005, p. 17). According to Steen and Van Bueren (2017), not only a problem, but also an idea can trigger the adoption of a LL approach.

Box 15.4 Example of triggers for a LL initiation.

Being situated in a peat area, the starting point of the Hegewarren LL was a mix of tangled problems: soil subsidence, CO_2 emissions, difficult water management, and nature destruction, for which solutions can only be explored through cross-sectoral and cross-level collaboration.

In the Essenburg Park project from Rotterdam, the trigger of the co-creation process was the neighbourhood residents' idea of transforming the old railway area into a publicly accessible park with a natural water retention area.

- **Attract others to work together**

In the initiation phase, a key role is played by the initiator, the person or organization who identifies a problem or comes up with an idea. The initiator is usually a public or private organization in the strategic and civil LL, and an individual or a group of individuals in the case of the grassroots LL. In this phase, the initiator makes the problem known to other potential key stakeholders—public actors, private actors, citizens or groups of citizens, and knowledge institutions—with the aim of gaining their support for a collaborative approach, for adopting a LL method, and for creating a partnership that has the capacity to set up the LL (see Box 15.5). The initiator needs to make sure that key stakeholders, which are usually also those that bring various kinds of resources to the LL in the later stages, show a high degree of commitment towards a 'LL way of working', which involves a high degree of openness, transparency, and trust. The persons or organizations that show interest in the initiative, even if not interested in being directly involved, can play the role of 'advocates' who support and spread the word about the initiative. When the support of key stakeholders for a LL approach is gained, the process moves to the preparation phase.

Box 15.5 Sparking connections with partners.

In the Essenburg Park case, the civic initiatives first gained the support of different neighbourhood actors (the Delfshaven borough, the Recreation and Sport department) and they later approached and gained the support of the coalition of political parties that wanted to create a new park in Rotterdam. Furthermore, they established informal interactions with the civil servants and the municipal councillors by inviting them to take a walk through the area to experience it.

In the Overdiepse polder project, the farmers from the polder first gained the support of the provincial deputy who manifested visionary leadership and decided to give them the space to come up with a different kind of solution from the one proposed by the government; further, it was essential that they had the support of an informal government group composed of high-level politicians, searching for exemplary projects that could put the new water policy into practice.

Phase 2: Preparation

- **Identify and select participants**

In the case of the grassroots LLs, which have a strong bottom-up nature and emerge from a particular community in response to a very local problem, the participants do not usually need to be 'recruited', but 'naturally' join the LL when they hear about it from their neighbours, colleagues, friends or family, and if they resonate with the problem/idea and feel that they can make a contribution.

By contrast, in the case of the strategic and civic LL types, the initiator needs to identify and assemble the network of potential LL participants. They do this together with the initial 'allies', who have "the vision, the energy, and the social skills to connect to diverse individuals and groups" (Krebs & Holley, 2004, p. 48) and act as 'webbers' (Heikkinen et al., 2007) or 'network weavers' (Hagman et al., 2018; Krebs & Holley, 2004). A way to identify the relevant web or network is to undertake a stakeholder analysis to identify the 'target communities', the stakeholders who are affected or have an interest in the problem (Gouillart & Hallett, 2015; Steen & Van Bueren, 2017).

The identification of the potential LL participants needs to be done by adopting an inclusive approach, therefore ensuring their diversity in terms of skills, knowledge, and resources; this is thought to be an essential condition for fostering creativity and innovation through interdisciplinary interaction. The identification of the potential LL participants can be done in various ways; one of these is by brainstorming about the different types of stakeholders and then grouping them in categories—as exemplified in Box 15.6.

Box 15.6 The identification of LL participants.

In the Hegewarren LL, the Province of Friesland, the waterboard and the Smallingerland municipality as initiators, undertook a stakeholder analysis and identified the stakeholders from the area, those from the vicinity areas, but also those that had an interest in the area. The identification was done by brainstorming about the different relevant actors for the area; these were then grouped into categories such as political actors, decision-makers, agriculture actors, recreation and tourism actors, water sector actors, nature conservation actors.

Particularly in the case of the strategic or civil LLs, after a list of potential LL participants is created, discussions about the participation space in relation to how many stakeholders can be included, which participants will be invited, what roles will they play, and what will they bring to the LL, can begin. An interesting technique to provoke such discussions and to make the step from the stakeholder analysis towards a participation strategy, is the 'rings of influence' model—see Fig. 15.2—which was also used in the Hegewarren LL case. It is a bulls-eye diagram that consists of four quadrants referring to four categories of actors—influencers, decision-makers, end-users, and suppliers—and of four concentric circles, each referring to the degree of the actors' involvement. This diagram can be used for three different types of analysis aimed at identifying the role that actors currently have, the role that they may want to have in the LL, and the role that the initiator would like them to have.

An important aspect to be taken into account is related to the number of participants. Although participants' diversity and inclusivity are desirable because it can enhance the 'collective creativity', a too large group of participants may lead to a less effective co-creation process. For this reason, filtering the potential participants is essential, since "getting the right people and the right chemistry is more important than getting the right idea" (Catmull & Wallace, 2014, p. 74).

- **Create the LL core – process design & management structure**

In addition to identifying the potential LL participants, the preparation phase is essential because then the initiator, who usually takes the role of the manager, together with key partners needs to identify the goals of the LL, identify the key resources and skills needed, create a working plan, envision the division of roles and responsibilities among key stakeholders, and design a management and communication system that will

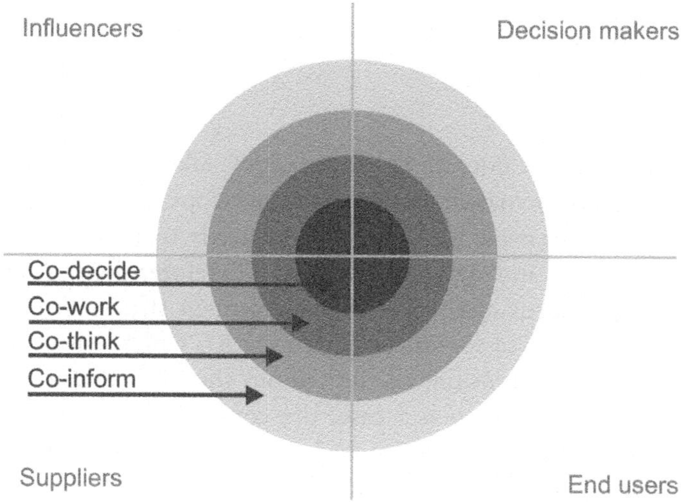

Fig. 15.2 Rings of influence model (*Source* Adapted with permission from www.publiec.nl)

allow the transparent spread of information and open dialogue within the LL. For all of this, the initiator and the key stakeholders need to dedicate enough time, because "whereas product design is self-evident in innovation processes, the design of the process is often forgotten, even though this activity proves crucial for the LL activities" (Steen & Van Bueren, 2017, p. 40). Based on the empirical research conducted, we observed that the design and management of LLs need to be flexible as they can be influenced by many factors, both internal and external—see Box 15.7.

Box 15.7 Flexibility in the face of external conditions – the case of the Hegewarren LL.
Flexibility needs to be a key element in the design and management of a LL, because there are many conditions, both internal and external, that can influence its evolution. For example, the planning of the Hegewarren LL started at the end of 2019, but the COVID-19 pandemic determined the transformation of the LL into a digital one, with exclusively online

meetings between the management 'layers' and with online co-creative workshops.

While in the grassroots LL the stakeholders take on roles and responsibilities and the management structure develops more organically; in the case of the strategic and civic LL, it needs to be designed in the preparation phase and develop onwards. Therefore, depending on the complexity and focus of the LL, its management structure may consist of different layers (see for example Box 15.8), which ensure the division of roles and responsibilities throughout the LL's existence (Steen & Van Bueren, 2017). In such a case, the communication arrangements between these different management layers is very important; they need to be constantly updated in line with developments in the LL. A key role in communications is played by the core team members, and especially by the initiator, who are part of all the management layers, therefore ensuring the dual-flow of information.

Box 15.8 Management structure.

In the Hegewarren LL, the province of Friesland, together with the waterboard and the Smallingerland municipality were initiators. The LL management structure consisted of a core team made of actors from the province of Friesland (the lead actor), an extended team made of the core team actors, and further actors from the province, from the waterboard, and the Smallingerland municipality, and an advisory team comprising LL and co-creation practitioners and researchers.

While the grassroots LL tends to have a very low degree of formalization, in the strategic or civic LL, where multiple public or private organizations:

> "are involved, there should be agreements in the form of 'contracts' that clearly specify the roles, tasks, and responsibilities are desirable as this brings clarity, raises institutional commitment and willingness to cooperate, whilst eliminating possible disagreements about responsibilities" (Rădulescu et al., 2020, p. 15).

In the case of strategic and civic LLs—which tend to be more techno-centric due to their more top-down, expert-led initiation (Garavaglia, 2020)—it is crucial for both the evolution and the outcomes of LLs to alleviate potential power asymmetries and to create a 'safe environment' for all participants. A way to deal with such aspects is by having the design and delivery of the co-creation activities to be carried out by an independent team of facilitators (Steen & Van Bueren, 2017). For this reason, in the preparation phase the management should 'bring in' such professionals, and provide them with enough information and creativity space to successfully craft the backbone for the co-creative design phase.

Phase 3: Co-Creative Design

The co-creative design phase is the central part of the LL methodology. The length of this phase may depend on the complexity of the problem or idea that triggered the LL's inception.

- Plan and design the co-creative sessions

The number of co-creative sessions depends largely on the scope of the LL, on the complexity of the problem, but also on the number of participants. However, a series of co-creative sessions usually starts with a kick-off meeting; this is essential for making the problem clear to all participants, for communicating pre-defined conditions, for discussing the co-creation process, and for collaboratively defining the ground rules that will guide the co-creation activities—see Box 15.9. Defining the ground rules collaboratively is very important because it helps to bring all the participants to agree on a set of shared values and modes of interaction.

Box 15.9 Co-creation workshops in the Overdiepse polder.

The farmers from the Overdiepse polder took part in a series of co-creative sessions for creating alternative plans for the polder. The co-creation sessions were designed and facilitated by an independent team of facilitators. The co-creation sessions had a results-oriented approach, so they were focused on doing, rather than on talking, and each focused on a specific subject: how to treat the people who wanted to leave the polder, the damage and compensations in case of high water levels.

The subsequent co-creative sessions can take many forms, such as workshops, design charettes or brainstorming sessions, in which stakeholders collaboratively and interactively come up with ideas, construct alternative scenarios, and engage in discussions about their potential benefits and challenges. Co-creation in water infrastructure planning and spatial development usually requires technical knowledge about the technical design of waterways and business models. For this reason it is good to create special sessions in which professional experts can offer detailed information to the participants. However, the complex technical aspects need to be synthetized and explained in plain language so that everyone can understand and follow the discussions. In the end, the co-creative design phase usually results in commonly ideated, designed, and supported alternative plans for solving the issue(s) that triggered the LL. This happens when the activities carried out in this phase transform potential conflicts or divergent perspectives of the participants into a joint and shared vision—see also Box 15.10.

Box 15.10 Co-creative design phase in the Hegewarren LL.

The start of the Hegewarren LL co-creative design phase was marked by a digital kick-off meeting (whose recording can be viewed here: https://www.youtube.com/watch?v=90uYa7Y_wa0) in which the goal of the LL and the way of working in a LL were explained, and questions could be asked. Based on invitations, but also on the reactions and interest shown by the participants that were present in the kick-off meeting, 17 participants, representing inhabitants and neighbours of the Hegewarren polder, and cross-regional stakeholders, were selected to take part in the co-creation process.

In the first workshop, the participants were presented to each other and they introduced their perspectives and interests regarding the area; in this way, 'local' knowledge and initial ideas about the area were collected, and each participant could get a grasp of the others' interests and perspectives. In the second workshop, the initial ideas were translated, with the help of a professional team of urban planners, into building blocks, which could then be used to construct five alternative scenarios. In the next workshops, ideas were further elaborated, combined and developed with the help of professional experts, architects and engineers, therefore leading to the formulation of five development scenarios for the area. In addition, the LL participants were offered more in-depth knowledge about relevant themes (e.g., water management, recreation and tourism, nature, agriculture) through a series of lectures given by experts.

Furthermore, an intermediary evaluation step was embedded towards the end of the co-creative design phase, therefore allowing for supplementary input and refinement of the future scenarios for the polder.

Communication was realized through a website (https://toekomsthegewarren.frl/), periodical newsletters and informal discussions between the participants and the facilitators and professional experts.

The co-creation sessions are usually (perceived as) intensive and can last 3–4 h. In planning the sessions, therefore, attention also needs to be paid to details such as the location and the layout of the room. In strategic or civil LLs, even when the initiator has enough meeting or conference rooms available at its headquarters, a neutral location is preferable so that power asymmetries are not further enhanced and a 'safe environment' is created. For the same reasons, supporting participants' equality throughout the process is essential and to this aim the setting of the co-creative sessions can play an important role. According to Haataja et al., (2018, p. 40), "a functional way to communicate equality is to position the participants in an open circle", maybe with everyone sitting at the same table, including the facilitators, therefore having no physical divisions between the participants.

- Perform 'temperature checks'

The co-creative sessions and their evolution can turn out to be unpredictable for both the facilitators and the participants. For this reason, the facilitators need to perform regular 'temperature checks' during the co-creative sessions to assess how the participants feel about the process and see if there are things that need to be done differently (monitoring). Such moments also create opportunities to enhance the feeling of trust between the participants and the facilitators. The 'temperature checks' can take the form of interim evaluation moments embedded in the co-creative design phase; this may lead to iterative loops and to the refinement of both the co-creation process and the co-created 'product'.

- Adopt a flexible attitude

Trying to facilitate and foster innovation and performing 'temperature checks' may bring uncertainty to the process. Consequently, while a thorough planning of the co-creation process must be in place, the facilitators, the experts and the participants must adopt a flexible attitude because activities in a LL do not follow a clearly defined path, and creativity comes with some degree of uncertainty. In addition, the initiator, the facilitator, the experts, as well as the key stakeholders, need to be highly sensitive to the evolution of the process, and be prepared to dedicate more time and resources to this phase if needed. At the same time, they need to openly communicate these aspects with the participants, especially because their participation in the LL is on a voluntary basis. Therefore, it is not only the 'product' of the LL that is co-created, but also the process.

- Communicate openly and transparently

Throughout this phase, but also throughout the entire LL process, open and clear communication between the LL participants, the management and the team of facilitators and experts is essential. For this reason, a communication system needs to be created and clearly made known to all those involved—for example, in the form of a website or periodical newsletters that keep track of the LL progress. Furthermore, enough opportunities need to be created for more informal, small-scale discussions, therefore ensuring that all participants' ideas and wishes are heard. Essential for communications in planning-related LLs is that the organizer and facilitators keep in mind that the participants are usually representatives of a much larger group—more formally (as representative of a community organization, NGO, etc.) or more informally (as a 'proxy' of the larger group of residents, farmers, etc.). Communication about the project and planning process, as well as formal participation processes, require careful attention so that engagement with the larger community and stakeholder groups evolves well. In addition, not all aspects discussed within the LL sessions can be communicated to the larger group. Therefore, agreements about the confidentiality of specific aspects need to be explicitly agreed on with the representatives at the start of the process. This is in order to prevent potential tensions and conflicts that may lead to mistrust and may spoil the creative mindset.

Phase 4: Evaluation and Link with Formal Decision-Making

"Evaluation is a core component of the LL approach" (Steen & Van Bueren, 2017, p. 66) that marks the end of the actual 'doing' in the LL. Despite its importance, evaluation is considered a very vulnerable part of the LL method; it usually receives less attention than the preceding co-creative design phase, and often it is not done (Verhoef & Bossert, 2019) because carrying out evaluations is seen as a challenging and time-consuming task. Evaluation is essential for reflecting not only on the 'product' of the LL, but also on the process. This helps those involved to internalize the experience of being part of a LL and transform it into a resource that can be used in similar future planning situations.

Furthermore, in LLs related to the planning domain, evaluation acts as a linking pin with the decision-making process that can lead either to the formal blending in of the LL 'product' and therefore to its development and implementation, or to its failure to gain political support.

Phase 5: Feedback

The feedback moment officially marks the end of a LL process. At this point, the team of facilitators, together with the initiators of the LL, need to arrange a last meeting with the LL participants to communicate what has happened with the LL's 'product' in the decision-making process. No matter what the formal planning decision is taken, the arguments, or any considerations that led to it, need to be clearly and transparently communicated to the participants. In this way, potential frustration or disappointment in the case of a 'negative' decision can be better dealt with and can be delimitated from the perceptions about the co-creation process per se. In this way, trust and enthusiasm for other similar processes is not diminished—preventing disillusionment and 'participation fatigue' among stakeholders (Esteves et al., 2012; Hamersma et al., 2018).

Reflections on the Use of Living Labs as a Planning Method

LLs are increasingly gaining attention in the planning domain, but their application comes with great challenges due to the nature of the field, where intricate dynamics play out, resulting from the interactions of the multiple levels, sectors, and actors involved. From our experience, this is especially visible in water infrastructure planning, where neither top-down nor bottom-up approaches are able to capture and respond to the complexity exposed by water—which "is not a single, discrete aspect of the environment. It is part of a greater interconnected whole; when one considers water, therefore, one must consider all that to which water is connected and related" (McGregor, 2021, p. 155). Therefore, while LLs as a planning tool are expected to highlight "the potential impact of collaborative interaction on the ability to foster new and innovative solutions to intractable problems" (Puerari et al., 2018, p. 804), in practice their application needs more reflection, especially because as we mentioned in the introduction, the application of LLs is often focused on technological-based innovation rather than socio-spatial issues. In the following part, based on our experiences in LLs in the planning domain, we try to indicate some key points that need attention when adopting a LL approach in planning.

First, there are many cases in planning where a LL is not a planning method chosen at the outset, but more a method that organically emerges as a reaction to the top-down initiative of an actor, situated on a higher level, which is perceived by the others as a threat—as with the cases of the Essenburg park or the Overdiepse polder. Looking back at the different cases, the initiation of a LL, both in terms of context and actors, defines the type of LL that it is going to be: strategic, civil or grassroots.

Furthermore, the initiation and evolution of a LL depend on a series of contextual conditions. For example, in the case of the Hegewarren LL, its initiation was triggered by the problems caused by the peat soil; however, it was also favoured because of a few contextual conditions, such as the new environmental law that demands more participation. The initiation was also favoured by the presence of motivated visionary leaders who are not afraid to take risks and are willing to experiment with new planning tools, such as the LL. Similarly, in the Overdiepse polder case that did not start as a LL but evolved into a co-creation process, the presence of motivated stakeholders with a pro-active attitude was essential for this evolution.

This leads us to our next point that is related to the actors involved and the roles they play. In strategic and civil LLs, the position of the public authorities is interesting as they not only initiate the process, but also act as patrons by supporting the innovation process, as webbers by selecting the LL participants, and as contributors by providing information throughout the process, therefore sometimes leading to confusion and distrust among the participants. Furthermore, a key role in a LL is played by the facilitator, whose task is to help the LL participants to understand their common objectives and perspectives, and to guide them to reach these objectives by offering them suitable ideation tools. To fulfil this role, the facilitator needs to manage the overall process, to lead the co-creative sessions, to establish the right conditions for the participants to feel safe to speak and express ideas and perspectives, to seek inclusive resolutions that work for all the participants, and to be prepared to react spontaneously to unforeseen changes. These kinds of unforeseen developments can lead to tensions, especially in strategic and civil LLs where flexibility may sometimes be at odds with the resource and time calculated approach of the initiator, or of the facilitator.

Another important aspect that needs to be kept in mind is that co-creation in a planning-related LL is not neutral: it is always developed in a political setting. Caution needs to be given to the evolution of the living labs, because they can become arenas of unequal expectations for various kinds of stakeholders, power games due to the influence and power of different actors, therefore leading to conflicts. Nevertheless, an important aspect that should not be forgotten is that the quality of the co-creation process is dependent on the history of relationships among stakeholders (Rădulescu et al., 2020). Therefore, LLs should be used carefully as they may easily become an umbrella for the same old practices due to more influential and powerful agendas and interests. However, they can also be a window of opportunity for re-designing and changing the present practices and approaches.

Finally, LLs as a planning approach provide opportunities for dealing with the challenges that the planning field faces by supporting experimentation, collaboration and learning. The non-linear, iterative nature of the creative process within a LL, marked by reflective and evaluative moments, provides the opportunity for continuous improvement through learning-by-doing and doing-by-learning, both in terms of process and outcomes within the boundaries of a specific initiative. Further, given the wide spectrum of stakeholders involved, LLs as a planning method may also be seen as a social learning opportunity. Nevertheless, when zooming out, LLs as a planning method provides the opportunity for organizational learning and even the diffusion of knowledge within the wider planning field, and may ultimately have an important contribution to sustainability transitions.

Recommendations for Using a LL as a Planning Method

Relying on the same traditional, siloed planning approaches will not get us too far. This is because it is evident that the current wicked challenges we are facing require collaborative and creative work across sectors and levels. In addition, creating or 'borrowing' concepts from other domains and using them in policy-making can be helpful, but is not sufficient.

Ultimately, in attempting to bring new, innovative, and creative solutions to light, the flashy and almost magical concepts of LLs need to be carefully put into practice. In anticipation of such a turn, the present chapter has given an overview and provided insights into what LLs are, why and how they can be effectively used in the planning field. Furthermore, based on both existing literature and our own experiences, we introduced LL as a five-staged methodology, with each step having its own characteristics. However, we acknowledge that there is no blueprint for such an interactive process, and that maximum flexibility for fine-tuning and adaptation must be accommodated because each process is unique and iteratively evolving. For this reason, we conclude by outlining a few recommendations that could be useful when considering the use of LLs as a planning method:

- **Involve all relevant stakeholders and be flexible regarding the scope:** It is important to adopt an inclusive, tailor-made approach when selecting the LL participants. This is because a diverse network of actors, with varied capabilities, skills and motivations, is a determinant for the emergence of creativity and innovation as a result of interdisciplinary interactions. In the planning domain, LLs are usually place-specific and their context is influenced by the interaction of multiple actors situated on different levels and scales. Therefore, when the initiator undertakes the initial stakeholder analysis, they need to think creatively, not only about the specific location of the LL but also about the larger scope needed to come to creative solutions that include multiple challenges. Nevertheless, a fine balance needs to be maintained between diversity and the number of participants so that the LL proves to be an efficient planning tool.
- **Let the LL grow organically:** While planning tends to be pre-defined, controlled and process-oriented, LLs as a planning method offer the opportunity for organic planning processes. To take advantage of this opportunity, one should restrain from assembling the list of LL participants solely according to the results of the stakeholder analysis, which is an institutionalized tool in planning practice. Instead, selected participants should discuss if further potential stakeholders should be brought into the process if they consider their stakes are

relevant. This leads to an organically grown LL that, in opposition to traditional planning processes that rely on a pre-defined and minutiously controlled approach, increases the diversity of LL stakeholders, helps to build trust, and opens up the role of the authorities.

- **Manage expectations:** As a LL is most often a parallel process to the 'official' planning and decision-making process, the initiator of the LL needs to be transparent about the goals of the LL and about its position within the (formal) planning process. They must make it clear from the beginning that the results and solutions/plans developed within the LL might, or might not, be taken up in the decision-making process, therefore eliminating potential frustrations.
- **Genuinely listen to the participants and continuously adapt the co-creation process:** It can be very easy to dismiss peoples' concerns or requests, arguing that they go beyond the scope or length of the process. This relates to the relevance of both expert knowledge of professionals and experiential knowledge of stakeholders. LLs often (implicitly) comprise science-society dialogues, where scientist experts might be reluctant to move beyond their own perspective of a particular issue. The potential strength of a LL approach is that it provides an interface for connecting expert and experiential types of knowledge. Therefore, throughout the process it is essential to try to understand where every piece of feedback comes from, to keep an open and flexible mindset, and try to sense the participants' needs.
- **Adopt an agile management approach:** LLs do not follow the same 'recipe' as traditional projects, so they do not need to—or should not—be run like one. Using phases as presented in Sect. 15.3 to plan the LL is good as it offers a perspective and a structure of the entire process. However, conducting a LL based on a rigid pre-defined plan and on a tight time and cost approach needs to be avoided. When doing an LL, major attention needs to be paid to the insights received from the various stakeholders and to their attitude and commitment to the process. This means that those running a LL need to be willing to adapt the process based on the participants' feedback; this can lead to small or even radical changes in the design of the process and its direction.

- **Do not default back to old approaches and roles:** Adopting new ways of doing things is especially hard when there is a tension between spurring innovation and creativity and quickly delivering concrete results. Furthermore, the adoption of new, experimental, and collaborative approaches in planning emphasizes new roles to be played by the involved stakeholders. Although adopting new approaches and roles might feel overburdening, and choosing the old ones or trying to incorporate them into the LL might be tempting, this will only defeat the initial purpose of adopting a LL as a planning method.
- **Do not be afraid of taking risks and possible failure:** Experimentation involves risk taking and this may lead to failure or partial success. However, even when a LL approach does not succeed in fostering innovative ideas, it can still be a source of learning in terms of process design.
- **LLs do not represent the holy grail for dealing with wicked problems:** In the examples presented in this chapter, adopting a LL approach had an influence on the planning practice. However, this may not always be the case. To be able to maximize the potential impact on planning policy and practice, it is important to clearly define and communicate the role of the LL in the planning process, to clarify its position in relation to the formal decision-making, and to explain the role of the authorities.
- **Do not focus on terminology, just keep it simple:** LLs are a buzzword, but so are urban living labs, design labs, city labs, fab labs; they are all experimental approaches that can be included in the larger category of real-world laboratories, which present numerous similarities and therefore result in being used interchangeably. Given this large diversity of similar concepts, it is important not to focus too much on terminology, but on deciding to adopt such approaches and letting them grow organically as interaction platforms.
- **Talk the language of the participants:** Using expert jargon communication is efficient between people from the same field, but in a LL, where participants' diversity and interaction are essential, it can lead to the exclusion of non-experts. Therefore, it is important to keep language simple, to try to eliminate jargon as much as possible to make all participants feel welcome, interested and willing to engage,

because this ultimately spurs creativity and fosters the development of innovative ideas.

- **Do not focus on reaching a compromise**, even though this might be tempting in such multifarious processes, with numerous actors that represent different interests. Instead, try to foster their interaction and the exchange of the diverse types of knowledge and experiences they embody. In the end, LLs are about creating the opportunity for a co-creative process, and not about forcing the development of solutions and reaching final planning decisions. For this reason, it is important to clearly demarcate the creative process and the formal decision-making and to constantly manage potential expectations about the outcome. Finally, to highlight this separation it is important that at the end of the LL the decision-makers give feedback about their decisions and the way these have been reached.

In the search for a sustainable future development of the Hegewarren polder, the province of Friesland adopted 'a living lab way of working'. This proved not to be an easy path as there is no 'magic recipe' when working with such new and innovative planning methods in a multi-scalar, multi-level, and multi-actor setting such as the planning of water infrastructure and spatial development. Despite the challenges and the temptation of falling back on old approaches, living labs certainly represent a relevant and growing practice in Dutch water infrastructure and spatial planning; they provide a valuable way of connecting local-scale and larger-scale planning issues and solutions.

Acknowledgements This research received funding from the European Union's Horizon 2020 research and innovation programme under the Marie Sklowdoska-Curie Grant Agreement No. 765389.

References

Bajgier, S. M., Maragah, H. D., Saccucci, M. S., Verzilli, A., & Prybutok, V. R. (1991). Introducing students to community operations research by using a city neighborhood as a living laboratory. *Operations Research, 39*(5), 701–709. https://doi.org/10.1287/opre.39.5.701

Ballon, P., Pierson, J., & Delaere, S. (2005). Test and experimentation platforms for broadband innovation: Examining European practice. *SSRN Electronic Journal*. https://doi.org/10.2139/ssrn.1331557

Bergvall-Kåreborn, B., & Ståhlbröst, A. (2009). Living lab: An open and citizen-centric approach for innovation. *International Journal of Innovation and Regional Development, 1*(4), 356. https://doi.org/10.1504/IJIRD.2009.022727

Bergvall-Kåreborn, B., Ihlström Eriksson, C., Ståhlbröst, A., & Svensson, J. (2009, December 6–9). *A milieu for innovation—defining living labs* (Paper presentation 2nd ISPIM Innovation Symposium, New York City, New York, United States). https://www.diva-portal.org/smash/get/diva2:1004774/FULLTEXT01.pdf

Catmull, E. E., & Wallace, A. (2014). *Creativity, Inc: Overcoming the unseen forces that stand in the way of true inspiration* (1st ed.). Random House.

Chesbrough, H. W. (2003). *Open innovation: The new imperative for creating and profiting from technology*. Harvard Business School Press.

de Roo, G., Hillier, J., & Wezemael, J. van (Eds.). (2012). *Complexity and planning: Systems, assemblages and simulations*. Ashgate.

Dell'Era, C., & Landoni, P. (2014). Living lab: A methodology between user-centred design and participatory design: Living lab. *Creativity and Innovation Management, 23*(2), 137–154. https://doi.org/10.1111/caim.12061

Dutilleul, B., Birrer, F., & Mensink, W. (2010). Unpacking European living labs: Analysing innovation's social dimensions. *Central European Journal of Public Policy, 4*(1), 60–85.

Eriksson, M., Niitamo, V.-P., & Kulkki, S. (2005). *State-of-the-art in utilizing living labs approach to user-centric ICT innovation-a European approach*. Luleå University of Technology.

Esteves, A. M., Franks, D., & Vanclay, F. (2012). Social impact assessment: The state of the art. *Impact Assessment and Project Appraisal, 30*(1), 34–42. https://doi.org/10.1080/14615517.2012.660356

European Commission. (2006). *The Helsinki manifesto*. European Commission. http://elivinglab.org/files/Helsinki_Manifesto_201106.pdf

European Network of Living Labs. (2015). *Introducing ENoLL and its living lab community* (1st ed.). https://issuu.com/enoll/docs/enoll-print

Evans, R., Guy, S., & Marvin, S. (1999). Making a difference: Sociology of scientific knowledge and urban energy policies. *Science, Technology, & Human Values, 24*(1), 105–131.

Følstad, A. (2008). Living labs for innovation and development of information and communication technology: A literature review. *Electronic Journal of Organizational Virtualness, 10*, 99–131.

Franz, Y. (2015). Designing social living labs in urban research. *Info, 17*(4), 53–66. https://doi.org/10.1108/info-01-2015-0008

Frissen, V., & van Lieshout, M. (2004). To user-centred innovation processes: The role of living labs. In TNO-ICT, Delft.

Garavaglia, L. (2020). Living labs: A tool for inclusive urban innovation. *Equilibri, speciale*, 133–144. https://doi.org/10.1406/98105

Geels, F. W. (2005). Processes and patterns in transitions and system innovations: Refining the co-evolutionary multi-level perspective. *Technological Forecasting and Social Change, 72*(6), 681–696. https://doi.org/10.1016/j.techfore.2004.08.014

Gouillart, F., & Hallett, T. (2015, Spring). Co-creation in government. *Stanford Social Innovation Review*. https://ssir.org/articles/entry/co_creation_in_government

Haataja, M., Hautamäki, A., Holm, E., Pulkkinen, K., & Suni, T. (2018). *Co-creation a guide to enhancing the collaboration between universities and companies* (No. 978–951–51–4096–8). University of Helsinki. https://blogs.helsinki.fi/andaction/files/2018/02/HY_Co_creation_web.pdf

Hagman, K., Hirvikoski, T., Wollstén, P., & Äyväri, A. (2018). *Handbook for co-creation*. City of Espoo. https://www.theseus.fi/handle/10024/159825

Hamersma, M., Heinen, E., Tillema, T., & Arts, J. (2018). Understanding resident satisfaction with involvement in highway planning: In-depth interviews during a highway planning process in the Netherlands. *Journal of Environmental Planning and Management, 61*(7), 1224–1249. https://doi.org/10.1080/09640568.2017.1339592

Heikkinen, M. T., Mainela, T., Still, J., & Tähtinen, J. (2007). Roles for managing in mobile service development nets. *Industrial Marketing Management, 36*(7), 909–925. https://doi.org/10.1016/j.indmarman.2007.05.014

von Hippel, E. (2005). *Democratizing innovation*. MIT Press.

IIIEE at Lund University. (2015, April 13). *GUST: Urban living labs* [Video]. https://www.youtube.com/watch?v=ITjSWVcWeiE&t=50s

Kemp, R., Parto, S., & Gibson, R. B. (2005). Governance for sustainable development: Moving from theory to practice. *International Journal of Sustainable Development, 8*(1/2), 12. https://doi.org/10.1504/IJSD.2005.007372

Krebs, V., & Holley, J. (2004). Building sustainable communities through social network development. *Nonprofit Quarterly, 11*(1), 46–53.

Lama, N., & Origin, A. (2006). *Innovation ECOSYSTEMS: Services engineering & living labs a dream to drive innovation?* EU/IST conference.

Leendertse, W., Langbroek, M., Arts, J., & Nijhuis, A. (2016). Generating spatial quality through co-creation: Experiences from the Blankenburgverbinding (The Netherlands). *Transportation Research Procedia, 14*, 402–411. https://doi.org/10.1016/j.trpro.2016.05.092

Leminen, S. (2015a). *Living labs as open innovation networks: Networks, roles and innovation outcomes* [Doctoral Dissertation, Department of Industrial Engineering and Management, Aalto University]. https://aaltodoc.aalto.fi/handle/123456789/17899

Leminen, S. (2015b). Q&A.What are living labs? *Technology Innovation Management Review, 5*(9), 29–35.

Leminen, S., Westerlund, M., & Nyström, A.-G. (2012). Living labs as open-innovation networks. *Technology Innovation Management Review, 2*(9), 6–11. Doi: https://doi.org/10.22215/timreview/602

Liedtke, C., Jolanta Welfens, M., Rohn, H., & Nordmann, J. (2012). LIVING LAB: User-driven innovation for sustainability. *International Journal of Sustainability in Higher Education, 13*(2), 106–118. https://doi.org/10.1108/14676371211211809

Malmberg, K., Vaittinen, I., Evans, P., Schuurman, D., Ståhlbröst, A., & Vervoort, K. (2017). *Living Lab Methodology Handbook*. https://doi.org/10.5281/ZENODO.1146321

Marvin, S., Bulkeley, H., Mai, L., McCormick, K., & Palgan, Y. V. (Eds.). (2018). *Urban living labs: Experimenting with city futures*. Routledge, an imprint of the Taylor & Francis Group.

McCormick, K., & Hartmann, C. (2017). *The Emerging landscape of urban living labs: Characteristics*. Lund University.

McGregor, D. (2021). First nations, traditional knowledge, and water ethics. In I. L. Stefanovic & Z. Adeel (Eds.), *Ethical water stewardship* (pp. 147–164). Springer International Publishing. Doi: https://doi.org/10.1007/978-3-030-49540-4

Molinari, F. (2011). Living labs as multi-stakeholder platforms for the egovernance of innovation. In E. Estevez & M. Janssen (Eds.), Proceedings of the 5th International Conference on Theory and Practice of Electronic Governance (ICEGOV 2011) (pp. 131–140). Association for Computing Machinery, New York, United States. https://doi.org/10.1145/2072069.2072092

Neef, M. R., Verweij, S., Gugerell, K., & Moen, P. L. R. (2017). *Wegwijs in living labs in infrastructuur en ruimtelijke planning: Een theoretische en empirische verkenning*. Rijksuniversiteit Groningen.

Pallot, M., Trousse, B., Senach, B., & Scapin, D. (2010, August). *Living lab research landscape: From user centred design and user experience towards user cocreation*. First European Summer School "Living Labs", Inria (ICT Usage Lab), Userlab, EsoceNet, Universcience, Paris, France. https://hal.inria.fr/inria-00612632/document

Pierson, J., & Lievens, B. (2005). Configuring living labs for a 'thick' understanding of innovation. *Ethnographic Praxis in Industry Conference Proceedings, 2005*(1), 114–127. https://doi.org/10.1111/j.1559-8918.2005.tb00012.x

Pollitt, C., & Hupe, P. (2011). Talking about government: The role of magic concepts. *Public Management Review, 13*(5), 641–658. https://doi.org/10.1080/14719037.2010.532963

Puerari, E., de Koning, J., von Wirth, T., Karré, P., Mulder, I., & Loorbach, D. (2018). Co-creation dynamics in urban living labs. *Sustainability, 10*(6), 1893. https://doi.org/10.3390/su10061893

Rădulescu, M. A., Leendertse, W., & Arts, J. (2020). Conditions for co-creation in infrastructure projects: Experiences from the overdiepse polder project (The Netherlands). *Sustainability, 12*(18), 7736. https://doi.org/10.3390/su12187736

Rosado, L., Hagy, S., Kalmykova, Y., Morrison, G., & Ostermeyer, Y. (2015). A living lab co-creation environment exemplifying Factor 10 improvements in a city district. *Journal of Urban Regeneration and Renewal, 8*(2), 171–185.

Sanders, L., & Simons, G. (2009). A social vision for value co-creation in design. *Open Source Business Resource*, December 2009, 27–34.

Schäpke, N., Stelzer, F., Caniglia, G., Bergmann, M., Wanner, M., Singer-Brodowski, M., Loorbach, D., Olsson, P., Baedeker, C., & Lang, D. J. (2018). Jointly experimenting for transformation? Shaping real-world laboratories by comparing them. *GAIA: Ecological Perspectives for Science and Society*, *27*(1), 85–96. Doi: https://doi.org/10.14512/gaia.27.S1.16

Ståhlbröst, A., & Holst, M. (2012). *The living lab methodology handbook*. Social Informatics at Luleå University of Technology and CDT.

Steen, K., & Van Bueren, E. (2017). *Urban living labs a living lab way of working*. Amsterdam Institute for Advanced Metropolitan Solutions (AMS).

Van den Brink, MA (2009) *Rijkswaterstaat on the horns of a dilemma*, Eburon.

Verhoef, L., & Bossert, M. (2019). *The university campus as living lab for sustainability: A practitioners guide and handbook*. Delft University of Technology, Hochschule für Technik Stuttgart. https://campusaslivinglab.org/wp-content/uploads/2019/06/new_RZ_Living_Lab_handbook_9.5.19.pdf

Voytenko, Y., McCormick, K., Evans, J., & Schliwa, G. (2016). Urban living labs for sustainability and low carbon cities in Europe: Towards a research agenda. *Journal of Cleaner Production, 123*, 45–54. https://doi.org/10.1016/j.jclepro.2015.08.053

Westerlund, M., & Leminen, S. (2011). Managing the challenges of becoming an open innovation company: Experiences from living labs. *Technology Innovation Management Review, 1*(1). http://timreview.ca/article/489

Woltjer, J. (2000). *Consensus planning : The relevance of communicative planning theory in dutch infrastructure development (dissertation)*. Ashgate.

Zivkovic, S. (2018). Systemic innovation labs: A lab for wicked problems. *Social Enterprise Journal, 14*(3), 348–366. https://doi.org/10.1108/SEJ-04-2018-0036

16

Supporting Institutional Transformations: Experimenting with Reflexive and Embodied Cross-Boundary Research

Gloria Giambartolomei, Alex Franklin, and Jana Fried

Introduction

In 2015, the Welsh Parliament introduced a ground-breaking new legislation to support an ambitious national political commitment to the principle of Sustainable Development: the 'Wellbeing of Future Generations (Wales) Act' (WBFGA, 2015). At the core of the Act are seven 'wellbeing goals' and five 'ways of working'. While the former are centred around achieving a 'healthier', 'prosperous', 'cohesive' and 'more

G. Giambartolomei (✉) · A. Franklin · J. Fried
Centre for Agroecology, Water and Resilience (CAWR), Coventry University, Coventry, UK
e-mail: giambarg@uni.coventry.ac.uk

A. Franklin
e-mail: ac0569@coventry.ac.uk

J. Fried
e-mail: ab9318@coventry.ac.uk

A. Franklin (ed.), *Co-Creativity and Engaged Scholarship*,
https://doi.org/10.1007/978-3-030-84248-2_16

equal' Wales, the latter enshrine 'collaboration', 'integration', 'prevention', 'long-term', and 'involvement' at the heart of all public-sector work. The wellbeing goals and the ways of working have since been embedded in the Environment (Wales) Act from 2016, and 'translated' into an overarching principle of sustainable management of the natural resources (SMNR) of Wales. In accordance with this legislation, natural resources must be used in a way and at a rate that "maintain and enhance the resilience of ecosystems and the benefits they provide" for present and future generations (National Assembly for Wales, 2016, p. 2). Although neither legislation explicitly talks about 'transformation', the implication is that to achieve compliance, all affected stakeholders need to engage in a deep and profound re-thinking of the ways we work to achieve human and ecological well-being.

Shortly after the introduction of the above legislation, a co-funded doctoral research project on collaborative forms of natural resource management was co-developed by a small transdisciplinary team of academics and (cross-divisional) civil servants within Welsh Government (WG). The aim was to enable an extended investigation of collaborative ways of working in a range of different SMNR settings (including both government-led and community-led initiatives). While a participatory action research (PAR) approach was always part of the original project design, it was only after the appointment of the doctoral researcher that PAR principles became as important in guiding the internal process of transdisciplinary working between the co-ordinating team, as they were in guiding the study of external SMNR initiatives. This chapter recounts the experience of the doctoral researcher (Giambartolomei, lead author) appointed to this project.

Specifically, we discuss Giambartolomei's experience of transdisciplinary collaboration through the methodological lens provided by blending the Formative Accompanying Research (Freeth, 2019) and the Embodied Researcher approach (Horlings et al., 2020). In addition, we also look at the role of creative methods and Theory U (Scharmer, 2018) in further promoting collaborative processes of meaning-making in transdisciplinary research settings; in particular, their role in enabling emotional and embodied ways of working to be forefronted. In doing so, this chapter contributes towards shedding more light on the dynamics as well as the challenges—personal, professional, and emotional—of adopting a collaborative way of working in the pursuit of institutional transformations towards sustainability.

Guided throughout by an understanding of policy actors "not just as rule-setting and rule-following beings, but relational agents who work out the substance of policy through interpersonal relationships and everyday transactions" (Lejano, 2020, p. 2), we offer a critical reflection on Giambartolomei's first-hand experience of co-experimenting alongside policy actors with alternative ways of working in the *spaces in between* the written publication and implementation of SMNR legislation and policy. We focus in particular on three of Giambartolomei's own formative moments in this shared journey. The narration of these three moments is based on evidence drawn from the multiple, intersecting pathways of experience and reflection that we encountered during her study. Although primarily recounted from her perspective, they encompass both 'inwards' (i.e., self-examining personal assumptions and mindsets) and 'outwards' (i.e., collectively examining structural and institutional barriers) dimensions of individual and collective group learning.

The chapter illustrates the role of emotional labour, vulnerability and energy in such co-experimental work and emphasizes the need for the practicing of *care* in building relationships of trust and collaboration, especially within the context of sustainability transformations (Moriggi et al., 2020). Despite being ever-present, such affective properties are rarely acknowledged as legitimate and relevant within governmental settings. Experimenting in the *spaces in between* written legislation and the relationships and practices occurring among actors tasked with implementing such legislation, we assert, provides a fertile ground for shaping and co-creating new and shared understandings. Furthermore, co-creative practices centred on embodiment and the experience of relationality, of *being and doing* with one another, are especially effective here; they bring to life the (dry and hollow) principles and provisions of legislation. At the same time, however, the critical challenges imposed by a neoliberal governmental approach that leaves very little time and space for people, especially civil servants, to be part of such experimental, co-creative spaces, must not be overlooked.

We conclude by emphasizing the importance of dedicating sufficient time and resources to enable a culture of care (Puig de la Bellacasa, 2017; Tronto, 2013) such that embodied and collaborative ways of working can be more fully supported and understood within governmental institutions. It is only by normalizing such ways of working outside transdisciplinary research projects, or during occasional, alternative moments of experimentation, that they can truly become far more integral to, and accepted within, everyday practice. It is, we argue, through such a process of care, through a collaborative and continuous process of aligning our heads, hearts, hands and feet, that we will be able to progress along the pathway of socially and ecologically just sustainability transformations, regardless of (and because of) our personal, professional or institutional starting points.

The Importance of Cultural Transformation for Wider Sustainability Transformations

Scholars in sustainability sciences emphasize the need for sustainability 'transformations' to ensure the survival of the human species, threatened by global systemic collapse triggered by anthropogenic climate change and biodiversity loss (among many other factors) (e.g., Blythe et al., 2018; Feola, 2015; O'Brien, 2012; Pelling, 2010; Pelling et al., 2015). The Intergovernmental Science-Policy Platform on Biodiversity and Ecosystems Services (IPBES) defines transformation as a "fundamental, system-wide change that includes consideration of technological, economic and social factors, including in terms of paradigms, goals or values" (IPBES, 2019, p. 7). Similarly, O'Brien (2012) defines transformations "as physical and/or qualitative changes in form, structure or meaning-making. It can also be understood as a psycho-social process involving the unleashing of human potential to commit, care, and effect change for a better life" (O'Brien, 2012, p. 670). Importantly, the type of transformations discussed by these authors are those labelled as 'deliberative and intentional', meaning that they are the results of chosen response paths—"in anticipation of collapse" (Pelling et al., 2015, p. 114)—rather than the unexpected or unintended outcomes of a process or event (O'Brien, 2012, p. 670).

There is wide agreement among social sustainability scientists that the transformative change we need is not *just* about technical solutions for which we merely need more scientific knowledge and evidence—the 'technical trap' (Nightingale et al., 2020). Instead, adopting a transformation-focused analytical lens means directly questioning and challenging the values, paradigms, norms, beliefs and assumptions that underpin and constantly exacerbate anthropogenic climate change, environmental destruction and socio-economic inequalities worldwide (Fazey et al., 2018; O'Brien, 2012). The concept of transformation requires that we address the root or structural causes of something considered a threat (e.g., climate change). This includes, for instance, investigating the existing social, cultural and economic relationships as well as power hierarchies that have or will trigger systems' failures (Pelling et al., 2015).

Fazey et al., (2018, p. 199) are explicit in their belief that "intentional transformative change is possible and that humanity is not entirely a slave to its past or current circumstances and trends". The concept of transformation as proposed by O'Brien and others (Fazey et al., 2018; O'Brien, 2012, 2013, 2018; O'Brien & Sygna, 2013; Pelling et al., 2015) draw attention to the necessary "shifts in the balance of power, rights and responsibilities in institutions, discourse and behaviour" (Pelling et al., 2015, p. 115) to overtly challenge the *status quo*, including those who largely and disproportionately benefit from current structures and systems of power (and oppression). Nevertheless, there is great uncertainty inside and outside academia around *what* exactly should change, and according to *whose* views and definitions of transformation (Feola, 2015; Nightingale et al., 2020; O'Brien, 2012). In fact, deliberative transformations hold at their core democratic participation and multiple forms of deliberation to define future (sustainability) pathways that have, so far, often failed to gain traction (Nightingale et al., 2020).

Among the biggest challenges to nurturing the conditions for deliberative transformations is the 'technical trap' mentioned above: in underplaying the critical role of "power and politics" in such processes, we also risk a great underestimation of the role of agency and the potential of people to move forward or hinder systemic change (O'Brien, 2018, p. 155). To overcome this 'trap' and the simplistic view of change attached to it, O'Brien and Sygna (2013) propose a heuristic device to understand transformations as a 'whole', made of three integrated and interconnected domains or spheres: the practical, the political and the personal (see Fig. 16.1).

The *practical* sphere includes technological innovations, infrastructures, and all those "specific actions, interventions, and strategies that directly contribute to a desired outcome" (O'Brien, 2018, pp. 155–156). The *political* sphere represents the systems and the structures (i.e., norms, rules, regulations, institutions, regimes, and incentives) that influence how systems are designed and governed (*Ibid.*). Finally, the

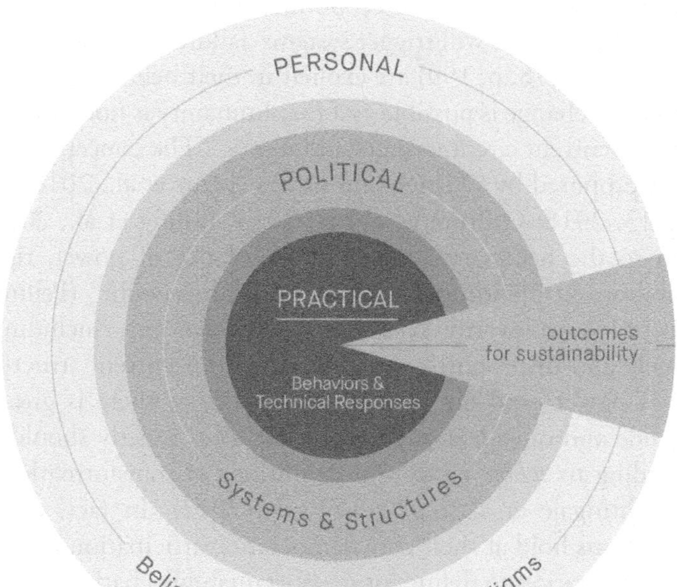

Fig. 16.1 The Three Spheres of Transformation (*Source* O'Brien and Sygna (2013))

personal sphere includes the beliefs, paradigms, assumptions, worldviews and values that individuals hold. These also represent the very factors that influence how systems and structures (i.e., the *political* sphere) are defined and ultimately changed (*Ibid.*).

The *personal sphere*, as visible in Fig. 16.1, is the overarching sphere that holds the other two dimensions together. It represents both individual and collectively-shared assumptions and understandings about the world, which shape how reality is perceived and socially constructed. This entails that it also "defines what is individually and collectively imaginable, desirable, viable and achievable" (O'Brien, 2018, p. 156). As O'Brien stresses, it would be "tempting to equate culture with the personal sphere" (*Ibid.*), when in fact, as she argues, culture is pervasive and transversal, cutting across all three spheres. Moreover, it is embedded and perceivable, especially in the interactions among these three domains.

The concept of culture is extremely wide and holds a myriad of (contested) meanings and understandings. Depending on the angles and contexts from which examine culture, "culture can mean anything from networks of meaning, to a way of life, to high culture and arts" (Soini & Dessein, 2016, p. 2) Without going into too many details of the scholarly discourse debating the meanings of and approaches to culture in the sustainability sciences, in this chapter, we draw on (Geoghegan et al., 2019) who investigate the issue of culture and climate changes from three perspectives. First, in terms of 'knowing' (cultural practices in past and contemporary scientific and epistemic communities); second as 'being' (embodied and lived experiences, emotional encounters and everyday practices); third, as 'doing' (concrete experiences of 'cultural work that pave the way for alternative social-ecological futures) (Geoghegan et al., 2019, p. 2). This three-pronged approach to the exploration of culture in climate change and sustainability transformations discourses confirms what has already been mentioned above: to culture is a *verb*, an ongoing, constantly evolving relational process that crosses and shapes the spheres of transformations, at both the individual and the collective level. But most of all, culture is about agency, and therefore a (much needed) cultural shift is about mobilizing collective imagination and agency, to envision, embody and realize alternative socio-ecological frames and futures, beyond the capitalist and neoliberal paradigm, beyond dyadic visions of the world and life (humans vs nature; body vs mind; individual vs collectivity; reason vs emotion; etc.) (Dieleman, 2017).

Based on the above reading, cultural transformations towards sustainability are, therefore, about re-imagining, re-envisioning things differently—us as a species, our relationships and whole systems. Hammond (2020, p. 3), conceives cultural transformations as "processes of individual and collective meaning-making as a way of broadening the society's imaginative space". Understood also as "necessarily dynamic, fluid, and heterogeneous" (Hammond, 2020, p. 8), cultural transformation is essentially a process in which we, individuals and societies, make and re-make culture through the co-creation of new *shared meanings*: "Climate facts arise from impersonal observation, whereas meanings emerge from embedded experience, and the environmental social sciences, arts and humanities are well-positioned to foster a more complex understanding of humanity's climate predicament" (Geoghegan et al., 2019, p. 3). As Geoghegan et al. (2019) suggest, there is an increasing recognition from across disciplines of the key role the social sciences, arts and humanities can play in helping policy-makers and communities to engage with fundamental 'cultural discussions' around the meanings, values and worldviews attached to terms such as sustainability, climate change, social and ecological well-being. Accordingly, the next section presents a methodological approach adopted by the first author as a way of engaging in such 'cultural discussions', through embodied, relational and caring practices.

Embodied and Accompanying Research

The methodological approach presented here is a creative blend of different theoretical concepts and a more spontaneous set of practices, at the core of which lie two distinct frameworks: the Formative Accompanying Research (Freeth, 2019) and the Embodied Researcher approach (Horlings et al., 2020). In this section, we summarize the key components of each framework, introducing also how a multifaceted theory of *care* reconciles these two approaches and underpins the whole research practice analyzed here.

Formative Accompanying Research (FAR) Framework

The Formative Accompanying Research (FAR) framework (Freeth, 2019), in essence, "is committed to promoting knowledge about collaboration while promoting the practice of collaboration" (Freeth & Vilsmaier, 2020, p. 58). At the core of the FAR approach lies a dynamic conception of the positionality of a (FAR) researcher: they can benefit from the *proximity* to their team or group, that allows them 'to experience the inner workings' involved in doing collaborative work, but also from the opportunity to 'move further away', to maintain an overview of the wider mechanisms of collaboration.

To navigate the blurring boundaries between the different roles that FAR researchers assume while working collaboratively in team or group settings, Freeth (2019) therefore distinguishes between three roles—*scientific researcher, team member* and *intervener*—and three related research orientations. The goal of the scientific researcher is ***learning about*** (the interdisciplinary team) and creating transferable knowledge; the team member ***learns with*** the team, alongside the team. Finally, the intervener ***learns for*** the team to support the advancement in terms of research outcomes. Although the context to which we apply this framework is different from that in which it originated (see Freeth & Vilsmaier, 2020), it nevertheless helps us explain and analyze "the idea of research positionality [as] *constituted in movement*, between outsider and insider roles" (Freeth, 2019, p. 54) in the context of cross-boundary collaborative research.

In understanding a researcher's positionality as a fluid, complex, and dynamic process, Freeth and Vilsmaier (2020) further identify three *balancing acts* and three related *practices*, to negotiate the paradoxes implicit to each balancing act. These balancing acts are needed to navigate the tensions that necessarily arise when moving between being an insider and outsider of the team. They argue that these acts are "a continuum, and that all positions along this continuum are possible and appropriate at different time" (2020, p. 62), with none of these positions existing independently, but only in relation to the others along the continuum. The first of such acts is balancing **participation** and **observation**, a well-documented tension described by a plethora of literature on ethnographic and participatory methodologies (e.g., Billo & Hiemstra, 2013; Cahill, 2007; Newton et al., 2012). Freeth and Vilsmaier (2020) propose a first accompanying practice of "**dynamic proximity**" to balance this tension and the paradox of being both an insider and participant, as well as an outsider and observer, along the same continuum. Keeping a dynamic proximity allows the researcher to be close enough to see finer details, but also to be able to step back, to hold a system view and see the 'whole-in-context' (Freeth & Vilsmaier, 2020, p. 62). By doing so, a dynamic proximity enables the researcher to provide the team (or group) with specific inputs for reflection and discussion. Finally, adopting dynamic proximity allows the researcher to be near enough "to perceive when the conditions are ripe for team-level learning", and distant enough "to avoid imposing a learning agenda" (Freeth & Vilsmaier, 2020, p. 63).

A FAR approach therefore helps the researcher to see the 'inner workings' and emotional labour of those involved in collaborative and interdisciplinary teams. Curiosity and care, the second balancing act introduced by Freeth and Vilsmaier (2020), shed further light on the emotional labour involved when digging deeper into certain (personal) matters of a group, towards which the researcher might be led by their curiosity. In fact, "curiosity and knowledge regarding the needs of an 'other' – human or not" (Puig de la Bellacasa, 2011, p. 98) are required for adequate care, which becomes "a doing necessary for significant relating" (*Ibid*). Recognizing the interdependence of all beings allows one to embrace the idea that "caring is not a romantic endeavour, nor an exclusive affair of motherly love, but a matter of earthly survival" (Rose, 1983). When curiosity, a basic characteristic of any researcher, meets care as a form of responsibility for the becoming of the object of the research (Puig de la Bellacasa, 2011), this creates the ground for a "careful curiosity (...) attuned to possible impacts of the research on others" (Freeth & Vilsmaier, 2020, p. 63). On this ground, the researcher, through a second accompanying practice of 'critical reflexivity', stays in inquiry mode, but at the same time is able to recognize appropriate times and conditions (or lack thereof) to dig deeper, and to challenge the others on uncomfortable territory. By being critically reflexive, the researcher therefore *accompanies* the group and its individual members (i.e., "walks in step with those being researched" (*Ibid.*)), while also taking responsibility for their own situatedness (normative positions and power exerted) within the research, all the while aware of the possible impacts on others.

Acknowledging that the interests and normative positions that a researcher holds do carry power within the research context, leads Freeth and Vilsmaier (2020) to the third balancing act with which researchers have to engage: the balancing between **impartiality** and **investment.** Given emphasis here is the fact that impartiality does not equal neutrality (i.e., no one is ever 'interests-free') but rather implies "being aware of interests but seeking to remain unbiased" (*Ibid*, p. 64). As the above discussion of the concept of care suggests, once we recognize the interdependence and fundamental relationality of all beings and things within the Earth system, *caring* becomes a doing, a practice necessary for survival, which implies caring for (i.e., maintaining) that web of relationships, and dealing with the vested interests and powers with which this web is imbued. As an "inevitable consequence of being in relationship" (Freeth & Vilsmaier, 2020, p. 59), we are partial and invested, especially when decisions taken within a group also necessarily impact on our role and work. A practice of "**embedded relationality**" allows us to balance, on the one hand, the need to overtly challenge certain interests by "claiming the power granted by an insider–outsider perspective" with, on the other, leaving the matter to the interpretation of the rest of the group. This often implies engaging in an exercise of enriching perspectives without having to necessarily achieve a compromise. Freeth and Vilsmaier remind us here of Haraway's understanding of 'embedded relationality': it produces "partial, locatable, critical knowledges sustaining the possibility of webs of connection called solidarity in politics and shared conversations in epistemology" (Haraway, 1991, p. 191 cited in Freeth and Vilsmaier (2020, p.64)).

Freeth and Vilsmaier (2020) finally propose three 'anchoring principles' for navigating the dynamic and fluid positionality at the core of the FAR approach: congruence, sensitivity and translucence (p. 64). However, the practices, the balancing acts and the whole experience that we present and discuss here have been mainly anchored to one overarching principle: care. As mentioned above, a (feminist) ethic of care stems from acknowledging the interdependent and relational nature of all things—first and foremost, of human and more-than-human lives. As Puig de la Bellacasa (2017) puts it, such interdependency is "the ontological state in which humans and countless other beings unavoidably live" (Puig de la Bellacasa, 2017, p. 4). More specifically to the context under study here, if we are to recognize the key role of relationships, relationality and interdependence in the context of *doing collaborative and transdisciplinary work,* then it is also worth remembering Fisher and Tronto's definition of care: "*a species activity that includes everything we do to maintain, continue, and repair our 'world' so that we can live in it as well as possible.* That world includes our bodies, ourselves and our environment, all of which we seek to interweave in a complex, life-sustaining web" (Fisher & Tronto, 1990, p. 40 original emphasis; also cited in Tronto, 1993, p. 103, 2013, p. 19).

Both Tronto and de la Bellacasa stress the intrinsic tensions and ambivalences attached to care as a three-dimensional concept made of maintenance work, affective engagement, and ethico-political involvement. Such an approach is far from an idealized, 'innocent' or essentialist conception of care as something necessarily and inherently 'feminine' or 'good' (Puig de la Bellacasa, 2017; Tronto, 2013). Instead, the doings and works of care aim to nurture an ongoing and hands-on process of re-imagining and re-creating 'as well as possible' relationships. It offers a way to ultimately re-claim care as a means to foster solidarities (Puig de la Bellacasa, 2017, p. 11), amidst unavoidable tensions and conflicts, while experimenting with more just ways of being and doing, of 'caring with' (Tronto, 2013) together. It is this understanding of the principle of care and its importance for doing collaborative and transdisciplinary research that leads us in turn to the concept of the embodied researcher.

The Embodied Researcher

The Embodied Researcher provides further 'grounding' of the practices characterizing the FAR framework within a place-based approach to sustainability research. Horlings et al. (2020) argue that researchers involved in place-based research "suspend the categorization of different roles" (e.g., reflective scientist, process facilitator, knowledge broker, change agent and self-reflexive scientist (see Wittmayer & Shapke, 2014, p. 488), engaging instead in transformative and situated research practices as '*embodied researchers*' (see Fig. 16.2). The embodied researcher is characterized by four elements: heart, hand, head and feet (Horlings et al., 2020, p. 479). This conceptualization portrays the researcher going into the field with their whole selves, adopting a reflexive approach inwards and outwards, also a key aspect of the FAR framework discussed above.

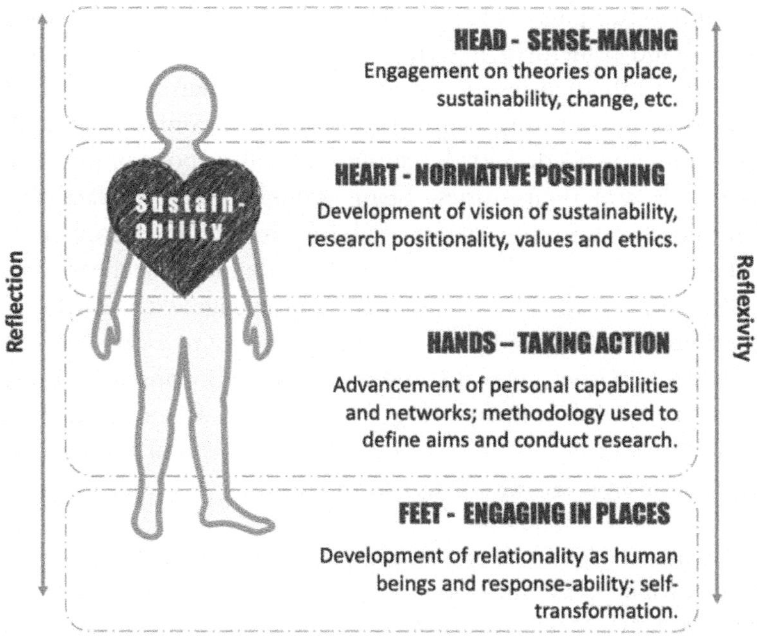

Fig. 16.2 The Embodied Researcher (*Source* Horlings et al. (2020))

Horlings et al. (2020) discuss how the embodied researcher practices self-reflexivity in the way they are aware of their own (evolving) positionality and normativity, and through their "responsibility and willingness to change" (*Ibid*). At the same time, importantly, they continuously consider and acknowledge the biases, values and positions held by the people involved in the research (beyond themselves); this ultimately informs a critical reflection on the research's dynamics, processes and data. As is also the case with FAR, neutrality is not an option; rather, being reflexively aware of our own partiality strengthens Haraway's idea that there is no contradiction between being objective and partial: "… a practice of objectivity that privileges contestation, deconstruction, passionate construction, webbed connections, and hope for transformation of knowledge and ways of seeing" (1988, pp. 584–585).

The embodied researcher's practices envisioned by Horlings et al. (2020) stem from a rooted normative stance (a deep wish to support change towards sustainability) present in their *heart*, which in turn acts as an "inner compass" (Horlings et al., 2020, p. 479). Conscious of the values and principles they stand for throughout the whole research process, the embodied researcher engages as a human being in the place and with its people, intertwining new personal connections with the communities involved, and developing ethical responsibilities towards those people and their stories. Being grounded in the place through commitment and a sense of responsibility towards the people is represented by the *feet*, as shown in the above figure. The heart and the feet allow the researcher to experiment and engage with situations and people through their hands and actions, according to a care-centred and process-based (rather than an outcome-oriented) approach. Engaging in research as a full human being, invested and aware of responsibilities, normative positions, roles, emotions and inner workings necessarily brings with it a process of self-transformation:

> "Self-transformation happens by engaging with critical theories related to sustainability and transformations (head), by reflecting upon one's own normative position as a researcher (heart), by experimenting with methods grounded on one's own values (hands) and by engaging in places

as a human being open to developing response-ability (feet)." (Horlings et al., 2020, p. 480)

Therefore, both the FAR and the embodied researcher frameworks share a vision of a self-reflexive, embedded, invested researcher that builds caring and creative practices, without ever compromising their analytical, critical, and enquiring attitude.

In the remainder of this chapter, the combination of the above two frameworks allows us to merge a descriptive and analytical set of characteristics of the embodied and transformative researcher, with a more dynamic (and open-ended/fluid) analysis of their practices. This supports a fuller account of what it takes—personally, emotionally and professionally—to be *care*-fully engaged in the pursuit of sustainability transformations in a transdisciplinary collaborative setting.

Co-creating Transdisciplinary Communities of Practice—A Journey

This section discusses three formative moments the doctoral researcher (lead author) experienced in the evolution of her transdisciplinary collaboration with the WG (and subsequently Natural Resource Wales (NRW, a pan-Wales 'public sponsored body')). These moments were selected since they were occasions when the implications and tensions of the embodied and accompanying research approach adopted by the first author emerged in all their complexity. In addition to critically reflecting on these three moments, we also further introduce the wider doctoral study context (2017–2021) in which they occurred. In doing so, we seek to produce an integrated account of how, in a transdisciplinary research setting, the personal and professional journey of a researcher can become interwoven with that of the non-academic partners in a mutual and relational dynamic process of co-evolution. Given the content of this section is founded upon the direct and first-hand experience of Giambartolomei, we therefore switch for this narrative account to the first person.

Woven throughout this narration are references to the multiple intersecting components and language of FAR, the Embodied Researcher and the overarching principle of care.

Embarking on a Transdisciplinary Journey

The initial stage of the collaboration (and of the PhD) was dedicated to *learning about* the context of Wales, its governance, policies and approaches to SMNR. It was also dedicated to getting to know the key individuals and organizations who work on the ground, including our partners at WG and NRW. Through my role of *scientific researcher* (Freeth & Vilsmaier, 2020) embedded in the collaborative project, I was able to dig into critical theoretical frameworks (e.g., political ecology, sustainability transformations) that could provide my partners with alternative or complementary approaches to those embraced in the Welsh legislation around SMNR.

Looking at SMNR through a distinctively political lens, critically reflective, for instance, of the structural socio-economic inequalities affecting access to and control over natural resources, enabled me to unpack and challenge taken-for-granted meanings, rooted in managerial and technocratic approaches to environmental governance. Given the need for pragmatism, such approaches often overlook the tensions and ambivalence of concept(s) such as 'sustainable' 'management' of 'natural resources' that can induce very different meanings, depending on *whose* assumptions and perspectives are privileged in the definitions. In accordance with the conceptual framework set out above, the *learning about*, therefore, went hand-in-hand with *learning with*. While providing my partners with new inputs from the literature and theoretical perspectives under study, I was gradually engaging my *head* and my *heart* in the research, *making sense* of such (new to me) theories, while also building my own understanding and normative position around the subject.

Four advisory meetings (i.e., namely the coordinating team of academic and WG collaborators) held during the first phase of the PhD (May 2017–April 2018), were focused on starting to explore together different assumptions, alternative ways of looking at the practical and political implications of pursuing 'collaboration' for 'SMNR'. Here, it is worth noting that these four advisory meetings were first conceptualized as meetings solely for guiding and advising the PhD researcher. However, they quickly transformed into a two-directional process, *learning about*, *learning for* and *learning with*. Within these meetings, I learned about the inner workings, achievements and challenges of implementing the SMNR principles. Together, we started to explore questions I raised that included, for example: *whose* definition of sustainability are we considering? *Who* is already sitting around these policy and collaborative tables? *Who* is missing from these conversations, but should be included? Is the currently followed managerial and a rather top-down approach only ticking the 'collaboration' box, or are we open and committed to truly transformative (inwards and outwards) practices and institutions that include in this conversation people who are not the 'usual suspects'? In posing these questions, I was trying to navigate and balance my *curiosity* about WG collaborators' own ideas and perspectives on these issues, which did not seem to be fully addressed by the current formulation of the policies and legislation, but all the while *maintaining a caring and 'safe' space* as much as possible. By doing so, I wanted to avoid anyone feeling (personally) attacked or criticized, but rather feel encouraged to reflect on current policies and practices from different points of view. In this initial phase, I therefore found myself engaging in the complex balancing act of *critical reflexivity*, through a practice of *care* that would allow me to nurture the space of safe collaboration we had started building (care as maintenance work) while also asking questions I profoundly care about (care as ethico-political involvement): e.g., where do we stand (as individuals, community members, citizens, policymakers, academics etc.) in the journey of Wales as a nation committed to social and ecological wellbeing for present and future generations?

After this initial phase and purposefully relocating to Cardiff in Spring 2018, a new and much richer phase of collaboration with WG began. My constant and physical presence in Wales facilitated our interactions, allowing a shift from formal and pre-organized quarterly meetings, to more spontaneous and frequent encounters. Moreover, I started to build new networks with people from various sectors and organizations involved in SMNR practices across the whole of Wales and, in so doing, also gradually enriched my knowledge of the people and the places at the core of my research. Concurrently, my sense of attachment and belonging to Wales was growing fast—parts of its landscape reminded me so much of my home village in Italy! The more I engaged with the people, the institutions and with its outstanding natural beauty, the more a sense of care (in an affective and emotional sense) and responsibility (as ethico-political involvement) was growing, defining my role as a researcher, and more simply as a human being, becoming fully committed to personally contribute—in whatever small way possible—to realizing human and ecological wellbeing in Wales. My *feet* were increasingly grounded and my *hands* more and more ready to take up an active role in advocating for a necessary 'cultural transformation' with and within governmental organizations. I felt fully present with myself and in my role as *embodied researcher*.

Formative Moment #1

In the first phase of my PhD, I was invited to a workshop organized as part of a joint programme put together by WG and NRW to further support the ongoing efforts to nurture a 'cultural transformation' within their organizations, towards a more collaborative and integrated way of working across departments and sectors (referred to hereafter as 'the WG-NRW joint programme'). The workshop was aimed at a variety of officers and stakeholders from community, public and third sector organizations involved in SMNR delivery, with a total of approximately 50 attendees. This being the first event I was invited to by my WG partners, I found myself still *learning about* and, unable to navigate FAR's spectrum of *dynamic proximity*. I felt mainly stuck in the (silent) observer's place, without being able to *participate* and articulate critical considerations. The observed conversations emphasized the need to find the right communication strategy to 'galvanize' people to learn to do things differently, in line with the new ways of working. Most of the debate, it appeared, was therefore about finding the right strategy to communicate the 'evidence' (mainly conceived as 'the facts' produced independently by scientific academic institutions) to those without access to, or not educated or interested enough to appreciate such evidence—"just avoid the high-level stuff" was one of the comments I recorded in my notes. The importance of blending and respecting different types of knowledge, experiences and perspectives was neglected in order to meet objectives of rapid and efficient delivery, through 'galvanizing' (uncompliant) people to do the right thing. I felt astonished by some conversations at the tables, and the frustration I experienced during this workshop required a lot of emotional labour on my part to contain potentially inappropriate reactions.

Personal and shared reflections with my WG and NRW collaborators after this workshop clearly suggested that more targeted work was urgently needed to bring officers and professionals working with SMNR initiatives into a space where they could collectively reflect on the meanings and implications of the new ways of working and NRP priorities in their everyday work. This would entail a far greater amount of time spent in conversation, listening deeply to one another—not always a common practice in these organizations, as reiterated by a number of individuals in both organizations on several occasions during my doctoral research. This opportunity to listen and interact with different people working in both organizations helped me to realize the direction in which I should focus my own contribution to the planning of future events of the WG-NRW joint programme.

Formative Moment #2

Subsequently, to reflect on the above workshop, WG and NRW collaborators organized a meeting where I was invited to provide my feedback and thoughts. When preparing for this meeting, I felt very strongly the need to convey messages about the "inside-out adaptation" (O'Brien, 2013) and the importance of challenging our own paradigms through the use of different media. I decided to *learn for* by gathering the most relevant theoretical and academic inputs for this discussion with WG and NRW in the form of a very rudimentary, imperfect, yet comprehensive illustration on flipchart paper (see Fig. 16.3). Creating this illustration, I felt deeply and fully embodied in my research while experimenting with

Fig. 16.3 Draft illustration of keywords (*Source* Constructed by author)

this unusual medium, imbued with a creative tension between 'letting all go' (on paper, with crayons, scissors and glue) and maintaining an extent of rigour and clarity of my messages (as usually expected of an academic and for interactions with governmental institutions).

This moment, which perhaps might sound like a minor, trivial detail, represented instead a turning point in my personal and professional journey, a step into my own process of self-transformation, as described by Horlings et al. (2020). Although at first I had never thought to show that illustration to my collaborators, I then in fact realized that I was the one proposing to *them* to *learn with* one another, embracing vulnerability and bringing our *whole selves* to our collaborative journey. Therefore, if we were hoping to acknowledge and challenge personal assumptions, paradigms and worldviews, I had to accept imperfection and perhaps also risk to appear ridiculous or a 'mere student' rather than a professional researcher. So, after this realization, I proudly went into the meeting with the drawing (Fig. 16.3). The reactions were generally positive and stimulated, as hoped for, a fruitful discussion and mutual learning. This experience and the trust and reciprocal respect already in place with my WG collaborators gave me strength and confidence to follow my intuition and experiment further with my own creativity.[1]

Formative Moment #3

Between the end of phase two and beginning of phase three of my PhD, WG and NRW collaborators invited me to a series of preparatory meetings aimed at planning and designing two new sets of pilot workshops, as part of the WG-NRW joint programme. It was especially my involvement with one of them that brought about my third formative moment in our collaborative journey. This workshop was co-designed with, and facilitated by, 'Emergence', an "evolving art, ceremonial and facilitation practice" with a "history in hosting transformative events and spaces for dialogue on issues of creativity and sustainability and change processes" (Emergence, 2020), and was led by Fern Smith and Phil Ralph.

The workshop focused on enhancing skills and capacity of practitioners across Wales to become *trusted intermediaries* and *change agents,* able to champion meaningful and transformative collaborative practices across sectors and organizations for the SMNR. To do so, we agreed on the need to learn about and practice deep-listening (to one another and to ourselves), and open and honest communication, from the shared basis of understanding and empathy. Using Theory-U (Scharmer, 2018) as a guiding framework for an embodied, practice-based learning (see also Pearson, this book), Phil and Fern proceeded to creatively guide participants in their 'journey along the U' to collectively develop the

[1] In addition, I had also built some confidence in using more creative and alternative means to express my reasoning (and myself more in general), thanks to my simultaneous involvement in the Project Skyline (a third sector-led feasibility study to assess the potential for community land transfer in the South Wales Valleys), that also formed part of my doctoral fieldwork. Through Project Skyline, I had the opportunity of taking part in and co-facilitating the engagement processes led by various artists' collectives in the involved communities. In those instances, I was able to experiment (alone, as well as collectively with other participants) with the power of art and artistic practices to bring ourselves out of our comfort zone and free up (mental) space for new thinking, new reflections and visions for action.

skills of 'learning-by-doing', through co-production, collaboration and prototyping new ideas.

At the core of this journey was the aim of discovering those *inner* and *structural* blind-spots of leadership, collaborative practices, and wider system change. To do so, space and time were created to purposefully look *inwards* and *outwards,* through a new pair of lenses (i.e., Theory U). From the outset, it was made clear by our facilitators that we would potentially be entering an uncomfortable space that would lead us to face vulnerability, uncertainty, fears and a sense of being lost amidst a process of *conscientization, "conscious raising"* (Freire, 1970) and empathetic self-awareness.

As Fig. 16.4 shows, 'going through the U' is an inner journey made of various steps and phases. Throughout the workshop, the way we experienced this was by switching between more individual reflections and collective sharing, either in groups of four to five people, or with the whole group of participants (around 20 each time). A key element of adopting Theory U was the focus on **embodiment**: the process of gradually unravelling the institutional (structural) barriers, as well as the inner ones, to fully embrace transformative change that requires an open will,

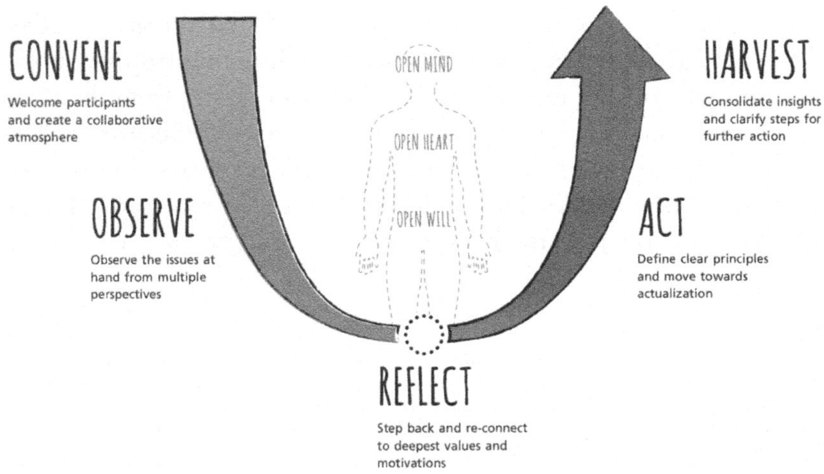

Fig. 16.4 The Theory U process of co-sensing and co-creating (*Source* Pearson et al. [2018] as adapted from Scharmer [2009])

an open heart and an open mind. Therefore, these three 'requirements' immediately put us, the participants, in a context where our professional hats were no longer relevant. Instead, we were asked to meet just as human beings. As some of the participants said in their feedback form, it was "powerful being just a person" although "getting rid of the expectations on my role" was considered challenging—sentiments with which I also fully concurred from my own experience.

In this process of *learning with* each other and taking time for *care*, we engaged our bodies, our hearts, and our hands in a dynamic relational process of sense-making: what does collaboration mean to us? What does 'deep-listening' mean? How do we do that? How can we learn to listen to ourselves and others, without interruption, leaving aside a judgemental attitude to embrace a welcoming and generative one? As one participant put it, "making sense of the mess in the way we did it" was perceived as highly valuable, but was also considered as challenging with the same participant reflecting: "[it was challenging] to make sense of the mess in my mind". Many exercises proposed by our facilitators helped us to reflect on these contradictions. As highlighted by participant feedback, one of the most appreciated parts of the workshop was a walk outside with one other person, whom we had (ideally) not met before, to share a formative moment of our lives with, while also practicing deep-listening. These walks and the request 'to have more of it' were among the most frequent answers to the "what was good about the workshop" question, as well as to the "what would you do differently" question, in the feedback forms.

Especially through the one-to-one walk, immersed in nature, present with ourselves, we had the opportunity to encounter each other, to feel connected, to feel being *in relation* with one another, as human beings. "Taking time out connecting with others", "sense of community you managed to create it!", but also "talking with people without having preconceptions of their views", "more connection and a different type of dialogue" and "meeting people and doing exciting, meaningful, and sustainable things" were among the participants' answers when asked what was good about the workshop. Being aware of our inherently relational nature, of being and doing together, implies being reflexive about the nature and dynamic of interdependence. A core part of the journey along the 'U' was in fact to fully embrace the truth that the one is not separate from the system. However, the more we recognized our interdependence, the more vulnerability, fear and (at times uncomfortable) intensity of emotions came along. For example, some of the participants reported the following as personally challenging: "Being vulnerable to others", "looking inwards", "being emotionally honest", "being uncomfortable, yet feeling safe". Observing and experiencing first-hand these dynamics through the lens of transdisciplinary research, helped me to further realize the complexity and arduousness of the emotional labour required for co-creative and collaborative working.

The workshop was organized twice, two weeks apart, with two different groups of participants, although I was the exception since I participated both times. This allowed me to experience the usually fluid and ambivalent role of the embodied and FAR-inspired researcher, always juggling between insider–outsider, participant-observer, impartial-invested roles, in a more distinctive way. During the first set of workshops, I fully embraced the role of *participant*, enthusiastically engaging with fellow participants in all the activities proposed by the facilitators. I thoroughly immersed myself, especially in the **self-reflexive process** core of the workshop. We crafted a space together in which we took the time to simultaneously reconnect, inwards and outwards, individually and collectively. The importance of time dedicated to nurturing reflexivity was mentioned by the majority of the participants in their feedback on what was good about the workshop: "loads of reflection—very much needed", "Time to go deep—nothing felt too rushed—helps drop down into reflective space", "time to recentre", "allowing time for reflection", "space and time for people, to let themselves out of their boxes".

At the same time, however, through the required emotional labour, moments of intense inner working drained much of my energy. I felt that my internal compass had become unbalanced: a propension towards only one side of the spectrum envisioned by the FAR framework brought me towards *participation, investment* and *care,* leaving no space or energy to counterbalance that instinctive need to be *just* a participant. The inner working, reflection, sharing and *learning with* the other participants resulted in a sense of loss of my usually ambivalent researcher's role. I was painfully letting myself into my own personal journey along the U: I noted in my journal "so hard to let it go, I feel very embarrassed and vulnerable". Nevertheless, I was aware that my normative positions and my strong will to contribute to that (cultural) change (at the *heart* and *feet* of my research practice) required me to fully embrace that vulnerability to be able to care, to co-create and to hold that space with others, that could let us all feel part of a wider community with common purpose. As one participant similarly experienced it: "interesting points on how to change—the need for pain and discomfort".

One example of such complex and difficult moments of vulnerability came towards the end of the first set of workshops. The final steps of the Theory U, as depicted in Fig. 16.4, required us to *crystalize* a vision, an intention, an idea, that was generated throughout the *presencing* phase (the uncomfortable and painful 'bottom of the U', the place from which we also generate and create 'the new') into a concrete and tangible *prototype*. This involved us listing and identifying a series of concrete actions to bring 'the new' to life, and make it real. When we were asked to make our own prototypes by the facilitators, I refused. I felt I was not ready yet to get out of 'the bottom of the U'. The painful but generative moment I was going through was not finished yet, I needed more time to process that pain and discomfort, before being able to 'prototype' my (new) intentions and vision. When we were asked to share our prototypes with the rest of the group, and I had to admit that I could not do it, it was embarrassing, but also liberating and empowering: through our shared experience and *learning with*, I had reached a sense of connection, safety and trust within that newly emerged 'community', that I felt confident and fine with being honest about my "failure".

Discussion

Earlier in this chapter, we proposed that *cultural shifts* are fundamental to systemic change towards sustainability; we are envisioning such cultural shifts as re-thinking, co-creating, and re-imagining new and alternative meanings and understandings of the world we want to live in and the people we want to be. The interactions (and reactions) that characterized some of the three formative moments reviewed in the preceding section show how artistic practices can facilitate such collaborative and creative processes of meaning-making. When looking at policy-making contexts as relational, Lejano (2020, p. 4) argues that "what is needed is closer, undivided attention to the workings, and the richness, of the relationships themselves" between the policy actors involved. Throughout this chapter, we have attempted to account for such richness and intensity by discussing the relational dynamics between participants, while making sense of principles, requirements and *ways of working*, established by the Welsh legislation in relation to SMNR.

The elements of *relationality* and *embodiment*, the *being* and *doing* together, have remained at the very centre of our account throughout. While we have elected to predominantly focus on the reflexive experience of the lead author, her experience, in turn, points to the wider relevance and potential of fore-fronting care when implementing policies and practices for sustainability transformations. Our experience therefore supports a conception of 'knowledge' and 'knowing' as multiple and distributed elements, which develop and evolve through people's relationships and practices. The Embodied Researcher and FAR frameworks have allowed us to critically analyze relationality and embodiment for creative and collaborative (research) practices.

Two crucial elements/challenges have emerged. First, the engagement in such creative and collaborative sense-making processes is extremely **time-, energy-** and **emotionally-intense**. Moreover, it has the potential to produce multiple and multi-faceted tensions within (as well as between) participants. In the case of the first author's own experience, such (emotional) intensity was exacerbated by the fact that she anchored her practice to the principle of care, which (as discussed above) is an ambiguous and multifaceted concept. In the instances in which she was not fully able to balance impartiality and investment, the intensity of her emotional involvement mixed with her ethico-political commitment to *practice* care, leaving her in an uncomfortable situation. As a result, she found it hard on such occasions to re-establish a 'safe distance', and a *dynamic proximity* between herself and the group of participants.

Second, the emotional work involved in such creative and collaborative processes goes hand-in-hand with the uncomfortable (but unavoidable) task of facing **vulnerability** and fears, triggered by *being* and *doing with* others. The experiences analyzed here stress the importance of meeting one another 'just' as human beings during shared, collaborative endeavours. From working with artists as professional facilitators we have learnt to experience vulnerability as a way of practicing and embracing interdependence and relationality. As highlighted by Tronto (2017, p. 32), we do go through a fundamental ontological shift, a fundamental rethinking of our very own nature, when we understand that "everything exists in relation to other things [...] people, other beings and the environment are interdependent", and that "all humans are vulnerable and fragile". At the same time, though, *knowing* that we (as the human species, and as Planet Earth inhabitants) are interdependent and vulnerable is not enough: the care involved in *being* and *doing together* dramatically helps us to fully embrace our very own condition.

Notably, however, the embodied and relational experiences and practices analyzed here require two fundamental elements: **time and mental space for *care*.** 'Time-out for thinking' was especially highlighted by many of the participants as the main good thing, for example, about the 'To the Moon and Back' workshop. A fundamental benefit of such a workshop for professionals from governmental organizations and other practitioners was indeed to carve out some time and space for themselves to engage in conversations, listening and reflecting alone and with others. Their everyday jobs, often filled with tight deadlines and narrowly-defined deliverables, rarely if ever allowed such engagement, and left little, if any, room for experimentation and possible failure. The professionals involved in the 'To the Moon and Back' workshops, once having overcome an initial reluctance to make time for them, found it 'refreshing' to have the opportunity to deeply engage with one another. Time and (mental) space to experience genuine collaboration and the sharing of stories about personal as well as professional lives, constitute the very base on which to build relationships of trust. Ultimately, these relationships underpin the whole legislative structure around SMNR in Wales.

An important question therefore arises from the discussion of our experience (Giambartolomei's especially) of collaborating with members of WG in a transdisciplinary research setting: how can we (academics) best support governmental organizations and the individuals within them, to institutionalize these alternative practices centred on embodied and relational experiences (e.g., deep listening, walking conversations, prototyping together), and mainstream them into the policy-making realm? How can governmental institutions enable such process-based practices to stimulate meaningful and creative collaboration, without exhausting or over-exploiting people's energy and motivation to get involved? We suggest that one place from which to start answering such questions is by asking these professionals directly, but also via more creative forms of inquiry: what support do *you* need in order to build stronger relationships and facilitate deeper collaboration throughout your—and our—ways of working? Space and time are certainly critical here, but given the potentially transformative nature of this question, undoubtedly numerous other forms of support will also continue to be required.

Conclusions

This chapter has discussed and analyzed the first-hand experience of a doctoral researcher in doing collaborative and transdisciplinary research with governmental institutions in Wales. Our account is grounded in an innovative methodological lens that, centred around the principle of care, combines the Formative Accompanying Research framework (elaborated by Freeth, 2019), and the Embodied Researcher approach (introduced by Horlings et al., 2020). Set in the context of the *spaces in-between* SMNR legislation and its implementation, we have applied these combined frameworks to analyzing fragments of multiple, intersecting journeys. These journeys include: the joint programme of the WG and NRW to improve their collaborative organizational culture; the experience and reactions of participants encountering this joint programme, each committed in their own ways, to realizing SMNR's principle on the ground; and the embodied trajectory of the doctoral researcher herself, attempting to *accompany* these professionals for a small part of their journeys, trying to listen as much as possible to their stories and needs, while also walking her own doctoral path. The conclusion drawn from this analysis is that such *spaces in-between* offer rich potential to co-creatively experiment with the relational and embodied dimensions of deeper, more meaningful and meaning-making forms of collaborative working. It is through *care*-full, iterative and reflexive experimentation that we can begin to better align our heads, hearts, hands and feet. Ultimately, such alignment is fundamental to co-creating a shared journey of socially and ecologically just sustainability transformations.

Acknowledgements The doctoral research on which this piece is based received funding from the Energy, Environment and Rural Affairs Department of Welsh Government and Coventry University. Agreement No. 5894.

References

Billo, E., & Hiemstra, N. (2013). Mediating messiness: Expanding ideas of flexibility, reflexivity, and embodiment in fieldwork. *Gender, Place and Culture: A Journal of Feminist Geography, 20*(3), 313–328. https://doi.org/10.1080/0966369X.2012.674929

Blythe, J., Silver, J., Evans, L., Armitage, D., Bennett, N. J., Moore, M.-L., … Brown, K. (2018). The dark side of transformation: Latent risks in contemporary sustainability discourse. *Antipode*, 0(0), 1–18. https://doi.org/10.1111/anti.12405

Cahill, C. (2007). The personal is political: Developing new subjectivities through participatory action research. *Gender, Place and Culture, 14*(3), 267–292. https://doi.org/10.1080/09663690701324904

Dieleman, H. (2017). Enchanting sustainability. From enlightened modernity towards embodiment and planetary consciousness. In S. Asikainen, C. Brites, K. Plebańczyk, L. R. Mijatović, & K. Soini (Eds.), *Culture-sustainability relation: Towards a transdisciplinary approach* (SoPhi, pp. 10–21). University of Jyväskylä, Department of Social Sciences and Philosophy.

Emergence. (2020). Emergence. Retrieved November 20, 2020, from https://emergence-uk.org/.

Fazey, I., Moug, P., Allen, S., Beckmann, K., Blackwood, D., Bonaventura, M., … Wolstenholme, R. (2018). Transformation in a changing climate: A research agenda. *Climate and Development*, 10(3), 197–217. https://doi.org/10.1080/17565529.2017.1301864

Feola, G. (2015). Societal transformation in response to global environmental change: A review of emerging concepts. *Ambio, 44*, 376–390. https://doi.org/10.1007/s13280-014-0582-z

Fisher, B., & Tronto, J. C. (1990). Towards a feminist theory of caring. In E. K. Abel & M. Nelson (Eds.), *Circles of care. Work and identity in women's lives*. SUNY Press.

Freeth, R. (2019). *Formative accompanying research with collaborative interdisciplinary teams*. Doctoral Thesis.

Freeth, R., & Vilsmaier, U. (2020). Researching collaborative interdisciplinary teams: Practices and principles for navigating researcher positionality. *Science and Technology Studies, 33*(3), 57–72.

Freire, P. (1970). *Pedagogy of the oppressed* (Peguin Cla). Penguin Random House.

Geoghegan, H., Arnall, A., & Feola, G. (2019). Climate and culture: Taking stock and moving forward. In G. Feola, H. Geoghegan, & A. Arnall (Eds.), *Climate and culture: Multidisciplinary perspectives on a warming world* (pp. 1–18). Cambridge University Press.

Hammond, M. (2020). Sustainability as a cultural transformation: The role of deliberative democracy. *Environmental Politics, 29*(1), 173–192. https://doi.org/10.1080/09644016.2019.1684731

Haraway, D. (1988). Situated knowledges: The science question in feminism and the privilege of partial perspective. *Feminist Studies, 14*(3), 575–599. https://doi.org/10.4324/9780203427415-40

Haraway, D. (1991). *Simians, cyborgs and women: The reinvention of nature*. Free Association Books.

Horlings, L. G., Nieto-Romero, M., Pisters, S., & Soini, K. (2020). Operationalising transformative sustainability science through place-based research: The role of researchers. *Sustainability Science, 15*, 467–484. https://doi.org/10.1007/s11625-019-00757-x

IPBES. (2019). *Decision IPBES-7/1: Rolling work programme of the intergovernmental science-policy platform on biodiversity and ecosystem services up to 2030*.

Lejano, R. P. (2020). Relationality: An alternative framework for analysing policy. *Journal of Public Policy*, 1–24. https://doi.org/10.1017/S0143814X20000057.

Moriggi, A., Soini, K., Franklin, A., & Roep, D. (2020). A care-based approach to transformative change: Ethically-informed practices, relational response-ability & emotional awareness. *Ethics, Policy, & Environment*, 1–18. https://doi.org/10.1080/21550085.2020.1848186.

National Assembly for Wales. Environment (Wales) Act (2016). Retrieved from http://extwprlegs1.fao.org/docs/pdf/uk154195.pdf.

Newton, J., Franklin, A., Middleton, J., & Marsden, T. (2012). (Re-)negotiating access: The politics of researching skills and knowledge for "sustainable communities." *Geoforum, 43*, 585–594. https://doi.org/10.1016/j.geoforum.2011.12.003

Nightingale, A. J., Eriksen, S., Taylor, M., Forsyth, T., Pelling, M., Newsham, A., ... Whitfield, S. (2020). Beyond technical fixes: Climate solutions and the great derangement. *Climate and Development*, 12(4), 343–352. https://doi.org/10.1080/17565529.2019.1624495.

O'Brien, K. (2012). Global environmental change II: From adaptation to deliberate transformation. *Progress in Human Geography*, 36(5), 1–10. Retrieved from http://web.b.ebscohost.com.rproxy.uwimona.edu.jm/ehost/pdfviewer/pdfviewer?vid=1&sid=818cc447-7d47-40e6-9930-51a187227 5a7%40sessionmgr103.

O'Brien, K. (2013). The courage to change: Adaptation from the inside-out. In S. C. Moser & M. T. Boykoff (Eds.), *Successful adaptation to climate change: Linking science and policy in a rapidly changing world* (pp. 306–320). Routledge.

O'Brien, K. (2018, April 1). Is the 1.5°C target possible? Exploring the three spheres of transformation. *Current Opinion in Environmental Sustainability*. Elsevier B.V. https://doi.org/10.1016/j.cosust.2018.04.010.

O'Brien, K., & Sygna, L. (2013). Responding to climate change: The three spheres of transformation. In *Proceedings of transformation in a changing climate* (pp. 16–23). Retrieved from www.cchange.no.

Pearson, K. R., Bäckman, M., Grenni, S., Moriggi, A., Pisters, S., & de Vrieze, A. (2018). *Arts-based methods for transformative engagement: A toolkit*. SUSPLACE.

Pelling, M. (2010). *Adaptation to climate change: From resilience to transformation*. Routledge.

Pelling, M., O'Brien, K., & Matyas, D. (2015). Adaptation and transformation. *Climatic Change, 133*(1), 113–127. https://doi.org/10.1007/s10584-014-1303-0

Puig de la Bellacasa, M. (2011). Matters of care in technoscience: Assembling neglected things. *Social Studies of Science, 41*(1), 85–106. https://doi.org/10.1177/0306312710380301

Puig de la Bellacasa, M. (2017). *Matters of care. Speculative ethics in more than human worlds*. University of Minnesota Press.

Rose, H. (1983). *Hand, brain, and heart: A feminist epistemology for the natural sciences. Source: Signs, Autumn* (Vol. 9).

Scharmer, C. O. (2009). *Theory U: Learning from the future as it emerges*. Berrett-Koehler Publishers.

Scharmer, O. (2018). *The essentials of theory U: Core principles and applications*. Berrett-Koehler Publishers.

Soini, K., & Dessein, J. (2016). Culture-sustainability relation: Towards a conceptual framework. *Sustainability, 8*(2), 167. https://doi.org/10.3390/su8020167

Tronto, J. C. (1993). *Moral boundaries: A political argument for an ethic of care*. Routledge.

Tronto, J. C. (2013). *Caring democracies. Markets, equality and justice*. New York University Press.

Tronto, J. C. (2017). There is an alternative: Homines curans and the limits of neoliberalism. *International Journal of Care and Caring, 1*(1), 27–43. https://doi.org/10.1332/239788217X14866281687583

Wittmayer, J. M., & Shapke, N. (2014). Action, research and participation: Roles of researchers in sustainability transitions. *Sustainability Science, 9*(4), 483–496. https://doi.org/10.1007/s11625-014-0258-4

17

How to Make Policy-Makers Care about "Wicked Problems" such as Biodiversity Loss?—The Case of a Policy Campaign

Agnes Zolyomi

Introduction

One of the most spectacular characteristics of this planet is its abundance of life with over 9 million plants, animals and fungi species, all of which provide the basis of human society and economy (Cardinale et al., 2012). Despite its fundamental role however, biodiversity is continuously damaged and decreased by human activities. The alarming loss of biodiversity, which threatens the mass extinction of over 1,000,000 species, does not only critically endanger the biosphere but our very own existence (IPBES, 2019). Despite the increasing scientific evidence, humanity's ultimate dependence on nature and its goods is seemingly not evident at the decision-making levels, resulting in a lack of policy prioritization of nature and its goods (Mace et al., 2018; Primmer et al., 2015;

A. Zolyomi (✉)
Coventry University, Coventry, UK
e-mail: agnes.zolyomi@coventry.ac.uk

A. Franklin (ed.), *Co-Creativity and Engaged Scholarship*,
https://doi.org/10.1007/978-3-030-84248-2_17

WWF, 2018). With the current business-as-usual scenario, both the rate of biodiversity's loss and the risks for humanity continue to grow.

Policy-makers are on top of the food chain in determining many aspects of our society and economy by paving the general directions in policy. They are one of the (if not the most important) key players to influence the steering of the political agenda in a certain direction. At the same time, they are also the key players that are rather difficult to reach or affect especially if one is not a member of a large industrial or other lobby power, but rather belongs to an under-represented group that usually gets more eye-rolling (which is mostly the case with the nature conservation sector). So, we are given an under-represented cause from an under-represented group.

As a result, in order to convince decision- and policy-makers to put nature and biodiversity higher on the relevant policy and decision-making agendas, we need unprecedented communication and further scientific efforts. Thanks to trailblazer NGOs, IPBES (Intergovernmental Science-Policy Platform on Biodiversity and Ecosystem Services) and other science-policy interfaces, such efforts have achieved promising results in recent years; however, numerous challenges are still to be faced. For one, communicating biodiversity and its loss is itself a grandiose task due to its complexity, and often indirect effects on one's life (e.g., we will most probably not feel the straightforward implications of the loss of tigers) (Kidd et al., 2019; Legagneux et al., 2018; Millner & Olivier, 2015; Sharman & Mlambo, 2012; Zaccai & Adams, 2012). However, communicating this to a special group of decision-makers at the highest level, who are targeted by myriads of messages from a vast number of often conflicting lobby groups and influencers, calls for all the creative and other resources and wits one can think of. If a communication piece is to stand out and to be heard, you need to have an actual message they are willing to listen to within an actual relevant policy process in an easily digestible, creative format that goes above their threshold level of uninterest. The challenge is that there is no exact secret or scientific recipe for how to do that—but there are indeed certain 'magic' ingredients.

In this chapter, I will discuss why biodiversity is particularly challenging to communicate, especially for policy-makers in democratic political systems that, for example, the EU represents. I will also showcase specific strategies for communication that have already proved successful and have been dispatched by the conservation sector. Further potential methods and tools stemming from behavioural economics will also be presented with the recommendation to consider them at a more elevated level when formulating future policy messages. Through my personal experience, I will explore the application of these scientific strategies in the specific conservation campaign I was involved in, which aimed to safeguard the pillars of the European Union's nature conservation policies from possible restructuring and watering down. I will furthermore reflect on the subsequent results and potential improvement of nature conservation and biodiversity messages to reach even more substantial policy impacts.

Biodiversity as a Diffuse Problem—And Other Inconvenient Truths

Apart from the everyday challenges of advocacy work in policy (e.g., short-termism, uncertainty and individual versus collective gains and losses), there is a further dimension to tackle complex issues such as climate change or biodiversity loss, the so-called diffuse or wicked problems (Millner & Olivier, 2015; Sharman & Mlambo, 2012; Zaccai & Adams, 2012). These problems are complex with no specific villain and victim, with a sense of remoteness of impacts and responsibility (Millner & Olivier, 2015) (again, it is sad that tigers will become extinct, but so what and how is this my fault anyway(?)). Additionally, stakeholders who manage or are interested in biodiversity are not only diverse and are therefore difficult to address and involve; the drivers of biodiversity losses are also multiple. In the context of the policy environment, stakeholders and interest groups are numerous, while their values and interests are often controversial and ambiguous (Sharman & Mlambo, 2012). Even more so, a solution to one aspect of a challenge can lead to probable further difficulties, inducing trade-offs and subsequent further complexity (Sharman & Mlambo, 2012). With the various interests and the lack of simple solutions to tackle the complex challenge in question, the issue is considered wicked (Rittel & Webber, 1973).

In particular, in terms of biodiversity, there is a limited understanding about societies' dependence on it, as well as of the sectorial and economic impacts of its loss (Zaccai & Adams, 2012). Confusion among policy-makers, as well as any particular individual, is further aided by the unclear definition of biodiversity and its loss, and accordingly, biodiversity policy messages and the various concepts in use (Legagneux et al., 2018; Sharman & Mlambo, 2012). In addition, personal ownership of responsibility for halting biodiversity loss is also incredibly low. The latter is partially due to the fact that neither the concepts of biodiversity and biodiversity loss, nor individual actions to tackle biodiversity loss, are adequately understood (Sharman & Mlambo, 2012)—or as a matter of fact, want to be understood.

The major drivers of biodiversity loss, which remain the over-exploitation of natural resources and agriculture (Maxwell et al., 2016), added to by inept governance (WWF, 2018), are related to our current world economy's and society's set-up. Biodiversity can be considered as the necessary victim of the ever-expanding global markets and human population (Maxwell et al., 2016) facing vested interests and political pressures to deliver on the economic scale (OECD, 2017). In addition, prioritizing nature can result in dreaded economic losses (e.g., enhancing environmental regulation can contribute to competitive disadvantage and higher costs that can affect jobs and growth (OECD, 2017)). In this arduous policy environment, it is therefore not too surprising to understand why communicating biodiversity is a demanding quest, and why certain tricks up one's sleeve are needed.

Ways and Strategies for Communicating Biodiversity Messages

To date, there is a very limited amount of literature about how to communicate biodiversity messages effectively (Kidd et al., 2019), let alone scientific knowledge about specifically transferring conservation-related messages targeting decision-makers. In a recent attempt to divulge how biodiversity is communicated, Kidd et al. (2019) conducted a systematic review with the result that ecosystem services and flagship species were most frequently the key frames within which authors transferred their messages. The familiarity principle (people tend to prefer those things that they can directly relate to or could be affiliated with (Reder and Ritter (1992)), risk perception, connection to nature or raising other emotions were also included in those articles as ways of communicating biodiversity. These communication methods and strategies will be briefly presented here. It is nonetheless important to note that these are only some examples of the probable tools; these can be further tailored by rhetoric analysis to choose what kind of concepts, terms, arguments, etc., can be used with different target audiences.

Communicating with Ecosystem Services

The ecosystem services concept was formed as a communication tool as early as the 1970s by Westman (1977) who described the value of services provided by nature (Bekessy et al., 2018). More focused work on nature services resulted in the term "ecosystem services" first described in the work of Ehrlich and Ehrlich (1981). The efforts to capture nature and its services in both ecological and financial terms were pushed by both ecologists and economists with the overall aim being to capture the immense dependence of human society and the economy on nature (Chaudhary et al., 2015). This concept was then further shaped and rose to international prominence with the renowned articles of Daily et al. (1997) and Costanza et al. (1997); these depicted the definitions of ecosystem services and provided an estimate of the monetary value of all ecosystem services in the world (over US$16–54 trillion annually). These studies called for further scientific debates and triggered myriads of additional research, especially in ecosystem services valuation (Costanza et al., 2017).

Chaudhary et al. (2015) provide a thorough overview of the history and development of the ecosystem services concept that also continued to gain ground in the international policy arena: the concept was adopted by the UN's Convention of Biological Diversity and the Millennium Development Goals in the early twenty-first century. Following its adoption by numerous multilateral agreements, a rapid uptake could be witnessed by various disciplines, reaching its peak in 2005 with the global Millennium Ecosystem Assessment (MEA) synthesis report. The report not only provided a warning overview of the degrading status of the ecosystems and their services, but also connected the world's major scientists in a global mission to contribute to saving nature. Additionally, the MEA provided frameworks and standards on definitions, which were the basis of many following initiatives and studies. The Economics of Ecosystem and Biodiversity (TEEB) was launched in 2007 as a response to a G8+5 nations' proposal to assess the economic benefits of ecosystem services and the associated economic loss. In the same year, the idea of Intergovernmental Science-Policy Platform on Biodiversity and Ecosystem Services (IPBES) was formulated with the aim of establishing a body similar to the International Panel on Climate Change (IPCC), specifically for biodiversity. The idea came into realization in 2010 with the overall goal being to communicate the importance of biodiversity and ecosystem services and to transfer scientific messages to policy-makers. Since this era, ecosystem services have gradually received prominence in additional (mostly environment and biodiversity relevant) global policies, including for instance, the Convention on Biological Diversity's Aichi Targets, Sustainable Development Goals and EU policies, and became the focus of thousands of scientific articles (Chaudhary et al., 2015).

Across its life course, the ecosystem services concept has attracted considerable criticism due to its anthropocentric and economic focus, which to some degree omits the intrinsic values of nature (Chaudhary et al., 2015). On the other hand, however, other arguments underpin the view that such a human-values-centred approach was needed for the concept to enable mainstreaming of the challenge and potential solutions among a wider range of stakeholders (Schröter et al., 2014). It is this very approach that is thought to enable nature to be depicted in a language that decision-makers in particular are able to grasp (Bekessy et al., 2018). The idea behind the valuation of ecosystem services ("putting a price tag on nature") is embodied in the notion that as a result of the valuation, decision-makers will understand the immense value of ecosystems and their services and therefore rational choices will be made to prioritize and protect them (Braat & de Groot, 2012; Costanza et al., 2017).

Not surprisingly, ecosystem services findings are also mainly prepared for these specific stakeholders (Bekessy et al., 2018), although to date, with limited evidence on their actual impacts (Martinez-Harms et al., 2015; Posner et al., 2016; Wright et al., 2017).

The international tendency for highlighting nature's services has also been picked up by the European Commission, who communicated the social benefits of Natura 2000 (e.g., ten Brink et al., 2013) with the hope that the vast benefits of the EU's protected area translated to socio-economic terms can serve as an eye-opener for decision-makers fixated on these figures. Ecosystem services also started to occupy significant roles in nature-relevant EU policy pieces; this includes the Biodiversity Strategy 2020 and its target 5, where member states were required to map and assess their ecosystem services and integrate them into their financial accounting system by 2020 (European Commission, 2019a). Furthermore, other major EU policies, including the Common Agricultural Policy and the Marine Strategy Framework Directive, also refer to ecosystem services (Schleyer et al., 2015).

Communicating with Flagship Species

In order to raise awareness about biodiversity, as well as raising funds for conservation, many strategies and campaigns use the flagship species approach. Here, organizations use an iconic and well-known species (e.g., panda, elephant, koala) most people are compassionate about and to which they attribute positive feelings. In this way, one can raise people's attention not only about specific issues that are in direct relevance with the flagship species, but also in more general terms about the environment (Schlagloth et al., 2018).

First described in 1988 by Mittermeier, the flagship species concept showed, via case studies, how particular species were used to successfully convey conservation messages to the general public. Flagship species do not necessarily have to be cute, cuddly or majestic; their "use" largely depends on the target audience of the campaign and their cultural, traditional or historical connection with the species (Schlagloth et al., 2018). As Jepson and Barua (2015) point out, employing flagship species can be categorized into three distinct groups: (1) providing a compelling moral background for policy work, (2) upscaling inter-institutional considerations, and (3) offering a good basis for justifying conservation efforts. However, while attracting attention to the importance of biodiversity, and often, biodiversity loss, this way of communication received critiques for putting too much focus on certain, mostly animal, species (Smith et al., 2012). On the other hand, some argue the flagship species approach is solely a way of communication used not only to raise funds successfully, but also for the conservation of those species, which may be less plausible to effectively advocate for (McGowan et al., 2020).

Others suggest extending the flagship approach to a flagship fleet by covering multiple species within one communication campaign to provide opportunities for less well-known or difficult-to-communicate species (Veríssimo et al., 2013). Although flagship species campaigns in many cases proved efficient and fruitful (e.g., in the case of the giant panda as the logo animal of one of the most successful NGOs), in other instances they failed to reach the desired outcome (e.g., in the case of the orangutan in a campaign addressing unsustainable palm oil, which did not entirely reach its target) (Jepson & Barua, 2015). There is limited knowledge on how exactly flagship species add to conservation outcomes and why certain species' relevant communication and advocacy actions became victorious, while others may achieve only modest outcomes (Jepson & Barua, 2015; Lundberg et al., 2019 and Veríssimo et al., 2013).

In the European Union context in relevance to campaigning for Natura 2000, and nature conservation in general, the large carnivores of Europe (brown bear, wolf, the European and Iberian lynx species, and the wolverine) as well as predatory birds have enjoyed a prominent role in campaigns (BirdLife, 2020; WWF, 2020). If we take a look at the specific campaign videos of BirdLife for instance advocating for strong nature conservation policies, mostly large mammals and birds steal the limelight with the occasional appearance of marine fauna (BirdLife, 2020), suggesting that these are the flagship species that are believed to attract people (see also Baimukhamedova, this book).

The Familiarity Principle and Targeting Emotions

The familiarity principle is the concept that, in decision-making, people tend to prefer something of which they have preliminary knowledge and are familiar with; this can play a key role in communicating biodiversity messages (Kidd et al., 2019). In a recent experiment with donation preferences for opportunities to fund various flagship species and familiar ecosystems, it was revealed that familiarity played the most important element when deciding whether to financially support a project (Lundberg et al., 2019). This concept is in close relation to reducing psychological distance in order to enable a closer connection between the target audience and the communicated entity, by emphasizing the relationship or the impacts of proximity to the audience (Kusmanoff et al., 2020). Accordingly, when compiling a communication or advocacy campaign, it is important that the specific stakeholder group addressed by the campaign can connect with the entity used, not only emotionally, but ideally geographically or conceptually as well.

Let us take a look again at the tiger case or the animal's habitats in Asia. Many would consider it a very unfortunate and sad event if more rainforests or birch forests disappeared, together with the tiger becoming extinct. However, without personal attachment or the familiarity feeling, it would most probably remain a distant calamity (such as the Australian wildfires eradicating koalas or the polar bear's disappearance—despite the flagship species). Now, consider the forests you have been visiting with your family for the summer holidays for 20 years being felled and the squirrels you watch every Saturday with your children disappearing for good. You would certainly feel a more significant loss due to your familiarity with the squirrels and would be more prone to take action against their disappearance.

At the heart of all this, of course, there are the emotions that steer our decisions in many, if not most, cases, contrary to the common belief that we decide rationally (Thaler, 2015). We can think about what drove us when we made our most recent donations. Whereas statistics and numbers may be believed to work well when selecting our options (the rational theory of decision-making states that we carefully consider our choices and choose the one that "best" serves our needs (Liebe & Preisendörfer, 2010)), evoking emotions is usually a more effective means to appeal to people, especially when we have such an arsenal of the cute and the cuddly. Such emotional messages can be built on compassion in a positive manner to ask for a contribution, or can take the shape of a negative message that is mostly built on fear (e.g., fear of the extinction of whales during the hunting season) (Kusmanoff et al., 2020). Understandably therefore, the European campaigns on nature conservation feature local species one can directly relate to and are familiar with. These trigger some specific emotions (awe, fear, affection) and maybe personal experience or memory, making these species for us more as a subject of concern in case of probable extinction. As we can see in the videos of BirdLife (2020) mentioned above, most species (brown bear, wolves, otter, deer, squirrels, seals, dolphins, various bird species) are known by European residents and there is a high chance they have already encountered them in person.

Amending Biodiversity Communication Messages with Social and Behavioural Economics Theories

Even though biodiversity professionals may not always draw scientific lessons from other disciplines—just as we could see in the specific case detailed below about the Fitness Check campaign for the EU nature policies review—economic, sociology, psychology, and communication studies can guide conservationists to reach the better results we desperately need. For instance, it was pointed out by Kusmanoff et al., (2016, 2020) that conservation messages often miss one of the initial steps of defining and targeting the precise audience, and the messages often land with people who are already supportive towards the pitch. Different audiences need different messages, simply because various groups hold distinct values and social norms. Typically, groups can be formed based on their motivations (e.g., motivated by self-interest, altruism or pro-environmental behaviour), and so these groups are prone to act differently to various messages. For instance, an ultimately dedicated environmental activist will be difficult to motivate purely by monetary benefits, whereas members of other groups may be driven by self-interest and, accordingly, will be more likely to act on knowledge of financial gains.

A major building block of a campaign should also be grounded on social norms as they are believed to be one of the determining factors of environmental behaviour (Ajzen, 1991; Cialdini et al., 1990; Farrow et al., 2017). Social norms define what is approved of in a given social context and what behaviour should not be conducted (for instance, if many of our friends are vegetarian and ardent animal right protectors, we are more likely to believe this is an important issue and act accordingly, e.g., donate to animal rights organizations, buying products that do not harm animals, turn to a vegetarian diet ourselves, etc.). By understanding behaviour and driving norms, messages can be framed in order to nudge the target audience (Kusmanoff et al., 2020). For instance, in terms of a biodiversity campaign, we can highlight that most people have already donated to the cause, or already signed the petition framing that it is the norm to do so (meaning you are strange not to sign it).

There are also numerous heuristics and biases that can be used in communicating biodiversity messages as they impact people's decision-making. The classical example of the framing effect is Kahneman and Tversky's (1979) prospect theory. Here, the authors refuted the rational choice theory element, the expected utility theory stating that people, instead of rationally and meticulously calculating their chances in uncertain situations, rather bid on the certain gain versus only a probable significantly larger value, while gamble eagerly to avoid losses. Accordingly, messages can be framed to highlight the benefits in avoiding losses, and use negative framing (which, however, is indeed challenging in the biodiversity loss context if it is competing with jobs and growth losses).

The endowment effect may also be worth considering as it states that a certain item owned by us will be considered as having a larger value than others in the market, as certain additional (e.g., emotional) values are contributed to it (Kahneman et al., 1990). This may be used in situations amplifying, for instance, a specific species' value for a specific area. A related concept, the scarcity heuristic (we tend to contribute more value to an item if it is scarce) may also be employed, especially considering that most species in communicating biodiversity are endangered or close to extinction (Kusmanoff et al., 2020). The confirmation bias (we tend to agree with those arguments that underpin our belief system (Nickerson, 1998)), together with the *status quo* bias (we tend to dislike changes in the system we are used to) (Samuelson & Zeckhauser, 1988)), also affect advocating for biodiversity—unfortunately, in an adverse way. Mostly, this is because it is unpleasant to hear that certain, big-scale changes are needed to overwrite the current business-as-usual scenario. To date, there are almost 200 identified heuristics and biases; these may be worth examining when compiling a campaign for biodiversity (Kusmanoff et al., 2020).

Communicating messages to influence decision-makers are, of course, embedded in a wide array of theories, far more than those listed above. These social and economic theories address variations of norms, beliefs, narratives or social biases and heuristics aiming to explain how a nuance of change in the communication may result in different outcomes. However, it is important to underline that at the time of working on our nature conservation campaign in 2014–2016, I had only very vague ideas of the science on communicating or advocating for biodiversity—despite being a member of an advocacy NGO for years. Yet, many of our strategies detailed below mirror well the currently emerging field of communicating biodiversity messages that borrows ideas from other well-established disciplines (Kidd et al., 2019). At the same time, it is overly evident that many advocacy and research messages should be further tailored in order to nudge decision-makers towards prioritizing biodiversity on a more extensive scale.

Communicating Biodiversity on the Frontline—A Case of Advocating for the European Union's Nature Policies

Overview of the Fitness Check Process and Relevant Actors

In 2014–2016, the conservation sector was preoccupied by the so-called Fitness Check process of the EU nature conservation policies (the Birds Directive and the Habitats Directive)[1]; which threatened the possible opening up, and consequently weakening of the nature legislation. Jeopardizing the EU's Nature Directives and together with it the EU-wide Natura 2000 network of protected areas, meant the century's war to prepare for in conservation, and the green NGO world I also worked in started to get ready for the fight.

At the EU level of conservation, the Natura 2000 network (defined by EU nature policies, the Birds and Habitats Directives) provides the basis for nature conservation covering almost 20% of the EU terrestrial area and over 6% of marine territory (European Commission, 2020). This network of protected areas includes most of the national parks in the EU, nationally protected sites and many nature reserves; it helps protect our natural heritage, the species and habitats that can only be found within the European Union area (European Commission, 2020). Natura 2000 is also very unique, being the single almost continent-wide, multiple-country overlapping area of protected sites, subsequently also needing EU-wide legislation to coordinate work in different national settings (European Commission, 2019b).

The Natura 2000 network is designated and managed based on two pieces of EU-level legislation, the Birds Directive and the Habitats

---

[1] Within the Regulatory Fitness and Performance Programme (REFIT) in 2012, the Commission wanted to ensure EU policies are smart, "fit for purpose", efficient and relevant to be pursued at the EU level. Within this scope, nature legislation also needed to be scrutinized by a fitness check process to assess the effectiveness, efficiency, coherence, relevance and EU added value and, if needed, to identify obsolete segments or excessive elements to feed into future policy considerations and amendments (European Commission, 2014).

Directive. They were drafted in 1979 (Birds Directive) and 1992 (Habitats Directive) to set the basis for an EU-wide network and establish its overall operational and management principle and processes (European Commission, 2019b). Natura 2000 is currently the basis of the nature conservation of the Union, being very much valued and cherished (although often criticized) by the conservation sector. Having an EU-wide network and relevant legislation means assurance, and in many cases, another pair of scrutinizing eyes on top of the national level. This signifies to the whole sector of nature conservationists that nature and protected areas are important at the EU level, and they can count on legal support and remedy. However, as such, to a whole set of other sectors it means a number of nuisances at various levels. For instance, Natura 2000 legislation calls for more rigorous scrutiny and assessment in terms of new investments affecting protected sites, while it also impacts on farmers, who often feel too strictly impeded by Natura 2000 restrictions on agricultural land. No wonder, therefore, that certain opposing sectors could see an opportunity, while nature conservationists were anxious when the so-called Fitness Check of the two directives were announced. Regular review of EU policies is a normal process as pieces of legislation with time and changing conditions often need a revisit to ensure the policy is still fit for purpose. However, knowing that many would be cheering for a weaker EU nature policy, nature conservationists were awaiting a probable desperate struggle.

In the case of the campaign addressing the Fitness Check of the Nature Directives, various stages and stakeholders needed to be tackled and addressed. First, the European Commission put together its evaluation study on whether the directives are fit for purpose, for which various evidence was needed. The evidence gathering was comprised of interest groups and online public consultations, national and scientific reports, followed by consultations on the Fitness Check results in 2015 (European Commission, 2014). Based on the information collected and provided, the Commission published its staff working document with the conclusion of the results in 2016 (European Commission, 2016). However, between the two phases of the information synthesizing study in March 2016 and publishing of the policy evaluation in the staff working paper in December 2016, over nine months had passed, and the announcement of the Fitness Check results was continuously postponed. The time was ticking, making the conservation sector anxious and uneasy of the probable results. To counteract any unfortunate events, further actors of the EU policy arena, including the European Parliament, the European Council, stakeholder groups and citizens, were additionally drawn upon to put additional pressure on the Commission until the publication of the staff working paper in December 2016. Eventually, the campaign yielded success; the document emerged stating that the Nature Directives are fit for purpose, and further actions are needed to enhance their implementation (European Commission, 2016).

In the above-described process, obviously all of the relevant actors and stakeholders needed to be addressed with different keywords and messages. Factual messages were needed to raise the interest or break down potential arguments mainly, at the official consultation scale. At the European Parliament level, for the Members of the Parliament who are directly elected by the public, it was inevitable to show that citizens support this initiative, or that it was in the interest of the public to pursue it. At the general public level, emotions needed to be awoken. Accordingly, various formulations of messages and numerous vessels, which are ideally compelling to the specific stakeholder groups, were selected and employed.

My Recollections of Our Campaign to Safeguard the European Union's Nature Directive

In the period 2014–2016, the international conservation NGO network I was involved in also found it quintessential to take part in the Fitness Check campaign as many of the NGO's national members from Central and Eastern Europe (CEE) worked directly with Natura 2000, and the main mission of the organization was to conserve biodiversity. The CEE region also took pride in saying that Natura 2000 gave the opportunity for the EU to preserve in the CEE region what no longer exists in Western Europe: nature. Additionally, with somewhat more pro-development and infrastructure favouring governments in many countries of the CEE region, weaker nature legislation could have resulted in imminent and irreversible loss of biodiversity. So in short, we had a lot at stake—and we needed a successful campaign.

When we started to prepare our specific communication campaign to safeguard the European Union's key nature conservation policies, a number of us were not very aware of the above scientific theories and strategies.[2] Nevertheless, we had a pre-concept: while we would aim to target decision-makers with facts (mostly on the socio-economic importance of nature), we would address the public by building on and raising their emotions to support nature. However, there were other factors of the campaign that also needed to be taken into account. For instance, the green NGO world needed to align its messages as it had to be shown to our key audience—EU decision-makers—that the whole conservation sector is behind the unified tagline that the Nature Directives should remain untouched. In addition to aligning our messages, it made sense to fall in line to a certain extent with the bigger NGOs (BirdLife, EEB, WWF), especially the 'Nature Alert' campaign, towards the public, as opposed to having several micro-campaigns by different organizations. However, it was also crucial that we reach our specific audience in the CEE and the CEE in Brussels (meaning decision-makers from this region), and add a special spice to the overall NGOs' campaign.

Accordingly, our planned campaign focused on the CEE region, and was ultimately two-fold. On one hand, facts and figures on Natura 2000 sites and its contribution to jobs and growth, as well as other socio-economic benefits, were prepared mostly for the decision-makers, who were deemed to make rational decisions (applying the rational choice theory as well as the ecosystem services concept). On the other hand, the other key element was the demonstration of public support, both by activating various stakeholders and the public (targeting emotions as well as additional biases). With both these components, we planned to have ample ammunition to show to the various EU decision-makers that

[2] We did, however, have communication professionals in-house, as well as cooperation with the network of the large international environmental NGOs, who are also experts in campaigning and who may have been aware of the above principles: I was definitely not among them.

the CEE was clear that the Nature Directives were central pieces of EU legislation and should not be touched.

To achieve this aim, we first built a detailed database of the groups we needed to address as final message receivers. We identified the final receivers as the EU decision-makers (including the European Commission, European Parliament, and European Council), who would ultimately steer and decide on the Fitness Check's outcome and, as a secondary group, the general public. The latter was deemed secondary only as it was planned that the general public would act more as a vehicle to transmit messages towards decision-makers, instead of being the final receiver. In addition to the general public, another transmitter was our group of CEE allies that we named "Friends of Natura 2000". We identified and sought to include these "friends", who came from various backgrounds (from artists through local businesses to farmers), and provided their names and personal messages and stories about supporting Natura 2000. This group of Natura 2000 allies acted in four ways: (1) it showed the wide support of the general population, (2) reduced psychological distance and evoked emotions (as politicians from that specific region or country or even field of work may relate more to their personal messages), (3) broke the *status quo* and confirmation bias and showed that not all farmers, businessmen, etc., who are thought to be ardent attackers of the directives, consider Natura 2000 as a necessary evil and hinderance of development, and (4) acted as an upscale platform to reach further stakeholders and the general public during communication. We featured 27 individuals and organizations (singer, tourist expert, wine maker, teacher, farmer, horticultural organization, etc.) from several countries, and could use their messages within the campaign.[3]

We built largely on the rational choice theory that, ideally, decision-makers consider socio-economic benefits of nature. When seeing nature's contribution to the then EU motto of "jobs and growth", they may think that they should persevere with the directives because of their vast socio-economic benefits. Therefore, when the Commission report was being prepared to show evidence about how the directives are fit for purpose,

[3] These messages and other elements of the campaign can be found here: http://www.ceeweb.org/work-areas/working-groups/natura-2000/welovenatura2000/

and during interservice consultation (when the directives were discussed among Directorates General (DG)), we provided a brochure to over 500 EU decision-makers on case studies about the contribution of the nature directives to socio-economic development in Central and Eastern Europe (strictly in brevity with short texts and beautiful photos). The brochure included for instance, how Natura 2000 sites contributed to the income of 2,300 families in Sighişoara-Târnava Mare, Romania, or how the restoration of river meanders resulted in flood risk reduction estimated at €500 per ha (Loppin & Kotulak, 2016).[4] In addition to this, in many of our advocacy letters to the European Parliament and various DGs to support the fact that nature directives are fit for purpose, we highlighted the factual messages on Natura 2000 contributions to the economy through showcasing ecosystem services (and flagship species, although the latter was mostly through visual materials of photos and videos). Our messages to decision-makers very much focused on the fact that European nature is unique and in peril. The responsibility of safeguarding our European natural heritage is in the hands of the decision-makers, which in a way can be seen as triggering the endowment effect and the scarcity heuristic. A specific Twitter campaign on the socio-economic benefits of Natura 2000 and its ecosystem services, together with messages from the Friends of Natura 2000 group from each CEE country, was also used for the campaign to call for contributions for the Commission's public consultation and for general support for the directives.

During these campaigns a specific landscape or a flagship species were displayed. To reduce the psychological distance, to build further on emotions and the familiarity principle, and to mobilize the general public (especially the youth that may be more familiar with producing short films), we also called for a short film contest specifically on the importance of Natura 2000. This short[5] film was deployed when we targeted decision-makers on Twitter and via emails addressing 779 Members of the European Parliament and other EU decision-makers asking them to

[4] The brochure is available: http://www.ceeweb.org/wp-content/uploads/2011/12/CEEweb_N2000_fact_sheet_ev.pdf.

[5] Too cool to be killed, watched over 4000 times - https://www.youtube.com/watch?v=0SKomG50Lwk.

demand the outcomes of the Fitness Check Publication (as it was continuously postponed), and to back up the Nature Directives. In this letter, we showed both the overwhelming support of the citizens triggering emotions and the socio-economic facts.[6]

Ultimately, our CEE campaign, as well as the overall EU-wide NGO campaign, proved victorious and it was deemed that the Nature Directives are fit for purpose; to be fully effective, however, further national prioritization and funding are required (European Commission, 2016). It was, however, difficult to grasp the exact recipe for the success. If you work on an international scale involving a number of different stakeholders and messages with a multitude of socio-economic and institutional factors in play, it is practically impossible to judge what exactly made an impact, what was the actual trigger, or how they formed a complex system to induce change. There could be literally thousands of momenta or actors or processes and their interplay that could have possibly sealed the fate of the Nature Directives. Still, in order to attempt to have some idea, at least of the probable impacts of your efforts, some steps may be advised. Receiving direct feedback, e.g., getting a phone call or email thanking you for your contribution (clearly, only helpful if it is evident that it is not an autoreply), or seeing CEE case studies as references in the evaluation, are extremely useful to track down any potential impactful measures. Continuous follow-up monitoring actions are also very much needed whether from direct contacts or via an Internet search or references made; you can inquire if your input was taken into account. In this regard, social media is a fantastic tool as it aids tracking of what had happened with your messages. For instance, when you notice that a CEE Member of Parliament Tweeted the short film you sent him calling for support for the Nature Directives, you know you hit home. Of course, sheer numbers also add to evidencing impact—although because the campaign at the EU scale was a joint effort between multiple organizations, it was difficult to know to what extent precisely. However, it is known that the Nature Directives related campaign triggered the largest

[6] The letter to the Members of the European Parliament can be found here: http://www.ceeweb.org/wp-content/uploads/2011/12/CEEweb-for-Biodiversitys-letter-to-MEPs-ask-for-the-fitness-check-results-on-Natura-20001.pdf.

public consultation in the EU's history with over half a million contributions (WWF, 2016), and by all means, we also contributed to this success, especially considering the CEE region.

Overall, when putting these actions under the auspices of social-economic and communication theories, our campaign was built on two areas: familiarity principle and emotional connections with public pressure evoking the responsibility of decision-makers to safeguard Europe's future by protecting its nature; and, the socio-economic contribution of Natura 2000 to appeal to decision-makers with the perceptions that they work under rational behaviour theories. Looking back with hindsight on the work of preparing and implementing our campaign for the Nature Directives, it is astonishing to see how much our ways of communication and methods used can be justified by the science behind communication or other social science disciplines, even if at the time we had no, or only limited knowledge about it. Many of our headlines were grounded on socio-economic facts about nature's contribution and therefore, rational decision-making theory, due to the belief that it is the only way to attract the other sectors and to diminish their arguments of Natura 2000 being an obstacle to development. In addition to this, we also used additional framings. We employed the familiarity principle and the endowment theory addressing, e.g., the Polish Members of Parliament with the socio-economic relevance of Polish Natura 2000 sites or Polish flagship species. We also used creative short videos on European nature produced by the public to appeal to their emotions (as well as to their senses by acting as their voters ask them).

On the other hand, however, we made relatively limited use of exploiting emotions and reframing Natura 2000 according to Kahneman and Tversky's Prospect Theory (1979) of highlighting losses and the potential value of this, amplifying fear and negative consequences. We could also have triggered other biases on a more elevated scale, including the confirmation bias with potentially more provocative campaigns. In addition, we could have dug deeper and tried to group our stakeholder groups based on their social norms and values and tailor messages accordingly—although of course, as with every NGO, we also struggled with limited resources. Overall, despite having adequate scientific knowledge of relevant social theories, we still exploited relatively well the possible biases and heuristics, as well as steering emotions. On the other hand, we massively built on facts and figures about the socio-economic relevance of Natura 2000 and ecosystem services with the firm belief that decision-makers will only make rational choices (now, I know better). It is also interesting to see how practitioners' knowledge can be apt even without the actual knowledge of scientific theories, but more importantly, how much practice and science can provide to each other, ultimately resulting in a more complex and conscientious plan of action.

Conclusions: Pry, Plan and Follow up

To date, various scientific theories and strategies stemming from them can aid the forming of more impactful messages targeting decision-makers. These theories and strategies may very well be employed in the context of biodiversity, even if the overall topic is often thought to be a challenging concept that may not resonate well with people at the top—either because of its diffuse and complex nature or because of vested and competing economic interests. If carefully crafted however, as can be seen in the above case, communicating rational arguments embedded in monetary values with ecosystem services, steering feelings of familiarity with flagship species, or showing overwhelming public support, can help in reaching the desired outcomes. Formulating messages, however quintessential, are only the beginning of the cumbersome process of communication and advocacy. The right vessel should be chosen, together with the right audience for which the messages are relevant and with which they can resonate. The most difficult challenge remains how to grasp and involve that audience, especially if they are in a special bubble. In this, meticulous planning and specific research on your audience's members and their background can assist. Discovering ways of how best to trigger the familiarity principle or showcase socio-economic messages, how to understand the ruling norms or even how to activate certain biases that push the right buttons, are the building blocks of forming impactful messages. Furthermore, following up on your tactics and potential results (e.g., by asking for direct feedback or monitoring indicators) can provide useful tips in finding out what worked best and what to employ and fine-tune for the next occasion.

References

Ainscough, J., de Vries, A., Metzger, M., Rounsevell, M., Schröter, M., Delbaere, B., de Groot, R. and Staes, J. (2019) Navigating pluralism: Understanding perceptions of the ecosystem services concept. *Ecosystem Services*, 36: 100892.https://doi.org/10.1016/j.ecoser.2019.01.004.

Ajzen, I. (1991). The theory of planned behavior. *Organizational Behaviour and Human Decision Processes, 50*, 179–211. https://doi.org/10.1016/0749-5978(91)90020-T

Bekessy, S. A., Runge, M. C., Kusmanoff, A. M., Keith, D. A., & Wintle, B. A. (2018). Ask not what nature can do for you: A critique of ecosystem services as a communication strategy. *Biological Conservation, 224*, 71–74. https://doi.org/10.1016/j.biocon.2018.05.017

BirdLife. (2020). Nature alert. https://www.birdlife.org/europe-and-central-asia/project/nature-alert.

Braat, L., & de Groot, R. (2012). The ecosystem services agenda: Bridging the worlds of natural science and economics, conservation and development, and public and private policy. *Ecosystem Services, 1*(1), 4–15. https://doi.org/10.1016/j.ecoser.2012.07.011

ten Brink, P., Bassi, S., Badura, T., Gantioler, S., Kettunen, M., Mazza, L., Hart, K., Rayment, M., Pieterse, E., Daly, E., Gerdes, H., Lago, M., Lang, S., Markandya, A., Nunes, P., Ding, H., Tinch, R., & Dickie, I. (2013). *The economic benefits of the Natura 2000 network*. Publications Office of the European Union.

Cardinale, B., Duffy, J., Gonzalez, A., Hooper, D., Perrings, C., Venail, P., Narwani, A., Tilman, D., Wardle, D., Kinzig, A., Daily, G., Loreau, M., Grace, J., Larigauderie, A., Srivastava, D., & Naeem, S. (2012). Biodiversity loss and its impact on humanity. *Nature, 486*, 59–67. https://doi.org/10.1038/nature11148

Chaudhary, S., McGregor, A., Houston, D., & Chettri, N. (2015). The evolution of ecosystem services: A time series and discourse-centered analysis. *Environmental Science and Policy, 54*, 25–34. https://doi.org/10.1016/j.envsci.2015.04.025

Cialdini, R. B., Reno, R. R., & Kallgren, C. A. (1990). A focus theory of normative conduct: Recycling the concept of norms to reduce littering in public places. *Journal of Personality and Social Psychology, 58*, 1015–1026. https://doi.org/10.1037/0022-3514.58.6.1015

Costanza, R., de Groot, R., Braat, L., Kubiszewski, I., Fioramonti, L., Sutton, P., Farber, S., & Grasso, M. (2017). Twenty years of ecosystem services: How far have we come and how far do we still need to go? *Ecosystem Services, 28*(2017), 1–16. https://doi.org/10.1016/j.ecoser.2017.09.008

Costanza, R., Arge, R., Groot, R., Farberk, S., Grasso, M., Hannon, B., Limburg, K., Naeem, S., O'Neill, R. V., Paruelo, J., Raskin, R., Sutton, P., & Belt, M. (1997). The value of the world's ecosystem services and natural capital. *Nature, 387*, 253–260. https://doi.org/10.1016/S0921-8009(98)00020-2

Daily, G. C., Alexander, S., Ehrlich, P. R., Goulder, L., Lubchenco, J., Matson, P. A., Mooney, H., Postel, S., Schneider, S. H., Tilman, D., & Woodwell, G. G. (1997). Ecosystem services: Benefits supplied to human societies. *Natural Ecosystems: Issues in Ecology, 1*(2), 1–18.

Ehrlich, P. R., & Ehrlich, A. H. (1981). *Extinction: The causes and consequences of the disappearance of species*. Random House.

European Commission. DG Environment. (2014). Fitness check mandate for nature legislation. https://ec.europa.eu/environment/nature/legislation/fitness_check/docs/Mandate%20for%20Nature%20Legislation.pdf.

European Commission. DG Environment. (2016). Fitness check of the birds and habitats directives. https://ec.europa.eu/environment/nature/legislation/fitness_check/index_en.htm.

European Commission. DG Environment. (2019a). Ecosystem services and green infrastructure. https://ec.europa.eu/environment/nature/ecosystems/index_en.htm.

European Commission. DG Environment. (2019b). Nature and biodiversity law. https://ec.europa.eu/environment/nature/legislation/index_en.htm.

European Commission. DG Environment. (2020). Natura 2000. https://ec.europa.eu/environment/nature/natura2000/index_en.htm.

Farrow, K., Grolleau, G., & Ibanez, L. (2017). Social norms and pro-environmental behavior: A review of the evidence. *Ecological Economics, 140*, 1–13. https://doi.org/10.1016/j.ecolecon.2017.04.017

Intergovernmental Science-Policy Platform on Biodiversity and Ecosystem Services (IPBES). (2019). Media release. *Nature's dangerous decline 'Unprecedented' species extinction rates 'Accelerating'*. https://www.ipbes.net/news/Media-Release-Global-Assessment.

Jepson, P., & Barua, M. (2015). A theory of flagship species action. *Conservation and Society, 13*(1), 95–104. https://doi.org/10.4103/0972-4923.161228

Kahneman, D., Knetsch, J. L., & Thaler, R. H. (1990). Experimental tests of the endowment effect and the Coase theorem. *Journal of Political Economy, 98*, 1325–1348. https://doi.org/10.1086/261737

Kahneman, D., & Tversky, A. (1979). Prospect theory: An analysis of decision under risk. *Econometrica*, 47(2): 263–291. https://doi.org/10.1142/9789814417358_0006

Kidd, L. R., Garrard, G. E., Bekessy, S. A., Mills, M., Camilleri, A. R., Fidler, F., Fielding, K. S., Gordon, A., Gregg, E. A., Kusmanoff, A. M., Louis, W., Moon, K., Robinson, J. A., Selinske, M. J., Shanahan, D., & Adams, V. M. (2019). Messaging matters: A systematic review of the conservation messaging literature. *Biological Conservation, 236*, 92–99. https://doi.org/10.1016/j.biocon.2019.05.020

Kusmanoff, A. M., Fidler, F., Gordon, A., Garrard, G. E., & Bekessy, S. A. (2020). Five lessons to guide more effective biodiversity conservation message framing. *Conservation Biology, 00*, 1–11. https://doi.org/10.1111/cobi.13482

Kusmanoff, A. M., Hardy, M. J., Fidler, F., Maffey, G., Raymond, C., Reed, M. S., & Bekessy, S. A. (2016). Framing the private land conservation conversation: Strategic framing of the benefits of conservation participation could increase landholder engagement. *Environmental Science & Policy, 61*, 124–128. https://doi.org/10.1016/j.envsci.2016.03.016

Legagneux, P., Casajus, N., Cazelles, K., Chevallier, C., Chevrinais, M., Guery, L., Jacquet, C., Jaffre, M., Naud, M., Noisette, F., Ropars, P., Vissault, S., Archambault, P., Bety, J., Bertaux, D., & Gravel, D. (2018). Our house is burning: Discrepancy in climate change vs. biodiversity coverage in the media as compared to scientific literature. *Frontiers in Ecology and Evolution*, 5: 175. https://doi.org/10.3389/fevo.2017.00175.

Liebe, U., & Preisendörfer, P. (2010). Rational choice theory and the environment: Variants, applications, and new trends. In M. Gross, H. Heinrichs (Eds.), *Environmental Sociology*. Springer.

Loppin, C., & Kotulak, M. (2016). Fact sheet—The socio-economic benefits of Natura 2000 in Central and Eastern Europe. http://www.ceeweb.org/wp-content/uploads/2011/12/CEEweb_N2000_fact_sheet_ev.pdf.

Lundberg, P., Vainio, A., Macmillan, D. C., Smith, R. J., Veríssimo, D., & Arponen, A. (2019). The effect of knowledge, species aesthetic appeal, familiarity and conservation need on willingness to donate. *Animal Conservation, 22*, 432–443. https://doi.org/10.1111/acv.12477

Mace, G. M., Barrett, M., Burgess, N. D., Cornell, S., Freeman, R., Grooten, M., & Purvis, A. (2018). Aiming higher to bend the curve of biodiversity loss. *Nature Sustainability, 1*, 448–451.

Martinez-Harms, M. J., Bryan, B. A., Balvanera, P., Law, E. A., Rhodes, J. R., Possingham, H. P., & Wilson, K. A. (2015). Making decision for managing ecosystem services. *Biological Conservation, 184*, 229–238. https://doi.org/10.1016/j.biocon.2015.01.024

Maxwell, S., Fuller, R., Brooks, T., Watson, J. (2016). Biodiversity: The ravages of guns, nets and bulldozers. *Nature*, 536. https://doi.org/10.1038/536143a.

McGowan, J., Beaumont, L. J., Smith, R. J., Chauvenet, A. L. M., Harcourt, R., Atkinson, S. C., Mittermeier, J. C., Esperon-Rodriguez, M., Baumgartner, J. B., Beattie, A., Dudaniec, R. Y., Grenyer, R., Nipperess, D. A., Stow, A., & Possingham, H. P. (2020). Conservation prioritization can resolve the flagship species conundrum. *Nature Communications*, 11, 1–7, Article 994. https://doi.org/10.1038/s41467-020-14554-z.

Millner, A., & Olivier, H. (2015). Beliefs, politics, and environmental policy. *Review of Environmental Economics and Policy*, 10(2). https://doi.org/10.1093/reep/rew010.

Nickerson, R. S. (1998). Confirmation bias: A ubiquitous phenomenon in many guises. *Review of General Psychology, 2*, 175–220. https://doi.org/10.1037/1089-2680.2.2.175

OECD. (2017). *The political economy of biodiversity policy reform.* OECD Publishing.

Posner, S. M., McKenzie, E., & Ricketts, T. H. (2016). Policy impacts of ecosystem services knowledge. *Proceedings of the National Academy of Sciences of the United States of America, 113*(7), 1760–1765. https://doi.org/10.1073/pnas.1502452113

Primmer, E., Termansen, M., Bredin, Y., Blicjarska, M., Garcia-Llorente, M., Berry, P., Jaaskelainen, T., Bela G., Fabok, V., Geamana, N., Harrison, P. A., Haslett, J. R., Cosor, G. L., & Andersen, A. H. K. (2015). Caught Between personal and collective values: Biodiversity conservation in European decision-making. *Environmental Policy and Governance*, 27(6), 588-604. https://doi.org/10.1002/eet.1763.

Raue, M., & Schneider, E. (2019). Psychological perspectives on perceived safety: Zero-risk bias, feelings and learned carelessness. In M. Raue, B. Streicher, E. Lermer (Eds.), *Perceived safety. Risk engineering.* Springer. https://doi.org/10.1007/978-3-030-11456-5_5.

Reder, L. M., & Ritter, F. E. (1992). What determines initial feeling of knowing? Familiarity with question terms, not with the answer. *Journal of Experimental Psychology: Learning, Memory & Cognition,* 18(3), 435–451. https://doi.org/10.1037/0278-7393.18.3.435.

Rittel, H. W. J., & Webber, M. M. (1973). Dilemmas in a general theory of planning. *Policy Sciences, 4*(2), 155–169. https://doi.org/10.1007/BF01405730

Samuelson, W., & Zeckhauser, R. (1988). Status quo bias in decision making. *Journal of Risk and Uncertainty, 1,* 7–59. https://doi.org/10.1007/BF00055564

Schlagloth, R., Santamaria, F., Golding, B., & Thomson, H. (2018). Why is it important to use flagship species in community education? The koala as a case study. *Animal Studies Journal*, 7(1), 127–148.

Schleyer, C., Görg, C., Hauck, J., & Winkler, K. J. (2015). Opportunities and challenges for mainstreaming the ecosystem services concept in the multi-level policy-making within the EU. *Ecosystem Services, 16*(2015), 174–181. https://doi.org/10.1016/j.ecoser.2015.10.014

Schröter, M., van der Zanden, E. H., van Oudenhoven, A. P. E., Remme, R. P., Serna- Chavez, H. M., de Groot, R. S., & Opdam, P. (2014). Ecosystem services as a contested concept: A synthesis of critique and counter-arguments. *Conservation Letter, 7,* 514–523. https://doi.org/10.1111/conl.12091

Sharman, M., & Mlambo, M. C. (2012). Wicked: The problem of biodiversity loss. *GAIA—Ecological Perspectives on Science and Society,* 21(4). https://doi.org/10.14512/gaia.21.4.10.

Smith, R. J., Veríssimo, D., Isaac, N. J. B., & Jones, K. E. (2012). Identifying Cinderella species: Uncovering mammals with conservation flagship appeal. *Conservation Letters, 00*(2012), 1–8. https://doi.org/10.1111/j.1755-263X.2012.00229.x

Thaler, R. H. (2015). *Misbehaving: The making of behavioral economics.* W.W. Norton & Co.

Veríssimo, D., Fraser, I. Girao, W., Campos, A. A., Smith, R. J., MacMillan, D. C. (2013). Evaluating conservation flagships and flagship fleets. *Conservation Letters,* May/June 2014, 7(3), 263–270. https://doi.org/10.1111/conl.12070.

Westman, W. E. (1977). How much are nature's services worth? *Science, 197*, 960–964. https://doi.org/10.1126/science.197.4307.960

Wright, W. C. C., Eppink, F. V., & Greenhalgh, S. (2017). Are ecosystem service studies presenting the right information for decision making? *Ecosystem Services, 25*(2017), 128–139. https://doi.org/10.1016/j.ecoser.2017.03.002

WWF. (2016). EU Nature Directives are 'Fit for Purpose', concludes European Commission expert study. https://www.wwf.eu/?272571/EU-Nature-Directives-are-Fit-for-Purpose-study.

WWF. (2018). *Living planet report-2018: Aiming higher*. In M. Grooten, & R. E. A. Almond (Eds.), WWF, Gland.

WWF. (2020). Large carnivores. https://wwf.panda.org/knowledge_hub/where_we_work/alps/our_solutions22222/large_carnivores/.

Zaccai, E., & Adams, W. M. (2012). How far are biodiversity loss and climate change similar as policy issues? *Environment, Development and Sustainability*, 14, 557–571 (2012). https://link.springer.com/article/10.1007/s10668-012-9344-x.

Index

A. Franklin (ed.), *Co-Creativity and Engaged Scholarship*,
https://doi.org/10.1007/978-3-030-84248-2

The manufacturer's authorised representative in the EU is Springer Nature Customer Service Centre GmbH, Europaplatz 3, 69115 Heidelberg, Germany. If you have any concerns regarding our products, please contact ProductSafety@springernature.com

Printed and bound by CPI Group (UK) Ltd, Croydon, CR0 4YY
30/09/2026
02225321-0015